T0340198

AMERICA INVENTS ACT PRIMER

AMERICA INVENTS
ACT PRIMER

SARAH HASFORD, ESQ.

Academic Press is an imprint of Elsevier
125 London Wall, London EC2Y 5AS, United Kingdom
525 B Street, Suite 1800, San Diego, CA 92101-4495, United States
50 Hampshire Street, 5th Floor, Cambridge, MA 02139, United States
The Boulevard, Langford Lane, Kidlington, Oxford OX5 1GB, United Kingdom

Notices
Knowledge and best practice in this field are constantly changing. As new research and experience broaden our understanding, changes in research methods, professional practices, or medical treatment may become necessary.

Practitioners and researchers must always rely on their own experience and knowledge in evaluating and using any information, methods, compounds, or experiments described herein. In using such information or methods they should be mindful of their own safety and the safety of others, including parties for whom they have a professional responsibility.

To the fullest extent of the law, neither the Publisher nor the authors, contributors, or editors, assume any liability for any injury and/or damage to persons or property as a matter of products liability, negligence or otherwise, or from any use or operation of any methods, products, instructions, or ideas contained in the material herein.

Library of Congress Cataloging-in-Publication Data
A catalog record for this book is available from the Library of Congress

British Library Cataloguing-in-Publication Data
A catalogue record for this book is available from the British Library

ISBN: 978-0-12-812096-5

For information on all Academic Press publications visit our website at
https://www.elsevier.com/books-and-journals

Working together
to grow libraries in
developing countries

ELSEVIER | Book Aid International

www.elsevier.com • www.bookaid.org

Publisher: John Fedor
Acquisition Editor: Katey Birtcher
Editorial Project Manager: Jill Cettel
Production Project Manager: Anitha Sivaraj
Designer: Mark Rogers

Typeset by Thomson Digital

In loving memory of my late Grandma Margaret who always inspired me to pursue further education.

CONTENTS

ACKNOWLEDGMENTS

Although I am listed as the sole author, this text would not have been possible without the contributions of many others. Most of all, I could not have undertaken, let alone finished this project without the unwavering support, encouragement and love of my husband, Justin. Special thanks also goes to our dear friend Dr. Andrew Thompson, who endured the arduous task of reviewing a draft of each chapter. Dr. Thompson provided valuable feedback, which has undoubtedly produced a better product for all readers. In addition, I would like to thank my friend Sandra (Sandy) Langley, who is cherished and adored more than she knows, for her enthusiastic encouragement and enduring confidence in me.

Appreciation also goes to the Federal Circuit Bar Journal, a publication of the Federal Circuit Bar Association, for my limited use of Joe Matal, *A Guide to the Legislative History of the America Invents Act: Part I of II*, 21 Fed. Cir. B.J. 435 (2012) and Joe Matal, *A Guide to the Legislative History of the America Invents Act: Part II of II*, 21 FED. CIR. B.J. 539 (2012). Additionally, I would like to thank AIPLA for allowing me to use Robert A. Armitage, *Understanding the America Invents Act*, 40 AIPLA Q.J. 1 (2012) and Keith M. Kupferschmid, *Prior User Rights: The Inventor's Lottery Ticket*, 21 AIPLA Q.J. 213 (1993). I acknowledge with thanks, as well, Ahmed J. Davis and Karolina Jesien of Fish & Richardson P.C. for graciously granting me permission to use their article entitled *The Balance of Power in Patent Law: Moving Towards Effectiveness in Addressing Patent Troll Concerns*, 22 FORDHAM INTELL. PROP. MEDIA & ENT. L.J. 835 (2012).

Lastly, I want to extend my sincerest gratitude to the team at Elsevier. I especially want to thank my Acquisitions Editor, Katey Birtcher, for taking a chance on me as a first-time author and for her patience and persistence in obtaining approval for this text. I also wish to thank and acknowledge my Editorial Project Manager, Jill Cetel, for her continuous encouragement during this project.

DISCLAIMER

The intent of this text is to educate readers on the subject matter that is discussed. Nothing in this text should be construed to create an attorney–client relationship between the author or publisher and anyone else. Furthermore, neither the publisher nor the author of this text is engaging in legal services or providing legal advice through this text. Each case is fact-specific, and the appropriate solution in any case will vary. This text therefore may not be relevant to any particular situation.

While every attempt was made to ensure that this text is accurate, errors or omissions may be contained therein, for which any liability is disclaimed. Furthermore, one of the only constants in the law is that it is constantly changing. Portions of this text may, therefore, become obsolete or incorrect in the future. Accordingly, this text should be used only as a general guide and citations should be cross-checked, where appropriate.

Neither the author nor the publisher have liability or responsibility to any person or entity with respect to any loss or damage caused or alleged to be caused directly or indirectly by this text or the information contained in this text.

The opinions expressed in this text are solely the personal opinions of the author in her individual capacity. Nothing in this text shall be attributable to the author in any representational capacity or to any other person or legal entity.

CHAPTER ONE

Introduction to the America Invents Act

Contents

I) INTRODUCTION

The America Invents Act ("AIA") is the common name for the Leahy-Smith America Invents Act, Pub. L. No. 112-29, 125 Stat. 284 that was signed into law by President Obama on September 16, 2011. The enactment of the AIA is the most significant change to U.S. patent law that has occurred since the 1952 Patent Act was passed. Indeed, by transitioning U.S. patent law from a "first-to-invent" to a "first-inventor-to-file" system, the AIA has significantly expanded the body of prior art that may be cited against claims in a patent or patent application, thereby bringing substantial changes to both patent prosecution and patent litigation practices. *Id.* at § 3. Furthermore, by providing new post-grant proceedings (i.e., inter partes reviews and post grant reviews) for challenging the validity of patents, the AIA has created an entirely new trial-like practice before the U.S. Patent and Trademark Office ("U.S. PTO"). *Id.* at §§ 6(a)-(f), 18. Additionally, by altering claims and defenses to patent infringement, such as willful infringement claims and the best mode and prior user rights defenses, the AIA has made several important changes to the way patent suits are brought and defended. *Id.* at §§ 5, 15, and 17. Beyond its changes to the practice of patent law, many provisions of the AIA also alter the manner in which the U.S. PTO operates. *See, e.g., id.* at §§ 7, 22, 23, 28, and 32. Accordingly, the AIA impacts virtually every area of patent law.

Because of the AIA's far-reaching effects, it is imperative that those who are involved in patent law gain a good understanding of the legislation. To help with this, a brief history of the legislation is provided in Section II. In

America Invents Act Primer. http://dx.doi.org/10.1016/B978-0-12-812096-5.00001-6

addition, Section III gives an overview of the most significant provisions[a] of the AIA, and Section IV lists free resources that are helpful for learning about, and better understanding, the AIA. Moreover, the following chapters provide in-depth discussions of several AIA provisions. In particular, whenever possible, the subsequent chapters address the following questions for each AIA provision discussed therein:

    i. How did the AIA provision change the law?

    ii. What legislative goals/objectives were addressed by the new law?

    iii. When did the change in the law take effect; i.e., what is the effective date of the AIA provision?

    iv. What are the practical implications of the new law?

Understanding the answers to these questions should help provide a deeper knowledge of the AIA than can be attained by merely reading through each AIA provision itself.

II) BRIEF HISTORY OF THE AIA

The bill that was eventually passed as the AIA was based on legislation that was originally proposed in 2005. *See, e.g.*, 157 Cong. Rec. S131, 2011 (statement of Sen. Leahy) ("The Patent Reform Act of 2011 is structured on legislation first introduced in the House by Chairman Smith and Mr. Berman in 2005."). Proponents of the bill had expressed concerns that the U.S. was not keeping up with the innovation efforts of other countries, such as China, and that part of the reason the U.S. was falling behind was its antiquated patent laws. *Id.* Indeed, China had been modernizing its patent laws in the years leading up to the enactment of the AIA, whereas U.S. patent law had not seen significant change since 1952. *Id.*

Just as important as the concerns over global competitiveness was the angst over the havoc that had been wreaked on many U.S. corporations by entities commonly known as "patent trolls."[b] *Id.*; *see also* Ahmed J. Davis & Karolina Jesien, *The Balance of Power in Patent Law: Moving Towards Effectiveness in Addressing Patent Troll Concerns*, 22 Fordham Intell. Prop. Media & Ent. L.J. 848 (2012) ("There can be little doubt that Congress had NPEs

---

[a]Please note that while this text attempts to be comprehensive in covering the AIA, it does not discuss every single provision of the new law. Indeed, this text primarily focuses on those AIA provisions that substantively affect patent law.

[b]Patent trolls are entities that do not themselves make or sell any products or services ▶ but instead exist for the sole purpose of acquiring patent rights that they then enforce against others who do sell products and/or services. *See* Ahmed J. Davis & Karolina Jesien,

in mind when it passed the Leahy-Smith America Invents Act (AIA) . . . Section 34 of the AIA expressly provides that '[t]he Comptroller General of the United States shall conduct a study of the consequences of litigation by [NPEs], or by patent assertion entities, related to patent claims made under title 35, United States Code, and regulations authorized by that title.'") (footnote omitted). In fact, in the years leading up to the AIA's passage, the number of suits brought by patent trolls had increased significantly. *See* Ahmed J. Davis & Karolina Jesien, *The Balance of Power in Patent Law: Moving Towards Effectiveness in Addressing Patent Troll Concerns*, 22 FORDHAM INTELL. PROP. MEDIA & ENT. L.J. 836 n.2 (2012) (citing *Litigations Over Time*, PATENT FREEDOM, http://www.patentfreedom.com/research-lot.html (last visited Feb. 25, 2012)) ("In 2001, NPEs brought approximately one hundred lawsuits targeting five hundred operating companies, while in 2010, the numbers increased to more than five hundred lawsuits targeting over 2,300 operating companies."). As a result, there was a growing desire to pass legislation that could curtail the activities of these entities to save businesses from spending millions of dollars in defending against patent suits, which often were based on dubious patents. *See, e.g.*, H.R. REP. NO. 112-98, at 54 (2011) ("A number of patent observers believe the issuance of poor business-method patents during the late 1990's through the early 2000's led to the patent 'troll' lawsuits that compelled the Committee to launch the patent reform project 6 years ago."); *Patent Trolls: Fact or Fiction?: Hearing before the Subcomm. on Courts, the Internet, and Intellectual Property of the H. Comm. on the Judiciary*, 109th Cong. 3 (2006) (statement of Rep. Berman, Member, House Comm. on the Judiciary) ("I have concerns about those

◄ *The Balance of Power in Patent Law: Moving Towards Effectiveness in Addressing Patent Troll Concerns*, 22 FORDHAM INTELL. PROP. MEDIA & ENT. L.J. 836 (2012) ("A patent troll is an entity that focuses solely on capitalizing on patent portfolios. The troll purchases or otherwise obtains patents from other companies for purposes of licensing and enforcing them, rather than practicing any inventions covered by those patents . . . A typical business model for [a troll] is to acquire patents that apply broadly across a particular industry (often business method patents), identify potential infringers, threaten litigation, and then either collect license fees from those entities or bring lawsuits against those that refuse to license.") (footnotes omitted); Xun Liu, *Joinder under the AIA: Shifting Non-Practicing Entity Patent Assertions away from Small Businesses*, 19 MICH. TELECOMM. & TECH. L. REV. 492 (2012) ("These [non-practicing] entities are also referred to as 'patent trolls' because they characteristically hold patents until a related product becomes profitable, then emerge to demand payments from unsuspecting companies, much like the trolls of folklore that ambush unsuspecting passersby.") (footnotes omitted). The politically correct term for "patent troll" is "non-practicing entity" or "NPE." As I believe that political correctness is a pernicious form of speech censorship, I refrain from using these terms except when they are used in quoted material.

who take advantage of the current patent system to the detriment of future innovations, whether called trolls, entrepreneurs or those that shall not be named. There is a significant problem if the patent being asserted is of questionable validity."); *id*. at page 1 (statement of Rep. Smith, Member, House Comm. on the Judiciary) ("This morning, the Subcommittee will conduct its seventh hearing on patent reform in the 109th Congress by exploring the much-maligned patent troll. We hope to define trolling behavior in the modern patent world, determine its degree of privilege in the patent system and explore legislative reforms to combat it if needed. Complaints about trolling heightened public interest in patent reform and led to the development of the legislative drafts that our Subcommittee has reviewed.").

According to legislators, the AIA would accomplish several major objectives. For instance, by transitioning to a first-inventor-to-file system, the AIA would create a more transparent, objective, and predictable patent system for patentees and harmonize the U.S. patent system with that of other patent offices across the world, which would in turn contribute to ongoing work-sharing processes. 157 Cong. Rec. S951, 2011 (statement of Sen. Hatch). Additionally, by establishing post-grant proceedings that could be used to bring administrative challenges to a patent's validity, the AIA would create cost-effective alternatives to formal litigation. *Id*. at S952 (statement of Sen. Grassley). Moreover, because the estoppel effects of the post-grant proceedings established by the AIA would be more extensive than the estoppel effects of pre-AIA procedures used to challenge a patent's validity (e.g., inter partes reexamination and ex parte reexamination), the AIA would decrease abusive patent litigation. *See id*. at S951-52 (statement of Sen. Grassley). Also, by providing third parties with an opportunity to submit prior art and other information related to a pending application for consideration by a Patent Examiner, the AIA would improve patent quality. *Id*. at S952 (statement of Sen. Grassley). Further, by ensuring that the U.S. PTO has sufficient funding, the AIA would speed up the patent application process and eliminate the backlog of over 700,000 pending patent applications that the U.S. PTO was experiencing at the time the AIA was passed. *Id*. at S951-52 (statements of Sen. Hatch and Sen. Grassley). By addressing these goals, many hoped that the U.S. could overcome some of the problems it faced with respect to its global competitiveness and the rampant litigation brought on by patent trolls.

The bill that ultimately became the AIA bounced around the Senate and House of Representatives for approximately six years before it was passed. Many might argue that it would have lingered in Congress even longer had the 2008–2009 downturn in the economy not occurred. Indeed, after the financial meltdown, legislators began promoting the patent reform legislation as a way to stimulate the U.S. economy. For instance, Senator Hatch explained:

> While we debate this important legislation, it is crucial that we keep the creation of jobs and economic prosperity at the forefront of our thoughts. After all, patents encourage technological advancement by providing incentives to invent, to invest in, and to disclose new technology. Now more than ever we must ensure efficiency and increased quality in the issuance of patents. This, in turn, will create an environment that fosters entrepreneurship and the creation of new jobs, thereby contributing to growth within all sectors of our economy.
>
> If we think about it, one single deployed patent has a ripple effect that works like this: A properly examined patent, promptly issued by the USPTO, creates jobs— jobs that are dedicated to developing and producing new products and services. Unfortunately, the current USPTO backlog now exceeds 700,000 applicants. The sheer volume of the patent applications not only reflects the vibrant, innovative spirit that has made America a worldwide innovative leader in science, education, and technology, but the patent backlog also represents dynamic economic growth waiting to be unleashed. We cannot afford to go down this path any longer. We need to take advantage of this opportunity to expand our economy.
>
> **157 Cong. Rec. S951, 2011**

Legislators argued that the AIA was "critical for [the United States'] economic growth if we are going to rebuild our economy and win the future" and that the legislation was needed to have a "patent system [that] promotes research and development, investment, job creation, and global competitiveness." *Id.* at S956 (statement of Sen. Bennet). Senator Leahy even characterized the bill as being "a key part to any jobs agenda." *Id.* at S948. Once the bill was perceived as being crucial to bolstering the U.S. economy, lawmakers on both sides of the aisle finally had the motivation they needed to get the bill passed. *See, e.g.,* Can the America Invents Act Deliver on Its Promise?, http://vklaw.com/2012/02/can-the-america-invents-act-deliver-on-its-promise/ (Feb. 16, 2012) ("Congress acted on [the AIA] under the premise that our patent system was the cause of the lagging U.S. economy.").

III) AIA-THE BIG PICTURE

It is easy to become overwhelmed with the AIA, particularly when getting into the details and history of each provision. Indeed, the vast majority of the 37 sections of the Act each contain several provisions that bring significant reform to how patent law was practiced just a few years ago. Grouping the key provisions of the legislation together in certain categories can, however, make understanding the AIA much more manageable because it shows how each individual provision fits into the bigger picture of patent law reform accomplished through the AIA.

The provisions of the AIA can be grouped together and categorized in several different ways. One common way to organize the AIA provisions is by the date on which they became effective. Unlike legislation that contains provisions that all become effective at one time, the provisions of the AIA became effective on a rolling basis. In general, the most minor changes to patent law were implemented first (i.e., on September 16, 2011, the date on which the AIA was enacted), and the most significant changes to patent law were implemented last (i.e., on March 16, 2013, the date that is 18 months after the AIA was enacted). The following chart illustrates this organizational model for many of the provisions of the AIA that will be discussed in this text:[c]

Provisions that Took Effect on September 16, 2011	Provisions that Took Effect on September 16, 2012	Provisions that Took Effect on March 16, 2013
• Change to Inter Partes Reexamination Threshold (AIA § 6) • Tax Strategies Deemed Within the Prior Art (AIA § 14) • Elimination of Best Mode as a Defense to Patent Infringement (AIA § 15) • Human Organism Prohibition (AIA § 33) • Virtual and False Marking (AIA § 16) • Venue Change from D.D.C. to E.D. VA for certain suits brought against the U.S. PTO (AIA § 9) • Establishment of micro-entity (AIA § 10) • Establishment of Pro Bono Program (AIA § 32) • Authorization of U.S. PTO's Fee-Setting Authority (AIA § 10) • Prior Commercial Use Defense (AIA § 5)	• Inventor Oath/Declaration Practices (AIA § 4) • Third Party Submission of Prior Art in a Patent Application (AIA § 8) • Supplemental Examination (AIA § 12) • Citation of Prior Art in a Patent File (AIA § 6) • Priority Examination for Important Technologies (AIA § 25) • Inter Partes Review (AIA § 6) • Post Grant Review (AIA § 6) • Transitional Program for Covered Business Method Patents (AIA § 18) • Patent Ombudsman for Small Business Program (AIA § 28)	• First-Inventor-To-File (AIA § 3) • Changes to the novelty requirements set forth in 35 U.S.C. § 102 (AIA § 3) • Changes to the non-obviousness requirement set forth in 35 U.S.C. § 103 (AIA § 3) • Derivation Proceedings (AIA § 3) • Repeal of Statutory Invention Registration (AIA § 3)

[c]The U.S. PTO's website gives an excellent overview of the effective dates of most, if not all, of the provisions of the AIA at http://www.uspto.gov/sites/default/files/aia_implementation/aia-effective-dates.pdf.

Another way to organize the provisions of the AIA is to categorize them by topic. This is the manner in which the chapters of this text have been organized. The following chart illustrates the specific topics by which the AIA provisions discussed in this text have been organized:

Post Grant Proceedings	Changes to Patent Infringement Claims and Defenses	Changes to Patentability Requirements
• Inter Partes Review (AIA § 6) • Post Grant Review (AIA § 6) • Transitional Program for Covered Business Method Patents (AIA § 18) • Supplemental Examination (AIA § 12) • Change to the Inter Partes Reexamination Threshold (AIA § 6) • Citation of Prior Art in a Patent File (AIA § 6)	• Elimination of Best Mode Defense (AIA § 15) • Prior Commercial Use Defense (AIA § 5) • Virtual and False Marking (AIA § 16) • Advice of Counsel (AIA § 17) • Joinder of Parties (AIA § 19)	• Amendments to 35 U.S.C. § 102 (AIA § 3) • Amendments to 35 U.S.C. § 103 (AIA § 3) • Tax Strategies Deemed within the Prior Art (AIA § 14) • Amendments to 35 U.S.C. § 112 (AIA § 4) • Ban on Patenting Human Organisms (AIA § 33)

Administrative Matters for the U.S. PTO	Procedural Changes to Patent Prosecution
• Creation of PTAB (AIA § 7) • Authorization of U.S. PTO's Fee-Setting Authority (AIA § 10) • Patent and Trademark Office Funding (AIA § 22) • Satellite Offices (AIA §§ 23 and 24) • Venue Change (AIA § 9) • Establishment of Micro Entity (AIA § 10)	• Transition to a "first-inventor-to-file" patent system (AIA § 3) • Derivation proceedings (AIA § 3) • Prioritized Examination (AIA § 11) • Third-Party Submissions (AIA § 8) • Inventor Oath/Declaration Practices (AIA § 4) • Repeal of the Statutory Invention Registration (AIA § 3) • Calculation of the 60-day period for Patent Term Extensions (AIA § 37)

IV) FREE RESOURCES THAT ARE HELPFUL FOR UNDERSTANDING THE AIA

In addition to the resources cited throughout the following chapters, following is a list of free resources that are particularly helpful for understanding the AIA:

1) **U.S. PTO's AIA webpage**: This page can be found at http://www.uspto. gov/patent/laws-and-regulations/america-invents-act-aia/resources. The

U.S. PTO's AIA webpage has several tabs that are very helpful for learning more about the AIA. The following tabs are particularly useful:

a. **"AIA Resources" tab**: This tab provides links to excerpts of the legislative history for the AIA (e.g., the Congressional Record and Committee Reports on the legislation).[d] In addition, under the "Legislation" heading, the tab provides a document that lists all of the effective dates for each AIA provision. Links to journal articles on the AIA (authored by Bob Armitage and Joe Matal) are listed underneath the "Miscellaneous" heading. These articles are invaluable for understanding the AIA.

b. **"Patent Examination" tab**: As the name suggests, this tab provides additional resources for those provisions of the AIA that affect patent examination (e.g., the provisions on Prioritized Examination, Preissuance Submissions, Supplemental Examination, and First-Inventor-to-File). In particular, links to the U.S. PTO's proposed and final rules, U.S. PTO forms, memos to Patent Examiners, and/or FAQs are provided for each of the AIA provisions contained within this tab.

c. **"Inter Partes Disputes" tab**: This tab provides resources for learning more about inter partes review, post grant review, transitional covered business method review and derivation proceedings. In particular, the tab provides links to the U.S. PTO's proposed and final rules, Trial Practice Guide, and/or FAQs for each of the aforementioned proceedings.

d. **"AIA Informational Videos" tab**: This tab contains links to helpful PowerPoint presentations that explain the AIA's changes to pre-AIA 35 U.S.C. § 102. Links to the PowerPoint presentations are listed under the "AIA Training PowerPoint Presentations" heading.

e. **"AIA Frequently Asked Questions" tab**: This tab allows users to view FAQs by topic (e.g., Best Mode, First Inventor to File, Preissuance Submissions). Review of the questions and answers can be extremely helpful when trying to understand many of the nuances of each AIA provision.

f. **"AIA Roadshow" tab**: This tab contains both a PowerPoint slideset and a video that are helpful for understanding the change from a "first-to-invent" to a "first-inventor-to-file" patent system.

---

[d]Note that for any citation such as "157 CONG. REC. S1333," the material cited can be accessed by clicking on "Congressional Record" at www.congress.gov and then entering the issue number and page. In the aforementioned citation, the issue number is "157" and the page is "S1333."

2) **Legislative History for the AIA**: The following website provides links for most, if not all, of the legislative history for the AIA: http://www.patentreform.info/legislative-history2.htm. Even though the U.S. PTO's AIA webpage provides links to the legislative history for the AIA, it does not appear to provide links to older portions of the legislative history (i.e., portions that date before 2011) that are provided by the aforementioned webpage. Furthermore, the Congressional debates on the AIA and statements about the AIA are scattered all throughout the Congressional Record from 2005 through 2011. Accordingly, this webpage is helpful because it provides more of the legislative history than is provided on the U.S. PTO's website and eliminates the need to search the entire Congressional Record for those portions that pertain to the AIA.

Despite all of the information that is provided in this text, it is important to remember that many provisions of the AIA have yet to be interpreted by the courts. As such, many uncertainties remain with regard to how the new law will be applied even though the AIA was passed over five years ago.

CHAPTER TWO

Transitioning U.S. Patent Law From a "First-to-Invent" to a "First-Inventor-to-File" System Under the AIA

Contents

I) INTRODUCTION

Of the 37 sections of the AIA, section 3 is arguably the most important and complex. Indeed, section 3 shifts the U.S. patent system from a "first-to-invent" to a "first-inventor-to-file" regime for determining patent priority. Leahy–Smith America Invents Act of 2011, Pub. L. No. 112-29, § 3, 125 Stat. 284, 285-293; *see also Madstad Eng'g, Inc. v. U.S. Patent & Trademark Office*, 756 F.3d 1366, 1368 (Fed. Cir. 2014) (citing AIA § 3, 125 Stat. at 285-294) ("On September 16, 2011, the President signed into law the

America Invents Act Primer. http://dx.doi.org/10.1016/B978-0-12-812096-5.00002-8

AIA. The AIA, *inter alia*, adopted the 'first-inventor-to-file' principle for determining priority among patents and patent applications."). In doing so, section 3 amends 35 U.S.C. §§ 102 and 103 so that the novelty and non-obviousness of a claimed invention are evaluated as of the invention's effective filing date instead of its date of invention. Leahy-Smith America Invents Act of 2011, Pub. L. No. 112-29, § 3, 125 Stat. 284, 285-288. Section 3 of the AIA also eliminates interference proceedings and establishes derivation proceedings. *Id.* at 288-293. The shift from the "first-to-invent" to the "first-inventor-to-file" regime, as well as the elimination of interference proceedings and the establishment of derivation proceedings, will be discussed in this chapter. Chapter 3 will then provide a more in-depth discussion of the AIA's amendments to pre-AIA 35 U.S.C. §§ 102 and 103.

II) PRE-AIA LAW: THE "FIRST-TO-INVENT" SYSTEM

To fully grasp the significance of the transition from the "first-to-invent" to the "first-inventor-to-file" patent system, it is helpful to first understand some of the fundamental aspects of the "first-to-invent" system. Put most simply, the "first-to-invent" system awards a patent to the first inventor of a claimed invention, regardless of whether that inventor filed his/her patent application first. By doing this, the "first-to-invent" system places more emphasis on a claimed invention's date of invention than its effective filing date. In certain instances, the "first-to-invent" system even allows patent applicants to rely on the date of invention as the priority date for a claimed invention for purposes of antedating activities or references that have effective prior art dates that precede the claimed invention's effective filing date.

Several U.S. patent practices have been developed to ensure that the first inventor of claimed subject matter is the only person who receives and maintains a patent on that subject matter and to allow patent applicants to establish their date of invention as the priority date for their claimed invention. Included in these practices are interference proceedings, certain patent litigation strategies, and a patent prosecution procedure that can be used by patent applicants to antedate prior art using their date of invention as the priority date for a claimed invention. These practices will be discussed in more detail throughout this section.

A. Interference Proceedings

Interference proceedings are utilized during patent prosecution to ensure that only the first inventor of a claimed invention is awarded a patent on that invention. MPEP § 2301 (9th ed. 2014). Indeed, if the same invention[a] is claimed by different inventive entities in two or more patent applications, or by different inventive entities in a patent and a patent application, an interference-in-fact exists and an interference proceeding can be instituted[b] to determine which inventive entity was the first to invent the commonly claimed invention. *Id.*; *see also* pre-AIA 35 U.S.C. § 135(a) ("Whenever an application is made for a patent which, in the opinion of the Director, would interfere with any pending application, or with any unexpired patent, an interference may be declared . . . ."). Under the "first-to-invent" system, only the inventive entity that is adjudged the first inventor is entitled to pursue a patent on the claimed invention.[c] *See* pre-AIA 35 U.S.C. § 102(g)(1) (explaining that a person is entitled to a patent unless

---

[a]The U.S. PTO and the case law define "same invention" to mean patentably indistinct inventions. MPEP § 2301.03 (9th ed. 2014); *see also Aelony v. Arni*, 547 F.2d 566, 570 (C.C.P.A. 1977) ("Moreover, we believe that there is ample precedent from this court for framing the test of interference in fact in terms of whether two sets of claims are patentably distinct from each other."). In other words, an interference exists if the subject matter of a claim of one party would, if prior art, have anticipated or rendered obvious the subject matter of a claim of the opposing party and vice versa. 37 C.F.R. § 41.203(a). Identical language in claims does not guarantee that they are drawn to the same invention. MPEP § 2301.03 (9th ed. 2014). Instead, every claim must be construed in light of the application in which it appears. 37 C.F.R. § 41.200(b). Thus, even determining whether an interference proceeding should be instituted (i.e., whether a claim of one inventor can be said to interfere with the claim of another inventor) is not a straightforward task.

[b]Before enactment of the AIA, interference proceedings were carried out by the Board of Patent Appeals and Interferences (i.e., the "BPAI"). The AIA, however, replaced the BPAI with the Patent Trial and Appeal Board (i.e., the "PTAB"). *See* Leahy-Smith America Invents Act of 2011, Pub. L. No. 112-29, § 7, 125 Stat. 284, 313-314 (amending pre-AIA 35 U.S.C. § 6 to replace the "Board of Patent Appeals and Interferences" with the "Patent Trial and Appeal Board"). Accordingly, any interference proceedings that are initiated on or after September 16, 2012 (i.e., the effective date of § 7 of the AIA) are heard by the PTAB. *Id.* at § 7(e), 125 Stat. 315.

[c]More specifically, the party who demonstrates that he/she was the first to reduce the claimed subject matter to practice will prevail in an interference proceeding unless the other party demonstrates that: (1) he/she was the first to conceive of the claimed subject matter and (2) he/she acted diligently to reduce that subject matter to practice from a time prior to when the first reducer to practice entered the field through the time that the first reducer to practice actually reduced the claimed subject matter to practice. MPEP § 2138.01(I) (9th ed. 2014). MPEP § 2138.01(II) (9th ed. 2014) depicts timelines that are particularly helpful for understanding which party will prevail in an interference proceeding. For the sake of brevity, the criteria for establishing diligence in reducing subject matter to practice is not discussed in this text. MPEP § 2138.06 (9th ed. 2014), however, gives an explanation of what constitutes diligence for purposes of establishing priority of invention.

"during the course of an interference conducted under [35 U.S.C. § 135] or [35 U.S.C. § 291], another inventor involved therein establishes, to the extent permitted in [35 U.S.C. § 104], that before such person's invention thereof the invention was made by such other inventor and not abandoned, suppressed, or concealed . . . .").

Determining who first invented claimed subject matter (i.e., who has "priority of invention") is not a straightforward task. Priority of invention is typically awarded to the first person to reduce a claimed invention to practice. *See Mahurkar v. C.R. Bard, Inc.*, 79 F.3d 1572, 1577 (Fed. Cir. 1996) (citing *Christie v. Seybold,* 55 F. 69, 76 (6th Cir. 1893)) ("In the United States, the person who first reduces an invention to practice is 'prima facie the first and true inventor.'"). The person who first conceives, and, in a mental sense, first invents, however, may date his patentable invention back to the time of its conception, if he connects the conception with its reduction to practice by reasonable diligence on his part, so that they are substantially one continuous act. *Id.*

Conception occurs when an "idea is so clearly defined in the inventor's mind that only ordinary skill would be necessary to reduce the invention to practice, without extensive research or experimentation," and the "inventor has a specific, settled idea, a particular solution to the problem at hand, not just a general goal or research plan he hopes to pursue." *Burroughs Wellcome Co. v. Barr Labs., Inc.*, 40 F.3d 1223, 1228 (Fed. Cir. 1994) (citation omitted). In other words, conception of an invention does not occur until an inventor has progressed beyond identifying a problem and setting out a research plan to solve that problem. Moreover, reduction to practice, which typically occurs after an invention has been conceived, can be accomplished in two different ways: (1) an invention may be <u>constructively</u> reduced to practice, or (2) an invention may be <u>actually</u> reduced to practice. MPEP § 2138.05 (9th ed. 2014). Constructive reduction to practice occurs when an inventor files a patent application that claims his/her invention and that patent application complies with the written description and enablement requirements of pre-AIA 35 U.S.C. § 112, ¶ 1. *Hyatt v. Boone,* 146 F.3d 1348, 1352 (Fed. Cir. 1998). Actual reduction to practice occurs when a party has constructed an embodiment of the invention, or performed a process that includes all elements of the claimed inventive process, and the embodiment or process operates for its intended purpose. *Eaton v. Evans,* 204 F.3d 1094, 1097 (Fed. Cir. 2000).

While the concepts of conception and reduction to practice of an invention may seem straightforward, proving when each of these events occurred for any given invention can be extremely challenging.[d] Indeed, proving when conception of a claimed invention occurred can be very difficult because conception is a purely mental exercise. As such, inventor testimony is usually required to evaluate when an inventor thought of certain elements of his invention and whether the inventor recognized and appreciated that what he had thought of was indeed inventive. *See Silvestri v. Grant,* 496 F.2d 593, 597 (C.C.P.A. 1974) ("[A]n accidental and unappreciated duplication of an invention does not defeat the patent right of one who, though later in time was the first to recognize that which constitutes the inventive subject matter."). Inventor testimony alone is not sufficient, however, to establish a date of conception for a given invention. Instead, proof of conception requires that the inventor also provides corroborating evidence that demonstrates that he/she somehow disclosed the invention to the extent necessary for those of ordinary skill to make that invention. *See Price v. Symsek,* 988 F.2d 1187, 1194 (Fed. Cir. 1993) ("As indicated above, an inventor's testimony, standing alone, is insufficient to prove conception—some form of corroboration must be shown.") (citation omitted). Corroborating evidence can come in the form of invention records (e.g., lab notebooks signed by a witness), presentation materials, or journal submissions. Because corroborating evidence is often required to establish the date of conception for an invention, most patent attorneys and patent agents advise their clients to maintain invention records for several years beyond the date on which their clients apply for a patent on any subject matter that is disclosed in such records. Maintenance of these records can, however, be incredibly burdensome for inventors and their employers, especially in cases where a company goes through a merger or dissolves.

Proof of either constructive or actual reduction to practice is also no easy feat. With regard to constructive reduction to practice, determining whether a patent application provides adequate written description and/or enablement for a claimed invention under pre-AIA 35 U.S.C. § 112, ¶ 1 can be extremely difficult and burdensome, particularly in the unpredictable arts of chemistry and biotechnology. *See, e.g.,* Margaret Sampson, *The Evolution*

---

[d]The act of filing a patent application can be sufficient to demonstrate that both conception and reduction to practice have occurred for the subject matter disclosed therein. *Hyatt,* 146 F.3d at 1352 (citation omitted). Proof of the date of invention for claimed subject matter is therefore not as challenging in instances where an applicant seeks to rely on the effective filing date of a patent application as the date on which conception and reduction to practice occurred.

of the Enablement and Written Description Requirements Under 35 U.S.C. § 112 in the Area of Biotechnology, 15 BERKELEY TECH. L.J. 1232, 1233 (2000) ("This Comment focuses on the current trend of the Federal Circuit to heighten both the enablement and written description requirements for biotechnological inventions under 35 U.S.C. § 112."). Expert testimony is often needed to evaluate whether a person of ordinary skill in the art would have been able to make and use a claimed invention based on what is disclosed in the specification of a patent application (i.e., to evaluate whether the claimed invention is enabled). Such testimony is also often needed to establish whether the inventor(s) listed on a patent application had possession of the claimed invention at the time the patent application was filed (i.e., whether there is sufficient written description for a claimed invention). And proving actual reduction to practice is just as difficult, if not more difficult, than proving that an invention was constructively reduced to practice. *See, e.g., Scott v. Finney,* 34 F.3d 1058, 1061-62 (Fed. Cir. 1994) (citing numerous cases that demonstrate that the stringency of testing required to show reduction to practice of an invention is directly proportional to the complexity of the invention and the problem it solved); MPEP § 2138.05(II) (9th ed. 2014) ("For an actual reduction to practice, the invention must have been sufficiently tested to demonstrate that it will work for its intended purpose, but it need not be in a commercially satisfactory stage of development.").

Finally, even if a party demonstrates that he/she was the first to invent claimed subject matter, he/she will not prevail in an interference proceeding if his/her opponent demonstrates that the claimed invention was abandoned, suppressed, and/or concealed before a patent application claiming the invention at issue was filed. Pre-AIA 35 U.S.C. § 102(g)(1). U.S. patent law recognizes two types of abandonment, suppression and concealment. The first type involves cases in which the inventor <u>deliberately</u> abandons, suppresses, or conceals his invention, and the second type involves cases in which a <u>legal inference</u> of abandonment, suppression, or concealment is drawn based on "too long" of a delay in filing a patent application. *Paulik v. Rizkalla,* 760 F.2d 1270, 1273 (Fed. Cir. 1985) (en banc).

Determining whether either type of abandonment, suppression, or concealment occurred before a patent application was filed is a formidable task. Indeed, demonstrating that an inventor deliberately abandoned, suppressed, and/or concealed his invention requires a showing of specific intent. *Peeler v. Miller,* 535 F.2d 647, 653-54 (C.C.P.A. 1976). There are also no bright line rules for determining when an inventor has taken too long to file a patent application such that a legal inference of abandonment, suppression, or concealment should be drawn. Instead, courts look at the facts and circumstances surrounding the first inventor's delay, as well as the reasonableness of the delay in filing a patent application, to determine whether such legal inference should be drawn. *See, e.g., Fujikawa v. Wattanasin*, 93 F.3d 1559, 1568 (Fed. Cir. 1996) (quoting *Young v. Dworkin*, 489 F.2d 1277, 1285 (C.C.P.A. 1974)) (explaining that the Federal Circuit "has not set strict time limits regarding the minimum and maximum periods necessary to establish an inference of suppression or concealment" but instead has "recognized that 'it is not the time elapsed that is the controlling factor but the total conduct of the first inventor.'").

Due to the complexities involved in proving conception, reduction to practice, abandonment, suppression, and concealment, interference proceedings are extremely time consuming, costly, and burdensome for the parties involved. *See* 157 CONG. REC. S1090, 2011 (statement of Sen. Leahy) ("In the outdated, current system, when more than one application claiming the same invention is filed, the priority of a right to a patent is decided through an 'interference' proceeding to determine which applicant can be declared to have invented the claimed invention first. This process is lengthy, complex, and can cost hundreds of thousands of dollars. Small inventors rarely, if ever, win interference proceedings."). In fact, one of the longest patent disputes in U.S. history involved an interference proceeding. *See Bard Peripheral Vascular, Inc. v. W.L. Gore & Associates, Inc.*, 670 F.3d 1171, 1175, 1177 (Fed. Cir. 2012) (noting that the present case had commenced in 1974 and that the patent at-issue in the case was previously the subject of an interference proceeding). Additionally, parties have little certainty in predicting who will prevail in an interference proceeding because the evidence used to establish dates of conception and reduction to practice (e.g., lab notebooks, invention disclosures, presentation materials) is rarely available to the public and can therefore only be discovered once the interference proceeding is well underway. Moreover, because the U.S. PTO will not grant a patent until the priority of invention is resolved, patent applicants who are involved in interference proceedings endure much longer delays in obtaining their patent rights than those who are not involved in such proceedings.

B. Patent Litigation Strategies That Are Unique to the "First-to-Invent" Patent System

In addition to interference proceedings, U.S. patent law provides certain litigation strategies that are unique to the "first-to-invent" patent system. Specifically, pre-AIA 35 U.S.C. § 291 allows the owner of a patent to file a civil suit against the owner of an allegedly interfering patent so that a federal court can evaluate the validity of such a patent. In addition, pre-AIA 35

U.S.C. § 102(g)(2) allows an alleged infringer to invalidate claims of a patent if he/she can demonstrate that before the patentee invented his/her claimed invention, "the invention was made in this country by another inventor who had not abandoned, suppressed, or concealed it." In other words, pre-AIA 35 U.S.C. § 102(g)(2) enables an alleged infringer to invalidate patent claims if he/she can show that the patentee was not the first inventor of the claim(s) asserted against him/her. These litigation strategies help ensure that only those who are the first inventors of claimed subject matter maintain patents on that subject matter.

A party who brings suit under pre-AIA 35 U.S.C. § 291, or raises a defense under pre-AIA 35 U.S.C. § 102(g)(2), must make the same showings as parties who are involved in an interference proceeding under pre-AIA 35 U.S.C. § 102(g)(1).[e] That is, in order to demonstrate that at least one of the claims of the patent is invalid, a party must demonstrate that the inventor(s) listed on the patent was/were not the first to invent the subject matter claimed therein by showing that either: (1) another party reduced the invention to practice first; or (2) another party was the first to conceive of the claimed subject matter and that other party acted diligently to reduce that subject matter to practice from a time prior to when the first reducer to practice entered the field through the time that the first reducer to practice actually reduced the claimed subject matter to practice. *See Solvay S.A. v. Honeywell Int'l Inc.*, 742 F.3d 998, 1000 (Fed. Cir. 2014) ("[35 U.S.C. § 102(g)(2)] provides, '[a] person shall be entitled to a patent unless . . . before such person's

---

[e]The main difference between challenging a patent under pre-AIA 35 U.S.C. § 291 as opposed to under pre-AIA 35 U.S.C. § 102(g)(2) is that, in certain instances, the burden for demonstrating invalidity under the former statute is lower than for demonstrating invalidity under the latter statute. In particular, a party to a suit brought under pre-AIA 35 U.S.C. § 291 need only prove priority of invention by a preponderance of the evidence if the patent applications upon which the patents involved in the proceeding were based were all co-pending when they were being examined by the U.S. PTO. If the allegedly interfering patent issued before the filing date of the application upon which the claimant's patent is based, however, the claimant in a suit filed under pre-AIA 35 U.S.C. § 291 must establish priority of invention by clear and convincing evidence. *Eli Lilly & Co. v. Aradigm Corp.*, 376 F.3d 1352, 1365 (Fed. Cir. 2004). In contrast to actions brought under pre-AIA 35 U.S.C. § 291, invalidity defenses brought under pre-AIA 35 U.S.C. § 102(g)(2) must always be established by clear and convincing evidence. *See Apotex USA, Inc. v. Merck & Co., Inc.*, 254 F.3d 1031, 1036 (Fed. Cir. 2001) ("[35 U.S.C. §] 282 applies with full force to a § 102(g) defense, and thus a party asserting invalidity under § 102(g) must prove facts by clear and convincing evidence establishing a prior invention that was not abandoned, suppressed, or concealed.").

invention thereof, the invention was made in this country by another inventor who had not abandoned, suppressed, or concealed it...' A patent is invalid under that section if the claimed invention was made in this country by another inventor before the patent's priority date."); *Medichem, S.A. v. Rolabo, S.L.*, 437 F.3d 1157, 1163 (Fed. Cir. 2006) ("As an aside, we wish to note that in parallel with the district court proceedings under 35 U.S.C. § 291, the Board of Patent Appeals and Interferences ('Board') has been considering essentially the same interference and priority issues pursuant to 35 U.S.C. § 135."). Accordingly, civil suits and defenses brought under pre-AIA 35 U.S.C. §§ 291 and 102(g)(2), respectively, can be just as complex, uncertain, costly, and burdensome as interference proceedings brought in the U.S. PTO.

C. Utilizing the Date of Invention as the Priority Date for a Claimed Invention

The last practice that is unique to the "first-to-invent" system concerns the procedure by which an inventor may rely on his/her date of invention as the priority date for claimed subject matter in order to antedate the effective prior art date of certain references or activities. For example, if an inventor's claims are rejected during prosecution[f] for lacking novelty,

---

[f]While this section focuses on the procedure for "swearing behind" references or activities during patent prosecution, it is important to understand that patent owners involved in litigation may also rely on their date of invention as the priority date for their claims if those claims are attacked using references or activities that qualify as prior art under pre-AIA 35 U.S.C. § 102(a) or (e). *See, e.g., Mahurkar,* 79 F.3d at 1577 ("Had Dr. Mahurkar not come forward with evidence of an earlier date of invention, the Cook catalog would have been anticipatory prior art under section 102(a) because Dr. Mahurkar's invention date would have been the filing date of his patent."). Similarly, patent owners may rely on the date of invention as the priority date if activities that qualify as prior art under pre-AIA 35 U.S.C. § 102(g), but that do not qualify as prior art under pre-AIA 35 U.S.C. § 102(b), are used to attack their patents during litigation. Because the date of invention can be utilized as the priority date for a claimed invention for purposes of defeating invalidity attacks based on pre-AIA 35 U.S.C. § 102(a) and (e) prior art, but not for purposes of defeating invalidity attacks based on pre-AIA 35 U.S.C. § 102(b) prior art, alleged infringers are far less likely to rely on pre-AIA 35 U.S.C. § 102(a) and (e) prior art in litigation. Indeed, an alleged infringer usually has no way of knowing the date of invention for any given patent claim until discovery is well underway during litigation. Thus, relying on pre-AIA 35 U.S.C. § 102(a) and (e) prior art can prove futile for an alleged infringer where the patent owner is able to antedate such prior art by relying on a date of invention that is earlier than the patent's effective filing date.

or for being obvious, over references[g] or activities that qualify as prior art only under pre-AIA 35 U.S.C. § 102(a) and/or (e), the inventor can "swear behind" the reference or activity by filing a declaration or affidavit under 37 C.F.R. § 1.131 that establishes that his/her date of invention precedes the effective prior art date of the reference or activity.[h] MPEP § 715(I)(A) and (B) (9th ed. 2014); 37 C.F.R. § 1.131. An inventor may similarly file a 37 C.F.R. § 1.131 declaration or affidavit to swear behind activities[i] that qualify as prior art under pre-AIA 35 U.S.C. § 102(g) (e.g., another person's reduction to practice of subject matter that is claimed in the inventor's patent application) so long as those activities do not also constitute prior art under pre-AIA 35 U.S.C. § 102(b). MPEP § 715(I)(C) (9th ed. 2014).

A declaration or affidavit filed under 37 C.F.R. § 1.131 must allege facts that are sufficient to establish that the claimed invention was reduced to practice before the effective prior art date of the reference or activity being cited as prior art against the claimed invention. 37 C.F.R. § 1.131(b). If prior reduction to practice cannot be established, the declaration or affidavit may alternatively allege facts that are sufficient to show that: (1) the invention was conceived before the effective prior art date of the reference or activity, and (2) the inventor was diligent in trying to reduce the invention to practice from a time period beginning on a date that is before the effective prior art date of the reference or activity and extending through

---

[g]This procedure for antedating references cannot be used to antedate a patent or patent application that allegedly claims the same invention that is claimed in another patent or patent application. MPEP § 715(II)(C) (9th ed. 2014). Indeed, only an interference proceeding can resolve priority of invention issues for two patent applications, or a patent application and an unexpired patent, that claim the same invention. MPEP § 715.05 (9th ed. 2014).

[h]Patent applicants cannot utilize declarations or affidavits under 37 C.F.R. § 1.131 to obviate lack of novelty or obviousness rejections that are based on activities or references that qualify as prior art under pre-AIA 35 U.S.C. § 102(b), (d), or (f). See MPEP § 715(I) and (II) (9th ed. 2014) (listing situations in which declarations and affidavits under 37 C.F.R. § 1.131 can and cannot be used).

[i]As mentioned previously, inventors cannot use declarations or affidavits filed under 37 C.F.R. § 1.131 to swear behind references (i.e., patent applications and patents) that qualify as prior art under pre-AIA 35 U.S.C. § 102(g) because only references that claim the same invention as another patent or patent application qualify as prior art under that statute and the only way to establish priority of invention over such references is through an interference proceeding. Pre-AIA 35 U.S.C. § 102(g)(1); MPEP §§ 715(II)(C) and 715.05 (9th ed. 2014).

the date that the invention was actually reduced to practice or the date on which the patent application at-issue was filed. *Id.* In other words, 37 C.F.R. § 1.131 declarations or affidavits must establish priority of invention by making the same showings that are required in an interference proceeding. The following timelines provide examples of fact scenarios that, if properly alleged in an affidavit or declaration filed under 37 C.F.R. § 1.131, would successfully antedate the aforementioned references or activities that would otherwise qualify as prior art under pre-AIA 35 U.S.C. § 102(a), (e), and/or (g).

Timeline #1: Antedating a Reference or Activity Based on Prior Reduction to Practice

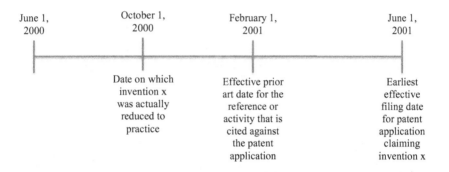

Timeline #2: Antedating a Reference or Activity Based on Prior Conception Coupled with Diligence in Reducing the Invention to Practice

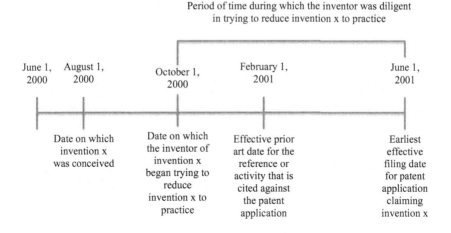

III) AIA § 3(i): REPLACEMENT OF INTERFERENCE PROCEEDINGS WITH DERIVATION PROCEEDINGS UNDER THE "FIRST-INVENTOR-TO-FILE" SYSTEM

The "first-inventor-to-file" system is fundamentally different from the "first-to-invent" patent system. For example, unlike the "first-to-invent" system, which generally awards a patent to the first person to invent a claimed invention, the "first-inventor-to-file" system generally awards a patent to the first inventor to file a patent application. By doing this, the "first-inventor-to-file" system places more emphasis on a claimed invention's effective filing date than its date of invention. Additionally, unlike the "first-to-invent" system, which allows patent applicants to rely on their date of invention as the priority date for their claimed invention for purposes of antedating certain activities or references that have effective prior art dates that precede the effective filing date of their patent application, only the effective filing date may be relied on as the priority date for claimed inventions in the "first-inventor-to-file" system. Thus, the practices discussed in Section II, which are only used to determine who invented a claimed invention first, are not relevant in the "first-inventor-to-file" system. *See* Leahy-Smith America Invents Act of 2011, Pub. L. No. 112-29, § 3, 125 Stat. 284, 285-293 (eliminating pre-AIA 35 U.S.C. § 102(g) and amending the U.S. patent statutes, including pre-AIA 35 U.S.C. §§ 135 and 291, to replace any references to interference proceedings with references to derivation proceedings). The following examples illustrate some of the differences between the "first-to-invent" and "first-inventor-to-file" systems:

> **Example #1**: Inventor A invents novel compound 1 on December 31, 2009 and files a patent application claiming that compound on January 1, 2010. Inventor B independently invents novel compound 1 on December 23, 2009 but does not file his application claiming that compound until January 29, 2010. Under the "first-to-invent" system, Inventor B would be entitled to the patent because even though inventor B filed his patent application after Inventor A, Inventor B was the first to invent novel compound 1. In contrast, under the "first-inventor-to-file" system, Inventor A would be entitled to the patent because Inventor A was the

first to file his patent application and he did not derive the invention from Inventor B.[j]

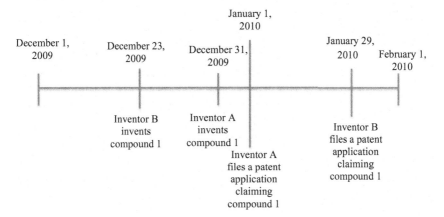

Example #2: Adam Smith invents novel compound 1 on January 1, 2011 and then files his first patent application claiming novel compound 1 on May 15, 2011. A journal article having an effective prior art date of April 20, 2011 discloses compound 1. Joy Jones, who is the sole author listed on the journal article, independently invented novel compound 1 (i.e., Joy did not derive compound 1 from Adam Smith). During examination of Adam Smith's patent application, a Patent Examiner finds Joy Jones' article and rejects Adam Smith's claim to compound 1 for lack of novelty over that article. Under the "first-to-invent" system, Adam Smith could overcome this rejection by filing a declaration or affidavit under 37 C.F.R. § 1.131 that establishes that he invented compound 1 on January 1, 2011, which is a date that precedes the effective prior

[j]Note that in this example, the first filer that was entitled to a patent under the "first-inventor-to-file" system was the *inventor* of the claimed invention and not just the first *person* to file the patent application. The legislative history for what became section 3 of the AIA makes it clear that the U.S. patent system was shifting to a "first-inventor-to-file" system and not merely a "first-person-to-file" system. *See, e.g.*, 157 Cong. Rec. S1092, 2011 (quoting a letter from the Intellectual Property Law Section of the American Bar Association) ("This current legislation, however, makes it clear that the award [of a patent] goes to the first inventor to file and not merely to the first person to file.").

art date of the journal article. By doing this, Adam Smith would "swear behind" the journal article such that it would no longer qualify as prior art under pre-AIA 35 U.S.C. § 102(a). Under the "first-inventor-to-file" system, however, Adam Smith could not rely on his date of invention to antedate Joy Jones' journal article. Accordingly, Joy Jones' journal article would prevent Adam Smith from obtaining a patent on compound 1.

The remainder of this section will discuss derivation proceedings, which replace interference proceedings in the post-AIA "first-inventor-to-file" patent system.

A. Derivation Proceedings

Notwithstanding the major differences between the "first-to-invent" and the "first-inventor-to-file" systems, one thing they both have in common is that under either system, only a person who *invented* claimed subject matter is entitled to a patent on that subject matter. Indeed, under the "first-to-invent" law, a claim could be invalidated during litigation, or rejected during prosecution, if an alleged infringer or Patent Examiner demonstrated that someone other than the inventor(s) listed on the patent or patent application was the true inventor of the subject matter claimed therein. *See* pre-AIA 35 U.S.C. § 102(f) (explaining that a person shall be entitled to a patent unless "he did not himself invent the subject matter sought to be patented . . . ."). Although pre-AIA 35 U.S.C. § 102(f) was eliminated by the AIA, the principle that only a true inventor is entitled to a patent remains true under the new law. *See, e.g.*, post-AIA 35 U.S.C. § 101 ("*Whoever* invents or discovers . . . may obtain a patent therefor, subject to the conditions and requirements of this title.") (emphasis added); Joseph D. Matal, *A Guide to the Legislative History of the America Invents Act: Part I of II*, 21 Fed. Cir. B.J. 435, 451 (2012) ("Some may think that, because § 102(f) has been repealed, there is no longer any legal requirement that a patent for an invention be obtained by the inventor. Not so. Both the Constitution and § 101 still specify that a patent may only be obtained by the person who engages in the act of inventing.").

In fact, the AIA established derivation proceedings to ensure that the first person to file a patent application is actually a true inventor of the claimed invention and to ensure that a person will not be able to obtain or maintain a patent for an invention that he did not actually invent. Post-AIA 35 U.S.C. §§ 135 and 291; 77 Fed. Reg. 56,068 (Sept. 11, 2012). Two types of derivation proceedings exist under the post-AIA law: one that can be brought at the U.S. PTO and one that can be brought in a district court. The following tables provide an overview of these proceedings

Type of Derivation Proceeding	Statutory Basis for Derivation Proceeding	How Derivation Proceeding is Initiated
Derivation Proceedings Brought in U.S. PTO	Post-AIA 35 U.S.C. § 135[k]	A patent applicant may file a petition to institute a derivation proceeding in the U.S. PTO. If the Director of the U.S. PTO determines that the petition demonstrates that the standards for instituting a derivation proceeding are met, the Director may institute such a proceeding. (Post-AIA 35 U.S.C. § 135(a)(1))[l]
Derivation Proceeding Brought in Federal District Court	Post-AIA 35 U.S.C. § 291[m]	The owner of a patent may file a civil complaint in a federal district court against the owner of another patent that claims the same invention[n] and has an earlier effective filing date, if the invention claimed in the other patent was derived from the inventor of the invention claimed in the patent owned by the person seeking relief. (Post-AIA 35 U.S.C. § 291(a))

[k]Before the AIA was passed, 35 U.S.C. § 135 was the statute that governed interference proceedings. As discussed earlier, interference proceedings were eliminated with the passage of the AIA. As such, post-AIA 35 U.S.C. § 135 sets forth the statutory basis for derivation proceedings.

[l]Note that post-AIA 35 U.S.C. § 135(a), as passed in AIA § 3(i), was amended by the Leahy-Smith America Invents Technical Corrections Act, which was passed on January 14, 2013. Pub. L. No. 112-274, § 1(k)(1), 126 Stat. 2456, 2457-58. The amendments to post-AIA 35 U.S.C. § 135(a) clarify the timing by which one must file a petition to institute a derivation proceeding and provide a definition of the term "earlier application" as used in post-AIA 35 U.S.C. § 135(a)(1). Id. The amendments to post-AIA 35 U.S.C. § 135(a) have the same effective date as AIA § 3(i), i.e., March 16, 2013. Id. at § 1(k)(2).

[m]Before the AIA was passed, 35 U.S.C. § 291 was the statute that governed interference actions brought in federal district courts. As discussed earlier, the AIA eliminated these types of civil suits. As a result, post-AIA 35 U.S.C. § 291 sets forth the statutory basis for bringing derivation complaints in federal district courts.

[n]The term "same invention" in post-AIA 35 U.S.C. § 291 has not yet been construed by the courts, but the U.S. PTO has interpreted the phrase to mean a "patentably indistinct invention." MPEP § 2301.03 (9th ed. 2014).

Type of Derivation Proceeding	Deadline for Bringing Derivation Proceeding	Opportunities for Appeal/ Review of PTAB's Decision
Derivation Proceedings Brought in U.S. PTO	The petition to institute a derivation proceeding may only be filed "during the 1-year period following the date on which the patent containing [a claim that is the same or substantially the same° as the petitioner's claim] was granted or the earlier application containing such claim was published, whichever is earlier." (Post-AIA 35 U.S.C. § 135(a)(2))	(1) The determination of whether to institute a derivation proceeding is final and cannot be appealed (Post-AIA 35 U.S.C. § 135(a)). (2) A party to a derivation proceeding may bring a civil action in the Eastern District of Virginia if that party is dissatisfied with the decision by the Patent Trial and Appeal Board. (Post-AIA 35 U.S.C. § 146). (3) Alternatively, a party may appeal the Patent Trial and Appeal Board's final decision in a derivation proceeding to the U.S. Court of Appeals for the Federal Circuit. Such appeal will be dismissed, however, if a party to the derivation proceeding files a civil action under post-AIA 35 U.S.C. § 146. (Post-AIA 35 U.S.C. § 141(d)).
Derivation Proceeding Brought in Federal District Court	An action must be filed "before the end of the 1-year period beginning on the date of the issuance of the first patent containing a claim to the allegedly derived invention and naming an individual alleged to have derived such invention as the inventor or joint inventor." (Post-AIA 35 U.S.C. § 291(b))	Parties to a derivation action in federal district court can appeal the district court's decision to the U.S. Court of Appeals for the Federal Circuit. (28 U.S.C. § 1295(a)(1))

°The phrase "same or substantially the same invention" has been construed by the U.S. PTO to mean "patentably indistinct." 77 Fed. Reg. 56,070 (Sept. 11, 2012). Moreover, because the phrase "same or substantially the same" was interpreted as meaning "patentably indistinct" in pre-AIA 35 U.S.C. § 135(b), it is likely the courts will interpret that phrase to have the same meaning in post-AIA 35 U.S.C. § 135(a)(2). *See Medichem, S.A. v. Rolabo, S.L.*, 353 F.3d 928, 934 (Fed. Cir. 2003) ("Though PTO regulations do not bind a district court, a district court defines 'the same or substantially the same subject matter' in the same manner as would the PTO under its own regulations — by using the two-way test."); 37 C.F.R. § 41.203 (explaining that under the two-way test, "[a]n interference exists if the subject matter of a claim of one party would, if prior art, have anticipated or rendered obvious the subject matter of a claim of the opposing party and vice versa."); *see also Midlantic Nat'l Bank* ▶

	Derivation Proceedings Brought in U.S. PTO	Derivation Proceedings Brought in District Court

Requirements for Bringing Derivation Proceeding	(1) An inventor seeking a derivation proceeding must file a patent application.[p]
	(2) The petition must show that the petitioner has at least one claim that is:
	  (i) the same or substantially the same as the respondent's claimed invention; and
	  (ii) the same or substantially the same as the invention disclosed to the respondent.
	(3) The petition must provide sufficient information to identify the application or patent for which the petitioner seeks a derivation proceeding.
	(4) The petition must demonstrate that a claimed invention was derived from an inventor named in the petitioner's application, and that the inventor from whom the invention was derived did not authorize the filing of the earliest application claiming such invention.
	(5) For each of the respondent's claims to the derived invention, the petition must:
	  (i) show why the claimed invention is the same or substantially the same as the invention disclosed to the respondent, and
	  (ii) identify how the claim is to be construed.
	(6) The petition must be supported by substantial evidence.[q]
	(Post-AIA 35 U.S.C. § 135(a)(1); 37 C.F.R. § 42.405; 37 C.F.R. § 42.402).
		Post-AIA 35 U.S.C. § 291 sets forth the following requirements for bringing a derivation action in federal court:
		(1) The patent belonging to the alleged deriver must claim the same invention as the patent belonging to the complainant;
		(2) the patent belonging to the alleged deriver must have an effective filing date that is earlier than the effective filing date of the complainant's patent; and
		(3) the invention claimed in the alleged deriver's patent must have been derived from the complainant.
		(Post-AIA 35 U.S.C. § 291(a))

◄ *v. New Jersey Dep't of Envt'l Protection*, 474 U.S. 494, 501 (1986) (citing *Edmonds v. Compagnie Generale Transatlantique*, 443 U.S. 256, 266-67 (1979)) ("The normal rule of statutory construction is that if Congress intends for legislation to change the interpretation of a judicially created concept, it makes that intent specific."); *Atlantic Cleaners & Dyers, Inc. v. U.S.*, 286 U.S. 427, 433 (1932) ("Undoubtedly, there is a natural presumption that identical words used in different parts of the same act are intended to have the same meaning.").

[p]Note that an inventor may copy an alleged deriver's application, make any changes to reflect accurately what the inventor invented, and provoke a derivation proceeding by filing a timely petition and fee. 77 Fed. Reg. 56,069 (Sept. 11, 2012).

[q]37 C.F.R. § 42.405(c) also provides that the derivation showing is not sufficient unless it is supported by at least one affidavit addressing communication of the invention to the alleged deriver and the alleged deriver's filing of a patent application without authorization.

	Derivation Proceedings Brought in U.S. PTO	Derivation Proceedings Brought in District Court
Determinations that Can be Made During the Derivation Proceeding	(1) Whether an inventor named in the earlier application derived the claimed invention from an inventor named in the petitioner's application; (2) whether an inventor named in the earlier application filed the earlier application without authorization from an inventor named in the petitioner's application; and (3) patentability issues (e.g., whether a claim is indefinite) where there is good cause to do so. (Post-AIA 35 U.S.C. § 135(b); 37 C.F.R. § 42.400(b)).	(1) Whether the patents involved in the derivation action claim "the same invention"; (2) whether the patent belonging to the alleged deriver has an earlier effective filing date than the complainant's patent; (3) whether the invention claimed in the alleged deriver's patent was derived from the complainant; and (4) whether the alleged deriver filed his/her patent without authorization from the inventor(s) named in the complainant's patent. (Post-AIA 35 U.S.C. § 291(a)).[r]
Remedies that Can Be Given During the Derivation Proceeding	(1) The PTAB can correct the naming of the inventor in any patent application or patent that is at issue in a derivation proceeding; (2) the PTAB can finally reject claims in a patent application where the claims in the patent application have been found to have been derived from another inventor; or (3) the PTAB can cancel claims in a patent where the claims of that patent have been found to have been derived from another inventor.[s] (Post-AIA 35 U.S.C. § 135(b) and (d)).	(1) a district court can invalidate any claim of a patent that has been found to have been derived from another inventor. (Post-AIA 35 U.S.C. § 291(a))

[r]Although not explicitly addressed in the statute, a district court could presumably construe the claims and determine patentability issues where a conclusion on such issues is necessary for determining whether the claimed invention in one patent was derived from one other than the named inventor(s). *See, e.g., Medichem, S.A.*, 353 F.3d at 933-34 (construing claims that were alleged to interfere with one another, and determining whether the claims were novel and obvious over one another, in order to determine whether the "same invention" was being claimed in two different patents).

[s]The cancellation of patent claims occurs only once no appeal or other review of the PTAB's decision is taken or had. Post-AIA 35 U.S.C. § 135(d).

Derivation proceedings are conducted by the PTAB and federal district courts to determine whether: (1) an inventor named in an earlier application, or in a patent having an earlier effective filing date, derived the claimed invention from an inventor who files a petition to institute a derivation proceeding in the U.S. PTO pursuant to post-AIA 35 U.S.C. § 135 or derived the claimed invention from a patent owner who files a civil complaint in a federal district court under post-AIA 35 U.S.C. § 291, and (2) the earlier application, or patent having an earlier effective filing date, claiming such invention was filed without authorization. If the U.S. PTO finds that someone derived an invention claimed in a patent or patent application from someone else, the U.S. PTO can: (1) correct the naming of the inventor(s) on the patent or patent application, (2) finally refuse patent application claims that are directed to the derived invention, or (3) cancel claims in a patent that are directed to the derived invention. Post-AIA 35 U.S.C. § 135(b) and (d). If a district court finds that the claims of an issued patent were derived from another inventor, the district court can invalidate those claims. Post-AIA 35 U.S.C. § 291(a).

Among other things, inventors seeking to institute a derivation proceeding in the U.S. PTO must demonstrate in their petition that: (1) the application or patent for which the derivation is being brought has at least one claim that is the same or substantially the same as the invention disclosed to the alleged deriver; (2) the inventor from whom the claimed invention was allegedly derived did not authorize the filing of the earlier patent application or patent claiming the derived invention; and (3) there is a basis for finding that an inventor named in an earlier patent application derived the claimed invention from an inventor named in the petition and, without authorization, filed the earlier application claiming such invention. 37 C.F.R. § 42.405(a) and (b). Similarly, a patent owner who brings a derivation action in district court must allege in the complaint that: (1) the patent belonging to the alleged deriver claims the same invention as the patent belonging to the complainant; (2) the patent belonging to the alleged deriver has an effective filing date that is earlier than the effective filing date of the complainant's patent; and (3) the invention claimed in the alleged deriver's patent was derived from the complainant. Post-AIA 35 U.S.C. § 291(a).

Patent applicants and patent owners alike have a very short window of time in which they are permitted to bring a derivation proceeding in either the U.S. PTO or in a district court. In fact, petitions to institute derivation proceedings in the U.S. PTO must be filed "during the 1-year period following the date on which the patent containing [a claim that is the same or substantially the same as the petitioner's claimed invention] was granted or the earlier application containing such claim was published, whichever is earlier."

Post-AIA 35 U.S.C. § 135(a)(2). In other words, the petition to institute a derivation proceeding must be filed within one year of the earlier of: (1) the issuance of an alleged deriver's patent or (2) the publication of the alleged deriver's application containing the allegedly derived claim(s), i.e., not within one year of when the claim of the party seeking to institute a derivation proceeding is published. 77 Fed. Reg. 56,070 (Sept. 11, 2012). Furthermore, a derivation action must be filed in a district court "before the end of the 1-year period beginning on the date of the issuance of the first patent containing a claim to the allegedly derived invention and naming an individual alleged to have derived such invention as the inventor or joint inventor." Post-AIA 35 U.S.C. § 291(b). That is, complainants have one year from the date on which an alleged deriver's patent issues to file suit to resolve the derivation issue in a district court.

Interestingly, even if the requirements for a petition to institute a derivation proceeding are met, the U.S. PTO is not required to institute a derivation proceeding. Indeed, post-AIA 35 U.S.C. § 135(a)(1) states that "[w]henever the Director determines that a [petition to institute a derivation proceeding] demonstrates that the standards for instituting a derivation proceeding are met, the Director *may* institute a derivation proceeding." (Emphasis added.) Furthermore, the U.S. PTO has already indicated that a derivation proceeding is unlikely to be instituted, even where the Director thinks the standard for instituting a derivation proceeding is met, if the petitioner's claim is not otherwise in condition for allowance. 77 Fed. Reg. 56,073 (Sept. 11, 2012). And the PTAB may decline to institute or continue a derivation proceeding where the patent applications, or patent and patent application, at-issue are commonly owned. 37 C.F.R. § 42.411. In these instances, the owner of the two applications, or the patent and application, is in the best position to resolve the derivation issue.

Although the AIA created a new derivation proceeding in the U.S. PTO, and amended pre-AIA 35 U.S.C. § 291 to allow patent owners to bring derivation complaints in federal district courts, the new legislation is unlikely to have any significant impact on U.S. patent system because derivation issues are unlikely to arise very often. In fact, according to one report, only six derivation petitions had been filed in the PTAB between March 16, 2013 (i.e., the date derivation proceedings became available) and November 24, 2014. Jones Day, First PTAB Decisions in Derivation Proceedings, http://www.jonesday.com/First-PTAB-Decisions-in-Derivation-Proceedings-12-05-2014/?RSS(true#_ftn1 (last visited July 27, 2016); *see also* Finnegan AIA Blog, Derivation in 2015: Still No Instituted Proceedings, http://www.aiablog.com/aia-did-you-know/derivation-proceedings-aia-did-you-know/derivation-in-2015-still-no-instituted-proceedings/ (explaining that only two petitions to institute a derivation proceeding were filed in 2015).

B. Comparison of Derivation and Interference Proceedings[t]

While interference and derivation proceedings may involve some of the same legal issues and evidentiary proofs, the two proceedings arise from the "first-to-invent" and "first-inventor-to-file" systems, respectfully, and are thus fundamentally different. Indeed, the most significant difference between the two proceedings is that an interference proceeding can be brought to determine who has priority of invention even if two inventors claim to have invented claimed subject matter independently, whereas a derivation proceeding can only be brought if there is a basis for finding that an inventor of an earlier-filed application derived an invention from the inventor named in the petition to institute the derivation proceeding, i.e., an inventor of a later-filed application. *See* post-AIA 35 U.S.C. § 135(a)(1) (explaining that a petition to institute a derivation proceeding "shall set forth with particularity the basis for finding that an individual named in an earlier application as the inventor or a joint inventor derived such invention from an individual named in the petitioner's application as the inventor or a joint inventor and, without authorization, the earlier application claiming such invention was filed."); 77 Fed. Reg. 56,068 (Sept. 11, 2012) ("Derivation proceedings were created to ensure that the first person to file the application is actually a true inventor . . . If a dispute arises as to which of two applicants is a true inventor (as opposed to who invented it first), it will be resolved through a derivation."); Eldora L. Ellison & Robert Greene Sterne, *Use of Interferences to Challenge Patents Before the USPTO*, The National Law Review, July 8, 2010, http://www.natlawreview.com/article/use-interferences-to-challenge-patents-uspto ("Occasionally, two or more inventors independently invent the same invention and file separate patent applications directed to such inventions, creating a dilemma for the [U.S. PTO], which must determine which inventor(s) is entitled to a patent. [1] When faced with two pending patent applications that claim the same invention, or when faced with a pending patent application and an issued patent that claim the same invention, the USPTO utilizes an interference proceeding to determine priority of invention.[2] Thus, interferences are, at their heart, primarily priority contests."). The fact that priority of invention between two independent inventors could be resolved by an interference proceeding under the pre-AIA patent system, but not by a derivation proceeding under the post-AIA patent system, highlights the fundamental difference between the "first-to-invent" and "first-inventor-to-file" patent systems. Indeed, the

---

[t]This section primarily addresses substantive differences between interference and derivation proceedings. Besides the differences mentioned here, there are also many procedural differences between the two proceedings.

"first-to-invent" system is ultimately and primarily concerned with awarding a patent to the *first inventor* to invent something, whereas the "first-inventor-to-file" system is ultimately and primarily concerned with awarding a patent to the *first true inventor to file a patent application* claiming his/her invention.

Because derivation proceedings are not merely concerned with who invented the claimed subject matter first, but are instead concerned with whether an earlier-filing inventor derived his/her claimed invention from a later-filing inventor, they require evidentiary showings that may be inessential to interference proceedings. In fact, a petition to institute a derivation proceeding must demonstrate that a claimed invention was derived from an inventor named in the petitioner's application, and that the inventor from whom the invention was derived did not authorize the filing of the earliest application claiming such invention. Post-AIA 37 C.F.R. § 42.405(b)(2). The petition to institute a derivation proceeding must also be supported by substantial evidence and must include at least one affidavit addressing communication of the derived invention and lack of authorization that, if unrebutted, would support a determination of derivation. Post-AIA 37 C.F.R. § 42.405(c). The showing of communication must also be corroborated. *Id.* No evidence regarding derivation of an invention, communication of an invention to another, or lack of authorization to file a patent application is required in an interference proceeding.

While there are significant differences between interference and derivation proceedings, the two proceedings share some common features. For instance, evidence of when a claimed invention was conceived and reduced to practice, i.e., evidence of the date of invention for claimed subject matter, will be relevant to both proceedings. *See* pre-AIA 35 U.S.C. § 102(g)(1) (explaining that a person is entitled to a patent unless "during the course of an interference conducted under section 135 or section 291, another inventor involved therein establishes, to the extent permitted in section 104, that before such person's invention thereof the invention was made by such other inventor and not abandoned, suppressed, or concealed . . . ."); MPEP § 2138.01(I) (9th ed. 2015) ("Subsection (g) of pre-AIA 35 U.S.C. 102 is the basis of interference practice for determining priority of invention between two parties."); MPEP § 2138 (9th ed. 2015) ("Pre-AIA 35 U.S.C. 102(g) issues such as conception, reduction to practice and diligence, while more commonly applied to interference matters, also arise in other contexts."); 77 Fed. Reg. 56,075 (Sept. 11, 2012) (citing 37 C.F.R. § 42.405(c)) ("Derivation requires both earlier conception by the party alleging derivation as well as communication of the conception. Thus, by requiring demonstration of derivation, the rule necessarily requires a showing of earlier conception as well as corroboration of that earlier conception."). Furthermore, neither an interference proceeding nor a derivation proceeding is likely to be instituted if the particular claims of the patent application that would be involved in such proceeding are not otherwise in condition for allowance. *See* 37 C.F.R. § 41.102 (explaining that an interference should rarely be suggested until examination is complete and there is at least one claim that: (1) is patentable but for a judgment in the contested case, and (2) would be involved in the contested case); 77 Fed. Reg. 56,075 (Sept. 11, 2012) ("A derivation is unlikely to be instituted, even where the Director thinks the standard for instituting a derivation proceeding is met, if the petitioner's claim is not otherwise in condition for allowance.").

Additionally, patent applicants have approximately one year after a claim is either published in a patent application publication or granted in an issued patent to provoke an interference or derivation proceeding. *See* pre-AIA 35 U.S.C. § 135(b)(1) (explaining that for interference purposes, a patent applicant may include in his/her patent application a claim that is the same as, or for the same or substantially the same subject matter as, a claim of an issued patent so long as the claim is included prior to one year from the date on which the patent was granted); pre-AIA 35 U.S.C. § 135(b)(2) (explaining that for interference purposes, a patent applicant may include in his/her patent application a claim that is the same as, or for the same or substantially the same subject matter as, a claim of an application published under section 122(b) so long as the claim is included before 1 year after the date on which the application is published); post-AIA 35 U.S.C. § 135(a)(2) ("A petition under this section with respect to an invention that is the same or substantially the same invention as a claim contained in a patent issued on an earlier application, or contained in an earlier application when published or deemed published under section 122(b), may not be filed unless such petition is filed during the 1-year period following the date on which the patent containing such claim was granted or the earlier application containing such claim was published, whichever is earlier."). Furthermore, the U.S. PTO only has jurisdiction to adjudicate interference and derivation proceedings as between two patent applications or between a patent and a patent application, but not between two issued patents. *See* pre-AIA 35 U.S.C. § 291 ("The owner of an interfering patent may have relief against the owner of another [patent] by civil action, and the court may adjudge the question of validity of any of the interfering patents, in whole or in part."); post-AIA 35 U.S.C. § 291(a) ("The owner of a patent may have relief by civil action against the owner of another patent that claims the same invention and has an earlier effective filing date, if the invention claimed in such other patent was derived from the inventor of the invention claimed in the patent owned by the person seeking relief under this section.").

C. Legislative Goals Addressed by the Establishment of Derivation Proceedings

Derivation proceedings were enacted to further the goal of shifting the U.S. patent laws to a "first-inventor-to-file" system. In earlier versions of the AIA, Congress had drafted the bill to shift the U.S. patent system to a pure "first-to-file" system. *See, e.g.,* 157 CONG. REC. S1096, 2011 (statement of Sen. Boxer) (explaining that "Section 2 of [an earlier version of the AIA] bill awards a patent to the first person to file, regardless of whether that person was the true inventor . . . ."). During the legislative process, however, there was concern that changing the law in this manner would directly conflict with the Constitution, which specifically awards patent rights to "inventors." *Id.* Thus, derivation proceedings, which provide a mechanism for determining the identity of the true inventor of a claimed invention, serve as a safeguard against awarding a patent to a person who is the first to file a patent application but who did not actually invent the subject matter claimed in his/her patent application.

D. Practical Implications of Derivation Proceedings

As mentioned in Section III.A., the chances that any inventor will be involved in a derivation proceeding are extremely low. Indeed, the number of derivation proceedings that are handled by the PTAB each year will likely be on the order of the number of interference proceedings that were handled by the BPAI each year. By one estimate, there were only about 50 such proceedings per year (out of approximately 480,000 patent applications filed) around the time the AIA was passed. 157 CONG. REC. S1095, 2011 (statement of Sen. Feinstein). That said, there are steps that inventors can take to reduce their chances of being involved in a derivation proceeding and to increase their likelihood of prevailing if they are ever involved in a derivation proceeding. These steps include:

    i. **Maintaining Good Research Records**: Inventors should be keeping regular records of what they are doing. The records can be kept in a lab notebook (either a hardcopy lab notebook or an electronic lab notebook). Records should be dated and signed by a witness in case there is ever a dispute as to when the inventor did something and what exactly he/she did.

    ii. **Filing Patent Applications Early**: Once an inventor has developed his/her invention to the point that he/she can meet the written description and enablement requirements of pre-AIA 35 U.S.C. § 112, ¶ 1, he/she should file a patent application without delay.[11] By doing so,

---

[11]Some practitioners advise inventors to file a provisional patent application even before their ideas are fully developed, such as at the earliest stages of conception. They argue that the ▶

there will be less opportunity for others to discover the invention (either from the inventor or on their own). Accordingly, derivation proceedings will be less likely to take place. Filing early will also minimize the body of prior art that can be used against the claims of a patent application.

iii. **Carefully Documenting the Terms of Any Collaborative Relationships**: It is not uncommon for scientists from different companies or universities to work together on a project. When this happens, it is imperative that the entities involved in the collaboration prepare a written agreement detailing who is doing what. If the relationship later turns sour, inventorship disputes often arise and it can be very difficult to corroborate accounts of what actually took place.[v]

IV) EFFECTIVE DATE OF THE AIA'S TRANSITION TO A "FIRST-INVENTOR-TO-FILE" SYSTEM AND THE IMPLEMENTATION OF DERIVATION PROCEEDINGS

The "first-inventor-to-file" system became effective 18 months after the America Invents Act was passed (i.e., on March 16, 2013). Leahy-Smith America Invents Act of 2011, Pub. L. No. 112-29, § 3(n)(1), 125 Stat. 284, 293. The new system applies to any patent application that has at least one claim having

---

◄ year-long period between the filing date of the provisional patent application and the deadline for filing the corresponding non-provisional patent application can be taken to fully flesh out the invention. The idea is that by doing this, the earliest possible priority for any ideas can be established. While this strategy may succeed in some of the more predictable arts (e.g., the mechanical arts), I tend to disagree with this approach for the unpredictable arts, such as the chemical and life sciences. The reason I disagree is that in order to establish that the filing date of anything claimed in a non-provisional application is that of the filing date of a provisional application upon which the non-provisional application is based, the provisional application must provide written description and enablement support for the invention claimed in the non-provisional application pursuant to pre-AIA 35 U.S.C. § 112, ¶ 1. If the provisional application does not fully flesh out the claimed invention of the non-provisional application, it is highly unlikely that it will meet the requirements of 35 U.S.C. § 112, ¶ 1 for that invention.

[v]To give you an anecdotal example, I was once involved in a litigation where a large company had collaborated with a professional at a university to synthesize new chemical compounds. As is usually the case, the relationship started out well with everyone getting along. A few years into the project, however, the honeymoon ended and the university professional became convinced that scientists at the large company were stealing his ideas. The relationship became very strained and eventually ended. Later on, when the large company brought an action to enforce its patent claiming one of the compounds that had been made around the time of the collaboration with the university professional, the university professional claimed that he had actually invented the compound. The large company was therefore faced with the task of not only defending its patent against the party it had originally sued but also against the university professional who maintained that he should have been listed as an inventor on that patent.

an effective filing date[w] that is on or after March 16, 2013. *Id.* at § 3(n)(1)(A) and (B). Complications arise, however, when a patent application contains at least one claim that has an effective filing date that is on or after March 16, 2013 and at least one claim that has an effective filing date that is before March 16, 2013. Accordingly, the following decision tree is provided to assist in determining what legal standards apply to a given patent application.

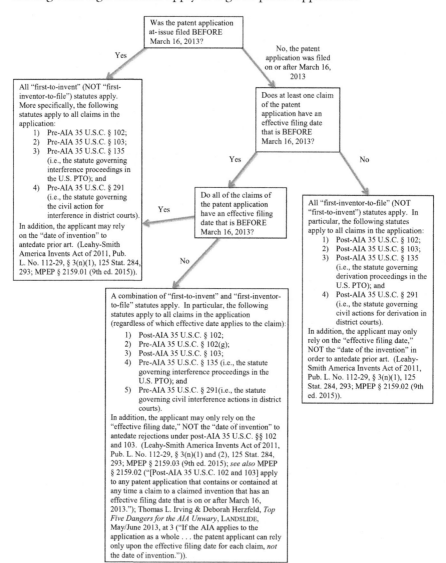

[w]Post-AIA 35 U.S.C. § 100(i)(1) states that "[t]he term 'effective filing date' for a claimed ▶ invention in a patent or application for patent means—

V) LEGISLATIVE GOALS ADDRESSED BY THE AIA'S TRANSITION TO THE FIRST-INVENTOR-TO-FILE SYSTEM[x]

While not an exhaustive list, some of the major goals of shifting the U.S. from a "first-to-invent" to a "first-inventor-to-file" patent system were to (1) create clear, objective, and transparent standards for determining patentability and patent validity issues, (2) provide more predictability and certainty for the patent system, (3) reduce legal costs, and (4) help U.S. companies and innovators compete globally.

First, shifting the U.S. from a "first-to-invent" to a "first-inventor-to-file" patent system should create clear, objective, and transparent standards for evaluating patentability during prosecution because unlike the "first-to-invent" system that allows inventors to antedate prior art references by relying on non-public evidence that shows an earlier date of invention for claimed subject matter, the "first-inventor-to-file" system allows prior art to be antedated only by disclosures and activities that are available to the public. *See* 157 CONG. REC. S1033, 2011 (statement of Sen. Coons) ("Under 'first to file,' [the] PTO's task of determining the priority of a patent application will be more straightforward because patent priority will depend on objective, public facts, rather than on secret files."). As explained by Senator Kyl:

> *To figure out if a patent is valid against prior art, all a manufacturer needs to do [under the "first-inventor-to-file" system] is look at the patent's filing date and figure out whether the inventor publicly disclosed the invention. If prior art disclosed the invention to the public before the filing date, or if the inventor [publicly] disclosed the invention within a year of filing but the prior art predates that disclosure, then the invention is invalid. If not, then the patent is valid against a prior art challenge.*

> **157 CONG. REC. S1105, 2011.**

The "first-inventor-to-file" system will also lead to greater transparency when assessing the validity of a patent in the litigation context. Indeed, Senator Kyl explained:

◀ A) if subparagraph (B) does not apply, the actual filing date of the patent or the application for the patent containing a claim to the invention; or

B) the filing date of the earliest application for which the patent or application is entitled, as to such invention, to a right of priority under section 119, 365(a), 365(b), 386(a), or 386(b) or to the benefit of an earlier filing date under section 120, 121, 365(c), or 386(c)."

[x]For a more comprehensive discussion of the legislative history on the U.S. patent law's transition to a "first-inventor-to-file" system, see Joe Matal, *A Guide to the Legislative History of the America Invents Act: Part I of II*, 21 Fed. Cir. B.J. 435 (2012).

Also, for businesses seeking legal certainty, our current system can be a nightmare. A company hoping to bring a new product to market in a particular field of technology has no way of knowing whether a competitor that belatedly sought the patent on its new product will succeed in securing a valid patent on the product. It all depends on the invention date the competitor will be able to prove relative to the [invention date] that the company developing the product can prove.

Given that both the product developer and competitor can rely on their own secret documents that the other side will not see until litigation over the patent commences, neither of these two parties can gain a clear picture of whether a patent is valid without years of litigation and millions of dollars of discovery and other litigation costs.

Id. at S5320.

In other words, the "first-inventor-to-file" system should be much more objective than the "first-to-invent" system because patents subject to the new law cannot be invalidated using non-public (i.e., secret) information (e.g., by alleged prior inventions of the subject matter claimed in the patent under pre-AIA 35 U.S.C. § 102(g) or pre-AIA 35 U.S.C. § 291). Instead, patent validity will be evaluated on the basis of objective, publicly available information (i.e., the effective filing date of patents or patent applications). *See* post-AIA 35 U.S.C. § 102(a)(1) and (2) (requiring that activities and references be dated before a claimed invention's effective filing date, instead of its date of invention, in order to constitute prior art).

Second, patent owners and patent applicants should also have more certainty and predictability with regard to validity issues under the "first-inventor-to-file" system. During the March 2011 debate on the AIA, Senator Coons explained:

Transition to first-to-file is an improvement over the current system because it provides increased predictability, certainty, and transparency. Patent priority will depend on the date of public disclosure and the effective filing date rather than on secret inventor notebooks, secret personal files which may or may not be admissible and often lead to long and contentious litigation, as the chairman mentioned in his floor comments as well.

This predictability, the predictability that the first-to-file system will bring, I believe will strengthen the hand of investors, inventors, and the public. All will know as soon as an application is filed whether it is likely to have priority over other patent applications.

157 Cong. Rec. S1179.

Third, as compared to the "first-to-invent" system, the legal costs associated with obtaining a patent should be reduced in certain instances under the "first-inventor-to-file" system. In particular, Senator Leahy explained that:

The first-inventor-to-file system will also reduce costs to patent applicants and the Patent Office. This, too, should help the small, independent inventor. In the outdated, current system, when more than one application claiming the same invention is filed, the priority of a right to a patent is decided through an "interference" proceeding to determine which applicant can be declared to have invented the claimed invention first. This process is lengthy, complex, and can cost hundreds of thousands of dollars. Small inventors rarely, if ever, win interference proceedings. In a first-inventor-to-file system, however, the filing date of the application is objective and easy to determine, resulting in a streamlined and less costly process.

157 Cong. Rec. S1090, 2011.

Fourth, by harmonizing the U.S. patent laws with the patent laws of other countries, the shift to a "first-inventor-to-file" patent system should help U.S. companies and innovators better compete in the global economy. *See* 157 Cong. Rec. S1033, 2011 (statement of Sen. Coons) ("By transitioning to a 'first to file' system, [the AIA] brings the U.S. into line with the rest of the world."). Indeed, during the debates leading to the passage of the AIA, Senator Leahy explained:

The transition to first-inventor-to-file is ultimately needed to help American companies and innovators compete globally. As business and competition increasingly operate on a worldwide scale, inventors have to file patent applications in both the United States and other countries for protection of their inventions. Since America's current outdated system differs from the first-inventor-to-file system used in other patent-issuing jurisdictions—all our competitors—it causes confusion and inefficiencies for American companies and innovators. Harmonization will benefit American inventors.

157 Cong. Rec. S1176, 2011.

Overall, because the "first-inventor-to-file" system no longer allows inventors and alleged infringers to rely on secret dates of invention as the priority date for a claimed invention, or on secret prior art to invalidate patent claims, the new system provides more transparency and predictability to U.S. patent law. The new system should also benefit companies that want to seek patent protection around the world because those companies will no longer have to comply with both "first-to-invent" and "first-inventor-to-file" patent systems.

VI) PRACTICAL IMPLICATIONS OF THE AIA'S TRANSITION FROM A "FIRST-TO-INVENT" TO A "FIRST-INVENTOR-TO-FILE" SYSTEM

The shift to a "first-inventor-to-file" patent system was one of the most controversial changes that was proposed in the America Invents Act legislation. The change in the law, however, has not significantly altered the manner in which attorneys advise their clients with regard to filing their patent applications. In fact, patent attorneys have long advised their clients to file patent applications on their inventions as soon as possible to minimize the prior art that can be used against them. Patent attorneys often say, "file early and file often." The advice to "file early" simply means that inventors should file their patent applications as soon as they have developed their technology enough to meet the written description and enablement requirements of pre-AIA 35 U.S.C. § 112, ¶ 1. And the advice to "file often" just means that inventors should file continuation applications (i.e., continuation, divisional or continuation-in-part applications) or new patent applications as they develop new aspects of their inventions or improve upon inventions that are disclosed in previous patent applications. Although it seems simple, the "file early, file often" saying remains sound advice after passage of the AIA.

CHAPTER THREE

The AIA's Impact on the Novelty and Non-Obviousness Patentability Requirements

Contents

I) INTRODUCTION

Two of the key requirements for patenting an invention in the U.S. are that the invention is novel, and not obvious, over the prior art. Under both the pre- and post-AIA law, 35 U.S.C. § 102 and 35 U.S.C. § 103 delineate the novelty and non-obviousness patentability requirements, respectively. Both pre- and post-AIA 35 U.S.C. § 102 also define what references and activities constitute prior art for purposes of showing that an invention lacks novelty and/or is obvious under 35 U.S.C. §§ 102 or 103. This chapter will explore how the AIA has modified these two key statutes.

II) NOVELTY REQUIREMENT OF 35 U.S.C. § 102

Pre-AIA 35 U.S.C. § 102 sets forth the novelty requirement for patenting a claimed invention in the U.S. Under pre-AIA 35 U.S.C. § 102, a claimed invention meets the novelty requirement for patentability unless it has been disclosed through one of the prior art activities, or in one of the references, that are listed in subsections (a)-(g) of the statute. Exceptions to these

America Invents Act Primer. http://dx.doi.org/10.1016/B978-0-12-812096-5.00003-X
Copyright © 2017 Sarah Hasford. Published by Elsevier Inc. All rights reserved.

prior art activities and references are also provided by the pre-AIA statute, wherein the activities or references are performed or created by the inventor within a certain grace period.

While most of the types of activities and references that can qualify as prior art remain the same after the AIA, the criteria for when those activities and references qualify as prior art, as well as inventors' grace periods for filing patent applications, have been significantly changed by the new patent legislation. The AIA altered 35 U.S.C. § 102 in four major ways with the most apparent change being the revision to the structure of the statute itself. In addition to the structural changes, the AIA modified pre-AIA 35 U.S.C. § 102 by: (1) easing the requirements for the activities and references set forth therein to qualify as prior art, (2) abridging the grace period for filing a patent application that inventors enjoyed under pre-AIA 35 U.S.C. § 102, and (3) eliminating certain "loss of right" provisions from pre-AIA 35 U.S.C. § 102. This chapter will provide an in-depth discussion of the post-AIA novelty requirements and will explore how those requirements differ from those of pre-AIA 35 U.S.C. § 102.

A. Structural Changes to 35 U.S.C. § 102

To begin to understand the many differences between the novelty requirements of pre- and post-AIA 35 U.S.C. § 102, it is helpful to first consider the structure in which each of the statutes is written. A comparison of the pre- and post-AIA statutes is shown as follows:

Pre-AIA 35 U.S.C. § 102	Post-AIA 35 U.S.C. § 102
A person shall be entitled to a patent unless – (a) the invention was known or used by others in this country, or patented or described in a printed publication in this or a foreign country, before the invention thereof by the applicant for patent, or (b) the invention was patented or described in a printed publication in this or a foreign country or in public use or on sale in this country, more than one year prior to the date of the application for patent in the United States, or	(a) NOVELTY; PRIOR ART.-- A person shall be entitled to a patent unless-- (1) the claimed invention was patented, described in a printed publication, or in public use, on sale, or otherwise available to the public before the effective filing date of the claimed invention; or (2) the claimed invention was described in a patent issued under section 151, or in an application for patent published or deemed published under section 122(b), in which the patent or application, as the case may be, names another inventor and was effectively filed before the effective filing date of the claimed invention.

(Continued on next page)

Pre-AIA 35 U.S.C. § 102	Post-AIA 35 U.S.C. § 102
(c) he has abandoned the invention, or (d) the invention was first patented or caused to be patented, or was the subject of an inventor's certificate, by the applicant or his legal representatives or assigns in a foreign country prior to the date of the application for patent in this country on an application for patent or inventor's certificate filed more than twelve months before the filing of the application in the United States, or (e) the invention was described in - (1) an application for patent, published under section 122(b), by another filed in the United States before the invention by the applicant for patent or (2) a patent granted on an application for patent by another filed in the United States before the invention by the applicant for patent, except that an international application filed under the treaty defined in section 351(a) shall have the effects for the purposes of this subsection of an application filed in the United States only if the international application designated the United States and was published under Article 21(2) of such treaty in the English language; or (f) he did not himself invent the subject matter sought to be patented, or	(b) EXCEPTIONS.-- (1) DISCLOSURES MADE 1 YEAR OR LESS BEFORE THE EFFECTIVE FILING DATE OF THE CLAIMED INVENTION.-- A disclosure made 1 year or less before the effective filing date of a claimed invention shall not be prior art to the claimed invention under subsection (a)(1) if-- (A) the disclosure was made by the inventor or joint inventor or by another who obtained the subject matter disclosed directly or indirectly from the inventor or a joint inventor; or (B) the subject matter disclosed had, before such disclosure, been publicly disclosed by the inventor or a joint inventor or another who obtained the subject matter disclosed directly or indirectly from the inventor or a joint inventor. (2) DISCLOSURES APPEARING IN APPLICATIONS AND PATENTS.-- A disclosure shall not be prior art to a claimed invention under subsection (a)(2) if-- (A) the subject matter disclosed was obtained directly or indirectly from the inventor or a joint inventor; (B) the subject matter disclosed had, before such subject matter was effectively filed under subsection (a)(2), been publicly disclosed by the inventor or a joint inventor or another who obtained the subject matter disclosed directly or indirectly from the inventor or a joint inventor; or (C) the subject matter disclosed and the claimed invention, not later than the effective filing date of the claimed invention, were owned by the same person or subject to an obligation of assignment to the same person.

(Continued on next page)

Pre-AIA 35 U.S.C. § 102	Post-AIA 35 U.S.C. § 102
(g) (1) during the course of an interference conducted under section 135 or section 291, another inventor involved therein establishes, to the extent permitted in section 104, that before such person's invention thereof the invention was made by such other inventor and not abandoned, suppressed, or concealed, or (2) before such person's invention thereof, the invention was made in this country by another inventor who had not abandoned, suppressed, or concealed it. In determining priority of invention under this subsection, there shall be considered not only the respective dates of conception and reduction to practice of the invention, but also the reasonable diligence of one who was first to conceive and last to reduce to practice, from a time prior to conception by the other.	(c) COMMON OWNERSHIP UNDER JOINT RESEARCH AGREEMENTS.-- Subject matter disclosed and a claimed invention shall be deemed to have been owned by the same person or subject to an obligation of assignment to the same person in applying the provisions of subsection (b)(2)(C) if-- (1) the subject matter disclosed was developed and the claimed invention was made by, or on behalf of, 1 or more parties to a joint research agreement that was in effect on or before the effective filing date of the claimed invention; (2) the claimed invention was made as a result of activities undertaken within the scope of the joint research agreement; and (3) the application for patent for the claimed invention discloses or is amended to disclose the names of the parties to the joint research agreement. (d) PATENTS AND PUBLISHED APPLICATIONS EFFECTIVE AS PRIOR ART.-- For purposes of determining whether a patent or application for patent is prior art to a claimed invention under subsection (a)(2), such patent or application shall be considered to have been effectively filed, with respect to any subject matter described in the patent or application-- (1) if paragraph (2) does not apply, as of the actual filing date of the patent or the application for patent; or

(Continued on next page)

Pre-AIA 35 U.S.C. § 102	Post-AIA 35 U.S.C. § 102
	(2) if the patent or application for patent is entitled to claim a right of priority under section 119, 365(a), or 365(b), or to claim the benefit of an earlier filing date under section 120, 121, or 365(c), based upon 1 or more prior filed applications for patent, as of the filing date of the earliest such application that describes the subject matter.

As shown previously, each of subsections (a)-(g) in pre-AIA 35 U.S.C. § 102 sets forth various prior art activities and references that, if found to have been dated before the date of application, i.e., the filing date, or date of invention of a claimed invention, can render a patent or patent application claim to that invention invalid or unpatentable for lack of novelty. For example, subsection (a) discusses when knowledge or use of a claimed invention constitutes prior art for purposes of assessing its novelty, while subsection (c) prohibits the patenting of a claimed invention where the inventor has abandoned that invention prior to filing his/her patent application. Furthermore, while the subsections of pre-AIA 35 U.S.C. § 102 do not explicitly set out exceptions to the prior art listed therein, such exceptions are implicitly provided through the requirements recited by the statute. For instance, one of the requirements of pre-AIA 35 U.S.C. § 102(b) prior art is that it must be dated more than one year prior to the filing date, i.e., "the date of the application for patent," of the application against which it is being cited. Thus, disclosures that are dated within one year of an application's filing date would not constitute prior art under this subsection.

In contrast to pre-AIA 35 U.S.C. § 102, wherein the categories of prior art are set forth in each of seven subsections, post-AIA 35 U.S.C. § 102 lists all prior art activities and references in one subsection—subsection (a). Post-AIA 35 U.S.C. § 102 also differs from the pre-AIA statute in that it explicitly lists out all of the exceptions to the prior art identified therein in post-AIA 35 U.S.C. § 102(b). The U.S. PTO has prepared the following slide that is helpful for understanding the relationship between subsections (a) and (b) of post-AIA 35 U.S.C. § 102:[a]

---

[a]This slide can be found at: http://www.uspto.gov/sites/default/files/aia_implementation/FITF_card.pdf.

AIA Statutory Framework

Prior Art 35 U.S.C. 102(a) (Basis for Rejection)		Exceptions 35 U.S.C. 102(b) (Not Basis for Rejection)	
102(a)(1) Disclosure with Prior Public Availability Date	102(b)(1)	(A) Grace Period Disclosure by Inventor or Obtained from Inventor	
		(B) Grace Period Intervening Disclosure by Third Party	
102(a)(2) U.S. Patent, U.S. Patent Application, and PCT Application with Prior Filing Date	102(b)(2)	(A) Disclosure Obtained from Inventor	
		(B) Intervening Disclosure by Third Party	
		(C) Commonly Owned Disclosure	

19

Yet another difference between pre- and post-AIA 35 U.S.C. § 102 is that the latter statute provides definitions for some of the terms used therein. For instance, post-AIA 35 U.S.C. § 102(c) explicitly defines the common ownership terms used in post-AIA 35 U.S.C. § 102(b)(2)(C). Additionally, pre- and post-AIA 35 U.S.C. § 102 differ in that the latter statute devotes an entire subsection to defining the effective prior art date for patents and published patent applications. See Post-AIA 35 U.S.C. § 102(d). Pre-AIA 35 U.S.C. § 102 provides guidance for the effective prior art date of patents and published patent applications in subsection (e), but this guidance is provided in the same subsection that sets forth the requirements for when those types of references qualify as prior art.

While the structural differences between pre- and post-AIA 35 U.S.C. § 102 may seem substantial, the most significant changes to the novelty requirement are actually seen in the substantive changes to the statute. These changes are discussed in the next three sections.

B. Expansion of the Realm of Prior Art Under Post-AIA 35 U.S.C. § 102

One of the most significant changes, if not the most significant change, the AIA made to U.S. patent law lies in the expansion of the body of prior art that is available for use against the claims of a patent or patent application under 35 U.S.C. § 102. Indeed, as compared to pre-AIA 35 U.S.C. § 102, post-AIA 35 U.S.C. § 102 generally makes it easier for certain activities and references[b] to qualify as prior art. The post-AIA statute does this by, for instance, encompassing more recent references and activities as prior art than pre-AIA 35 U.S.C. § 102 and by eliminating certain requirements that had to be met for activities and references to qualify as prior art under pre-AIA 35 U.S.C. § 102. As a result, far more activities and references may qualify as prior art under the post-AIA statute than under the pre-AIA statute.

Post-AIA 35 U.S.C. § 102 defines prior art in two parts of subsection (a): 102(a)(1) and 102(a)(2). The exceptions to the prior art set forth in these two subsections are then provided in two parts of subsection (b). More specifically, the exceptions to the prior art set forth in post-AIA 35 U.S.C. § 102(a)(1) are provided in post-AIA 35 U.S.C. § 102(b)(1), while the exceptions to the prior art set forth in post-AIA 35 U.S.C. § 102(a)(2) are provided in post-AIA 35 U.S.C. § 102(b)(2). A discussion of how the AIA has changed the conditions for when certain activities and references qualify as prior art under 35 U.S.C. § 102 is provided in the following sections.

i. Prior Art Under Post-AIA 35 U.S.C. § 102(a)(1)

Post-AIA 35 U.S.C. § 102(a)(1) defines when patenting, disclosures in printed publications, public uses, sales, and public availability of an invention will constitute prior art to a claimed invention. The following subsections explain how this new statute expands the scope of the prior art that was available under pre-AIA 35 U.S.C. § 102 and discusses the exceptions to the post-AIA 35 U.S.C. § 102(a)(1) prior art that are set forth in post-AIA 35 U.S.C. § 102(b)(1).

a. Public Use of a Claimed Invention[c]

The following table provides an overview of both the pre- and post-AIA statutes that govern when a public use of a claimed invention constitutes prior art.

---

[b]The changes to activities that used to qualify as prior art under pre-AIA 35 U.S.C. § 102(g) (i.e., previous invention by another) are not addressed in this chapter because they are discussed in Chapter 2.

[c]Under pre-AIA 35 U.S.C. § 102, use of a claimed invention and knowledge of a claimed invention were two prior art activities that were often discussed together. *See, e.g.*, MPEP § 2132(I) (9th ed. 2014). Knowledge, however, is not expressly addressed in this section because post-AIA 35 U.S.C. § 102 does not explicitly list knowledge as its own category of prior art. Nevertheless, public knowledge of a claimed invention likely still qualifies as prior art under the "otherwise available to the public" language that was introduced in post-AIA 35 U.S.C. § 102(a)(1). This new category of prior art is discussed in more detail in Section II.B.i.c.

Relevant Statutory Language Governing When the Use of a Claimed Invention Constitutes Prior Art Under Pre-AIA Law	Conditions That Must be Met for the Use of a Claimed Invention to Constitute Prior Art Under the Pre-AIA Statute	Relevant Statutory Language Governing When the Use of a Claimed Invention Constitutes Prior Art Under Post-AIA Law	Conditions that Must be Met for Use of a Claimed Invention to Constitute Prior Art Under the Post-AIA Statute	Uses of a Claimed Invention That Do NOT Constitute Prior Art Under Post-AIA Law
"[a] person shall be entitled to a patent unless . . . (a) the invention was . . . used by others in this country . . . before the invention thereof by the applicant for a patent . . ." (Pre-AIA 35 U.S.C. § 102(a)) "[a] person shall be entitled to a patent unless . . . (b) the invention was . . . in public use . . . in this country, more than one year prior to the date of the application for patent in the United States . . . ," (Pre-AIA 35 U.S.C. § 102(b))	1) Under both pre-AIA 35 U.S.C. § 102(a) and (b), the use of the claimed invention must have been **public.** (MPEP § 2132(I)(A) (9th ed. 2014)) 2) Under both pre-AIA 35 U.S.C. § 102(a) and (b), the use of the claimed invention must occur **in the United States.** (*See* pre–AIA 35 U.S.C. § 102(a) and (b) (explaining that in order to qualify as prior art, the invention must have been used by others "in this country").) 3) Under pre-AIA 35 U.S.C. § 102(a), the use of the claimed invention must have been by someone **other than the inventor** (i.e., a person or group of persons that is different from the inventor or inventors who are listed on the patent or patent application). (*See* pre–AIA 35 U.S.C. § 102(a) (explaining that a person shall be entitled to a patent unless the invention was used "by others").)	"[a] person shall be entitled to a patent unless --(1) the claimed invention was . . . in public use . . . before the effective filing date of the claimed invention." (Post-AIA 35 U.S.C. § 102(a)(1))	1) The use of the claimed invention must be **public.** (*See* Post-AIA 35 U.S.C. § 102(a) (1) (reciting "public use" as opposed to just "use")) 2) The use of the claimed invention must have occurred **before the effective filing date of the claimed invention.** (Post-AIA 35 U.S.C. § 102(a) (1))	1) A use of a claimed invention that occurs one year or less before the effective filing date of a claimed invention is not prior art if the use was by the inventor or joint inventor or by another who obtained the subject matter disclosed directly or indirectly from the inventor or joint inventor. (Post-AIA 35 U.S.C. § 102(b)(1)(A))

4) Under pre–AIA 35 U.S.C. § 102(a), the use of the claimed invention must occur **before the date on which the claimed invention was invented**, i.e., "the date of invention." (*See* pre–AIA 35 U.S.C. § 102(a) (explaining that a person shall be entitled to a patent unless the invention was used by others "before the invention thereof by the applicant for a patent").)

5) For purposes of pre–AIA 35 U.S.C. § 102(b), the use must occur **more than one year prior to the filing date** of the patent or patent application against which it is being cited as prior art. (*See* pre–AIA 35 U.S.C. § 102(b) (explaining that a person shall be entitled to a patent unless the invention was in public use "more than one year prior to the date of application for patent in the United States").)

2) A use of a claimed invention that occurs one year or less before the effective filing date of a claimed invention is not prior art if, the subject matter used had, before such use, been publicly disclosed by the inventor or a joint inventor or another who obtained the subject matter disclosed directly or indirectly from the inventor or a joint inventor. (Post–AIA 35 U.S.C. § 102(b)(1)(B))

As shown in the previous table, post-AIA 35 U.S.C. § 102(a)(1) makes it significantly easier for a public use of a claimed invention to qualify as prior art by: (1) eliminating some of the conditions that were required for such prior art under pre-AIA 35 U.S.C. § 102(a) and (b), and (2) deferring the deadline by which a public use must occur in order to qualify as prior art.

First, post-AIA 35 U.S.C. § 102(a)(1) eliminates some of the conditions that were required of public uses under pre-AIA 35 U.S.C. § 102(a) and (b). In particular, post-AIA 35 U.S.C. § 102(a)(1) eliminates the requirement that a public use must occur in the U.S., which is set forth in both pre-AIA 35 U.S.C. § 102(a) and (b). Indeed, by omitting the phrase "in this country" that appeared in pre-AIA 35 U.S.C. § 102(a) and (b), post-AIA 35 U.S.C. § 102(a)(1) does not require that the public use of a claimed invention occur in the United States to qualify as prior art. Instead, a public use that occurs anywhere in the world can be used as prior art against a claimed invention under post-AIA 35 U.S.C. § 102(a)(1).

Post-AIA 35 U.S.C. § 102(a)(1) also expands the scope of public uses that qualified as prior art under pre-AIA 35 U.S.C. § 102(a) by disposing of the requirement that a public use must have been by someone other than the inventor, i.e., by a person or group of persons that differs from the inventive entity listed on the patent or patent application. More specifically, by omitting the language "by others" that was used in pre-AIA 35 U.S.C. § 102(a), post-AIA 35 U.S.C. § 102(a)(1) permits an inventor's own public use of a claimed invention to constitute prior art.[d]

Second, more public uses are potentially available as prior art under the new law because post-AIA 35 U.S.C. § 102(a)(1) defers the deadline by which a public use of a claimed invention must occur in order for that activity to constitute prior art. As shown in the previous table, in order to qualify as prior art, pre-AIA 35 U.S.C. § 102(a) requires a public use to occur before the **date of invention** of claimed subject matter,[e] and pre-AIA

---

[d]There are exceptions to when an inventor's own actions constitute prior art under the post-AIA law. *See* post-AIA 35 U.S.C. § 102(b)(1)(A) and (B). These are discussed in more detail in Section II.B.i.e.

[e]As discussed in Chapter 2, one way inventors can overcome a rejection under pre-AIA 35 U.S.C. § 102(a) is to file a declaration or affidavit under 37 C.F.R. § 1.131. MPEP § 706.02(b)(2) (9th ed. 2014). A successful declaration or affidavit alleges sufficient facts to demonstrate that the date of invention for claimed subject matter is earlier than the effective date of the prior art cited against the patent application. Once an earlier date of invention is established, a Patent Examiner will withdraw the rejection under pre-AIA 35 U.S.C. § 102(a). Because the post-AIA novelty requirement focuses on the effective filing date (and not the date of invention) of a claimed invention, however, applicants will not be permitted to use 37 C.F.R. § 1.131 declarations to overcome rejections under post-AIA 35 U.S.C. § 102(a)(1). *See* MPEP § 715 (9th ed. 2014) (explaining that 37 C.F.R. § 1.131 is not applicable to patents and patent applications that are subject to the post-AIA novelty requirements).

35 U.S.C. § 102(b) requires such use to occur **more than one year before the date on which an application for a patent is filed**. Post-AIA 35 U.S.C. § 102(a)(1), however, merely requires a public use to occur **before the effective filing date** of the claimed invention. With regard to pre-AIA 35 U.S.C. § 102(a), the new statute shifts the focus from the **date of invention** to the **effective filing date** of a claimed invention. And in reference to pre-AIA 35 U.S.C. § 102(b), the post-AIA statute eliminates the requirement that a public use occur **more than one year** before the filing date of a patent or patent application. The **date of invention** for claimed subject matter virtually always precedes the **effective filing date** of any patent or patent application claiming that subject matter by several months because it often takes at least that much time to draft a patent application once an inventor discloses his/her invention to a patent attorney or patent agent. Thus, by postponing the deadline by which a public use of a claimed invention must occur (i.e., from the **date of invention** to the **effective filing date**) in order for that use to constitute prior art, post-AIA 35 U.S.C. § 102(a)(1) has the potential to include more public uses as prior art than did pre-AIA 35 U.S.C. § 102(a) and (b).

In sum, post-AIA 35 U.S.C. § 102(a)(1) has the potential to expand the body of "public use" prior art that was available under pre-AIA 35 U.S.C. § 102(a) and (b)[f] by: (1) allowing public uses of a claimed invention that occur anywhere in the world to constitute prior art, (2) allowing an inventor's own public use of a claimed invention to constitute prior art, and

---

[f]There is one way in which post-AIA 35 U.S.C. § 102(a)(1) might narrow the scope of prior art public uses that were available under pre-AIA 35 U.S.C. § 102(b). In particular, pre-AIA 35 U.S.C. § 102(b) permitted certain secret uses by an inventor to be cited as prior art against that inventor's claimed invention. *See Metallizing Eng'g Co., Inc. v. Kenyon Bearing & Auto Parts Co., Inc.*, 153 F.2d 516, 517-18, 520 (2d Cir. 1946) (holding that an inventor had forfeited his right to a patent on a metallizing process where the metallizing process had been used more than one year before the inventor applied for a patent application because although the use of the metallizing process was done in secret, and the process had not become known by the public, the main purpose of the inventor's use of the process was commercial); *see also Invitrogen Corp. v. Biocrest Mfg., L.P.*, 424 F.3d 1374, 1382 (Fed. Cir. 2005) ("In [*Metallizing Engineering*], the patentee used a secret process to recondition worn metal parts for its customers before the critical date. The Second Circuit correctly held 'that it is a condition upon an inventor's right to a patent that he shall not exploit his discovery competitively after it is ready for patenting; he must content himself with either secrecy, or [a patent].'") (alteration in original). The U.S. PTO, however, has interpreted post-AIA 35 U.S.C. § 102(a)(1) as only encompassing uses that are publicly accessible based on that statute's inclusion of the language "or otherwise available to the public." MPEP § 2152.02(c) (9th ed. 2014). Thus, if federal courts interpret post-AIA 35 U.S.C. § 102(a)(1) in the same way as the U.S. PTO, the scope of prior use prior art will be narrowed in that it will not include secret prior uses such as those identified in *Metallizing Engineering*.

(3) allowing public uses that occur any time before the effective filing date of a claimed invention to constitute prior art.

b. Patenting a Claimed Invention or Describing the Claimed Invention in a Printed Publication

The following tables provide an overview of both the pre- and post-AIA statutes that govern patent[g] and printed publication[h] prior art.

Relevant Statutory Language Governing When the Description of a Claimed Invention in a Patent or Printed Publication Constitutes Prior Art Under Pre-AIA Law	Conditions that Must be Met for Patenting or Description of a Claimed Invention in a Printed Publication to Constitute Prior Art Under Pre-AIA Statute
"A person shall be entitled to a patent unless - (a) the invention was … patented or described in a printed publication in this or a foreign country, before the invention thereof by the applicant for patent …." (Pre-AIA 35 U.S.C. § 102(a))	1) the claimed invention must have been patented (i.e., claimed in a patent) or described in a printed publication anywhere in the world; 2) the patent or printed publication must have been issued or published before the date of invention for the claims at-issue; and 3) the authorship or inventorship listed on the patent or printed publication must be different from the inventive entity listed on the patent application at-issue. (MPEP § 2132 (9th ed. 2014))

(Continued on next page)

---

[g]Note that pre-AIA 35 U.S.C. § 102(a) and (b) and post-AIA 35 U.S.C. § 102(a)(1) govern patents as prior art when the patent issues <u>before</u> the date of invention (for pre-AIA statutes), or the effective filing date (for post-AIA statutes), of a claimed invention. In contrast, pre-AIA 35 U.S.C. § 102(e) and post-AIA 35 U.S.C. § 102(a)(2) govern patents as prior art when the patent issues <u>after</u> the effective filing date of the claimed invention but the patent application from which the patent issued was filed <u>before</u> the effective filing date. 78 Fed. Reg. 11,074 (Feb. 14, 2013).

[h]Printed publications include patents and published patent applications. MPEP § 2152 (9th ed. 2014). Thus, patents or published patent applications that describe a claimed invention can constitute prior art under post-AIA 35 U.S.C. § 102(a)(1).

Relevant Statutory Language Governing When the Description of a Claimed Invention in a Patent or Printed Publication Constitutes Prior Art Under Pre-AIA Law	Conditions that Must be Met for Patenting or Description of a Claimed Invention in a Printed Publication to Constitute Prior Art Under Pre-AIA Statute
"A person shall be entitled to a patent unless . . . (b) the invention was patented or described in a printed publication in this or a foreign country . . . more than one year prior to the date of the application for patent in the United States . . . ." (Pre-AIA 35 U.S.C. § 102(b))	1) the claimed invention must have been patented (i.e., claimed in a patent) or described in a printed publication anywhere in the world; and 2) the issue or publication date of the patent or printed publication must be more than one year earlier than the filing date of the patent or patent application at-issue. (MPEP § 2133 (9th ed. 2014))
"A person shall be entitled to a patent unless . . . (d) the invention was first patented or caused to be patented, or was the subject of an inventor's certificate, by the applicant or his legal representatives or assigns in a foreign country prior to the date of the application for patent in this country on an application for patent or inventor's certificate filed more than twelve months before the filing of the application in the United States . . . ." (Pre-AIA 35 U.S.C. § 102(d))	1) the foreign patent application describing the claimed invention must have been filed more than 12 months before the effective filing date of the U.S. patent application at-issue; 2) the foreign application describing the claimed invention must have been filed by the same applicant (or his/her legal representatives or assigns) as the patent application that was filed in the United States; 3) the foreign patent or inventor's certificate that describes the claimed invention must have been actually granted before the filing date of the U.S. application at-issue; and 4) the invention that was patented in the foreign patent or inventor's certificate must be the same invention for which a patent is being sought in the U.S. application. (MPEP § 2135 (9th ed. 2014))

Relevant Statutory Language Governing When Patenting or Describing a Claimed Invention in a Printed Publication Constitutes Prior Art Under Post-AIA Law	Conditions that Must be Met for Patenting or a Description of a Claimed Invention in a Printed Publication to Constitute Prior Art Under Post-AIA Law	Patenting or Disclosures of a Claimed Invention in a Printed Publication that Do NOT Constitute Prior Art Under Post-AIA Law
"A person shall be entitled to a patent unless—(1) the claimed invention was patented, described in a printed publication . . . or otherwise available to the public before the effective filing date of the claimed invention . . . ." (Post-AIA 35 U.S.C. § 102(a)(1))	1) The patent or printed publication that describes the claimed invention must have been issued or published before the effective filing date for the claimed invention at issue. (Post-AIA 35 U.S.C. § 102(a)(1)) Note regarding the date patented: The effective date of a patent for purposes of determining whether the patent qualifies as prior art under the post-AIA law is its grant date (unless the patent was secret, in which case the effective prior art date of the patent is the date on which it became publicly available). (MPEP § 2152.02(a) (9th ed. 2014))	1) Any patenting or disclosure of a claimed invention in a printed publication that occurs one year or less before the effective filing date for a claimed invention is not prior art if the patenting or disclosure was by the inventor or joint inventor or by another who obtained the subject matter disclosed directly or indirectly from the inventor or joint inventor. (Post-AIA 35 U.S.C. § 102(b)(1)(A)) 2) Any patenting or disclosure of a claimed invention in a printed publication that occurs one year or less before the effective filing date for a claimed invention is not prior art if, before such patenting or disclosure, the claimed invention had been publicly disclosed by the inventor or a joint inventor or another who obtained the subject matter disclosed directly or indirectly from the inventor or a joint inventor. (Post-AIA 35 U.S.C. § 102(b)(1)(B))

Because post-AIA 35 U.S.C. § 102(a)(1) defers the deadline by which patenting or a disclosure of a claimed invention in a printed publication must occur in order to constitute prior art, the new statute has the potential to expand the scope of patent and printed publication prior art that was available under pre-AIA 35 U.S.C. § 102(a) and (b). Indeed, as shown in the previous table, in order to qualify as prior art, pre-AIA 35 U.S.C. § 102(a) requires that patenting or a disclosure of a claimed invention in a printed publication occur before the **date of invention** of claimed subject matter, and pre-AIA 35 U.S.C. § 102(b) requires that such patenting or disclosure occur **more than a year before the date on which an application for a patent is filed**. Post-AIA 35 U.S.C. § 102(a)(1), however, merely requires that patenting or a disclosure of a claimed invention in a printed publication occur **before the effective filing date** of the claimed invention. The new statute therefore shifts the focus from the **date of invention** to the **effective filing date** of a claimed invention and eliminates the requirement that patenting or disclosure in a printed publication occur **more than one year**[i] before the filing date of a patent or patent application. By postponing the deadline by which patenting or a disclosure of a claimed invention must occur, therefore, post-AIA 35 U.S.C. § 102(a)(1) has the potential to include more patent and printed publication prior art than was encompassed by pre-AIA 35 U.S.C. § 102(a) and (b).

Post-AIA 35 U.S.C. § 102(a)(1) also broadens the scope of patent and printed publication prior art that was available under pre-AIA 35 U.S.C. § 102(a) by disposing of the requirement that the inventorship or authorship listed on a patent or printed publication be different from that of the claimed invention. MPEP § 2152.02(f) (9th ed. 2014). As shown in the previous table, however, post-AIA 35 U.S.C. § 102(b)(1) excludes from post-AIA 35 U.S.C. § 102(a)(1) prior art certain patents and printed publications that are dated within one year of the effective filing date of a claimed invention.

In addition, post-AIA 35 U.S.C. § 102(a)(1) expands the scope of patent and published patent application prior art by attributing an earlier effective prior art date to certain patents and printed publications than was given to such references under pre-AIA 35 U.S.C. § 102(a) and (b). Indeed, under

---

[i]Note that post-AIA 35 U.S.C. § 102(b)(1)(A) and (B) provide certain exceptions to post-AIA 35 U.S.C. § 102(a)(1) so that certain patents or printed publications that have an effective prior art date that is within one year of the effective filing date of the claimed invention will not qualify as prior art under that statute. These exceptions are discussed in further detail in Section II.B.i.e.

the pre-AIA law, U.S. patents and published patent applications could only be used as prior art as of their earliest effective U.S. filing date. MPEP § 2154.01(b) (9th ed. 2014). Accordingly, under the pre-AIA law, if a Patent Examiner wanted to cite a U.S. patent application that was filed in the U.S. on August 2, 2012 and claimed priority to an Australian patent application that was filed on August 3, 2011, the Examiner could only use the August 2, 2012 date to establish that the patent application was prior art to a pending U.S. application. The post-AIA law, however, allows these documents to be used as prior art as of their earliest effective U.S. or foreign filing dates. Post-AIA 35 U.S.C. § 102(d); MPEP § 2152.01 (9th ed. 2014). Accordingly, in the previous example, the Examiner could use the August 3, 2011 date to establish that the patent application was prior art to a pending U.S. application. Thus, post-AIA 35 U.S.C. § 102(a)(1) has the potential to encompass far more patent and published patent application prior art than was included under pre-AIA 35 U.S.C. § 102(a) or (b).

Finally, as compared to pre-AIA 35 U.S.C. § 102(d), post-AIA 35 U.S.C. § 102(a)(1) actually narrows the scope of foreign patent prior art that is available. In particular, when an invention is described in a patent, it may only constitute prior art under pre-AIA 35 U.S.C. § 102(a) and (b) if that patent is made available to the public.[j] *See In re Ekenstam*, 256 F.2d 321, 325 (C.C.P.A. 1958) ("For the reasons given, we are of the opinion, both on principle and authority, that the word 'patented' as used in 35 U.S.C. § 102(a) and (b) is limited to patents which are available to the public."). In contrast, when a patent applicant patents his/her invention in a foreign country, that patent can bar the patenting of his/her claimed invention in

---

[j] Also recall that pre-AIA 35 U.S.C. § 102(b) bars the patenting of an invention where that invention was patented or described in a printed publication that has an issue or publication date that is more than one year prior to the effective filing date of the U.S. patent application at issue. In contrast, pre-AIA 35 U.S.C. § 102(d) bars the patenting of an invention where that invention was described in a foreign patent application that was filed more than 12 months before the corresponding U.S. patent application was filed and the foreign patent stemming from that patent application was granted prior to the effective filing date of the corresponding U.S. patent application. Unlike a patent that serves as prior art under pre-AIA 35 U.S.C. § 102(b), a foreign patent cited as prior art under pre-AIA 35 U.S.C. § 102(d) need not have been granted more than one year before the corresponding U.S. patent application was filed (only the foreign patent application from which the foreign patent issued to have been filed more than one year before the corresponding U.S. patent application was filed). Of course if the foreign patent application was published more than 12 months prior to the filing of the corresponding U.S. application, the foreign application itself could be cited as prior art (i.e., as a printed publication) under pre-AIA 35 U.S.C. § 102(b).

a U.S. patent application under pre-AIA 35 U.S.C. § 102(d) even if the foreign patent was not made available to the public. *See In re Kathawala,* 9 F.3d 942, 946 (Fed. Cir. 1993) ("[I]t is irrelevant under *section 102(d)* whether the Spanish patent was publicly available prior to the U.S. filing date. Rather, the Board correctly concluded that an invention is 'patented' in a foreign country under *section 102(d)* when the patentee's rights under the patent become fixed.") (emphasis in original). The U.S. PTO currently interprets the term "patented" in post-AIA 35 U.S.C. § 102(a)(1) to have the same meaning it had under pre-AIA 35 U.S.C. § 102(a) and (b). MPEP § 2152.02(a) (9th ed. 2014). Thus, foreign patents that are not available to the public (i.e., secret patents) cannot currently be cited against a claimed invention under post-AIA 35 U.S.C. § 102(a)(1).

Overall, post-AIA 35 U.S.C. § 102(a)(1) has the potential to expand the body of patent and printed publication prior art that was available under pre-AIA 35 U.S.C. § 102(a) and (b) by: (1) deferring the deadline by which patenting of subject matter, or disclosure of subject matter in a printed publication, must occur in order for such subject matter to constitute prior art to a claimed invention, and (2) providing an earlier effective prior art date for patents and printed publications (i.e., published patent applications) that are entitled to a right of priority, or the benefit of, a foreign filing date. By disposing of the requirement that the inventorship or authorship listed on a patent or printed publication be different from that of the claimed invention, post-AIA 35 U.S.C. § 102(a)(1) also has the potential to expand the body of patent and printed publication prior art that was available under pre-AIA 35 U.S.C. § 102(a). On the other hand, post-AIA 35 U.S.C. § 102(a)(1) narrows the scope of foreign patent prior art that was previously available under pre-AIA 35 U.S.C. § 102(d) by requiring that foreign patents be available to the public before they can be cited as prior art.

c. Claimed Inventions That are "Otherwise Available to the Public"

One of the most controversial changes the AIA made to pre-AIA 35 U.S.C. § 102 was the introduction of the phrase "or otherwise available to the public" at the end of post-AIA 35 U.S.C. § 102(a)(1). For ease of reference, the full text of post-AIA 35 U.S.C. § 102(a)(1) is provided as follows:

> *(a) NOVELTY; PRIOR ART.-- A person shall be entitled to a patent unless-- (1) the claimed invention was patented, described in a printed publication, or in public use, on sale, or otherwise available to the public before the effective filing date of the claimed invention . . . .*

The phrase "or otherwise available to the public" has been the subject of many disputes because the courts have not yet interpreted its meaning. Some practitioners think that the phrase creates its own catch-all category of prior art—i.e., a category of prior art that encompasses activities and references that render an invention "available to the public" even if those activities or references are not explicitly listed in post-AIA 35 U.S.C. § 102(a)(1)—while others believe that the phrase modifies all of the activities and references that are listed before it (e.g., "on sale" or "patented") such that those activities must be "available to the public" in order to constitute prior art under the new law.

One of the reasons that the interpretation of the new phrase is so important is that pre-AIA 35 U.S.C. § 102 allows certain non-public (i.e., secret) activities to qualify as prior art. For instance, a secret sale or offer for sale of an invention that occurred more than one year prior to the date of application for a patent could, under certain circumstances, be used as prior art against claims of that application under pre-AIA 35 U.S.C. § 102(b). MPEP § 2133.03(b)(III)(A) (9th ed. 2014) ("Unlike questions of public use, there is no requirement [under pre-AIA 35 U.S.C. § 102(b)] that 'on sale' activity be 'public.'"). Accordingly, if the correct interpretation of the "or otherwise available to the public" language is that it modifies all of the activities preceding it, the so-called "secret" prior art of pre-AIA 35 U.S.C. § 102 will cease to exist under the new law.

While we do not yet know how federal courts will interpret the new statutory language, the U.S. PTO has already interpreted the phrase "or otherwise available to the public" as both creating a "catch-all" category of prior art and as modifying the activities listed before it.[k] In fact,

---

[k]The U.S. PTO is not the final authority on the interpretation of patent statutes. *See, e.g., Ethicon, Inc. v. Quigg*, 849 F.2d 1422, 1425 (Fed. Cir. 1988) (quoting *FEC v. Democratic Senatorial Campaign Comm.*, 454 U.S. 27, 32 (1981)) ("[T]he courts are the final authority on the issues of statutory construction. They must reject administrative constructions, whether reached by adjudication or by rulemaking, that are inconsistent with the statutory mandate or that frustrate the policy Congress sought to implement."); *In re Gibbs*, 437 F.2d 486, 491 (C.C.P.A. 1971) ("[C]learly this court is not bound by a ruling of a Commissioner of Patents . . . regardless of how ingrained his ruling may have become in Patent Office practice."). Accordingly, if a federal court disagrees with the U.S. PTO's interpretation of post-AIA 35 U.S.C. § 102(a)(1), the court's interpretation will control.

MPEP § 2152.02(e) specifically categorizes the "or otherwise available to the public" language as a "catch-all" provision, which defines a new additional category of potential prior art not provided for in pre-AIA 35 U.S.C. § 102. The MPEP goes on to explain that this "catch-all" provision permits decision makers (i.e., Patent Examiners and Administrative Patent Judges) to focus on whether the disclosure was "available to the public," rather than whether a disclosure constitutes a "printed publication" or falls within another category of prior art that is explicitly listed in post-AIA 35 U.S.C. § 102(a)(1). *Id.* According to the MPEP, a claimed invention might have become available to the public in, for example, a student thesis in a university library, a poster display or other information disseminated at a scientific meeting, a patent application or patent that has been laid open, a document that is electronically posted on the internet, or a commercial transaction that does not constitute a sale as that term is interpreted under post-AIA 35 U.S.C. § 102(a)(1). *Id.* Even though these references and activities do not fall within those activities and references that are explicitly listed in post-AIA 35 U.S.C. § 102(a)(1), they may still constitute prior art under the "otherwise available to the public" provision of post-AIA 35 U.S.C. § 102(a)(1), provided that the claimed invention is made sufficiently available to the public.

The U.S. PTO has also indicated that it views the phrase "or otherwise available to the public" as modifying at least the "on sale" activity of post-AIA 35 U.S.C. § 102(a)(1). MPEP § 2152.02(d) (9th ed. 2014); *see also In re Enhanced Security Research, LLC*, 739 F.3d 1347, 1354 (Fed. Cir. 2014) (appearing to read the phrase "otherwise available to the public" as modifying the prior art categories listed in post-AIA 35 U.S.C. § 102(a)(1) by explaining that "[w]hether a document qualifies as a 'printed publication' that is 'available to the public' for the purposes of *35 U.S.C. § 102(a)(1)* is a question of law based on underlying findings of fact.") (emphasis in original). This interpretation is discussed in further detail in the following section.

d. Selling a Claimed Invention

An overview of both the pre- and post-AIA statutes that govern when the sale of a claimed invention constitutes prior art is provided in the following table:

Relevant Statutory Language Governing When the Sale of a Claimed Invention Constitutes Prior Art Under Pre-AIA Law	Conditions that Must be Met for the Sale of a Claimed Invention to Constitute Prior Art Under the Pre-AIA Statute	Relevant Statutory Language Governing When the Sale of a Claimed Invention Constitutes Prior Art Under Post-AIA Law	Conditions that Must be Met for Sale of a Claimed Invention to Constitute Prior Art Under the Post-AIA Statute	Sales of a Claimed Invention that Do NOT Constitute Prior Art Under Post-AIA Law
"A person shall be entitled to a patent unless . . . (b) the invention was . . . on sale in this country, more than one year prior to the date of the application for patent in the United States . . . ." (Pre-AIA 35 U.S.C. § 102(b))	1) The sale of the claimed invention must have occurred **in the United States**. (*See* pre-AIA 35 U.S.C. § 102(b) (explaining that in order to qualify as prior art, the invention must have been on sale "in this country").) 2) The sale of the claimed invention must have occurred **more than one year** prior to the filing date of the U.S. patent application at-issue. (*See* pre-AIA 35 U.S.C. § 102(b) (explaining that in order to qualify as prior art, the invention must have been on sale "more than one year prior to the date of the application for patent in the United States").)	"A person shall be entitled to a patent unless— (1) the claimed invention was . . . on sale . . . before the effective filing date of the claimed invention . . . ." (Post-AIA 35 U.S.C. § 102(a)(1))	1) The sale of the claimed invention must have occurred before the **effective filing date** of the claimed invention. (Post-AIA 35 U.S.C. § 102(a)(1))	1) A sale of a claimed invention that occurs one year or less before the effective filing date of a claimed invention is not prior art if the sale was by the inventor or joint inventor or by another who obtained the invention that was sold directly or indirectly from the inventor or joint inventor. (Post-AIA 35 U.S.C. § 102(b)(1)(A)) 2) A sale of a claimed invention that occurs one year or less before the effective filing date of a claimed invention is not prior art if, the invention that was sold had, before such sale, been publicly disclosed by the inventor or a joint inventor or another who obtained the disclosed subject matter directly or indirectly from the inventor or a joint inventor. (Post-AIA 35 U.S.C. § 102(b)(1)(B))

In comparison to pre-AIA 35 U.S.C. § 102(b), post-AIA 35 U.S.C. § 102(a)(1) expands the types of sales and offers for sales[1] that may constitute prior art to a claimed invention by: (1) eliminating the requirement that a sale or offer for sale of a claimed invention must occur in the United States; and (2) eliminating the requirement that a sale or offer for sale must occur more than one year prior to the date of the application for the patent at issue. Under the new law, a sale of a claimed invention that occurs anywhere in the world, and at any time prior to the effective filing date of the claimed invention, can be used as prior art against a claimed invention.

Depending on how the courts interpret post-AIA 35 U.S.C. § 102(a)(1), the statute could also narrow the types of sales and offers for sale that constitute prior art by requiring such sales and offers to be "available to the public." Prior to the AIA, the courts interpreted 35 U.S.C. § 102(b) to permit the use of certain non-public (i.e., secret) sales and offers for sale as prior art. *See, e.g., Hobbs v. Atomic Energy Comm'n*, 451 F.2d 849, 860 (5th Cir. 1971) (finding that any conditions of secrecy which may have existed at the time the patented invention at issue was placed "on sale" were irrelevant to whether that sale barred the patentability of the invention under pre-AIA 35 U.S.C. § 102(b)). Post-AIA 35 U.S.C. § 102(a)(1), however, contains the phrase "or otherwise available to the public." As mentioned earlier, the U.S. PTO has already interpreted this phrase as modifying the "on sale" activity listed in post-AIA 35 U.S.C. § 102(a)(1). MPEP § 2152.02(d) (9th ed. 2014). Accordingly, the U.S. PTO does not currently treat secret sales or secret offers for sale as prior art to claimed inventions that are subject to post-AIA 35 U.S.C. § 102 because those sales and offers for sale are not "available to the public." Secret sales and offers for sale could, however, still qualify as prior art under the post-AIA law if a federal court finds that the "or otherwise available to the public" language of post-AIA § 102(a)(1) does not modify the term "on sale" and/or if a federal court finds that "on sale" under the post-AIA law retains the same meaning that it had under the pre-AIA law.

In sum, post-AIA 35 U.S.C. § 102(a)(1) has the potential to expand the body of sales and offers for sale that were available as prior art under pre-AIA 35 U.S.C. § 102(b) by: (1) allowing sales and offers for sale of a claimed invention that occur anywhere in the world to constitute prior art, and (2) allowing sales and offers for sale that occur any time before the effective filing date of a claimed invention to constitute prior art. On the other hand, depending on how courts interpret the statute, post-AIA 35 U.S.C. § 102(a)(1) may narrow the body of sales and offers for sale that were available as prior art under

---

[1] Although the statute only lists "sales" as being a prior art activity, that term has long been interpreted to include certain offers for sale. MPEP § 2133.03(b) (9th ed. 2014).

pre-AIA 35 U.S.C. § 102(b) by requiring that such sales and offers for sale be "available to the public" in order to constitute prior art.[m]

e. Exceptions to Post-AIA 35 U.S.C. § 102(a)(1) Prior Art

In certain instances, a claimed invention might be disclosed in a reference, or through an activity, that is mentioned in post-AIA 35 U.S.C. § 102(a)(1) but the reference or activity will not actually constitute prior art because it will meet one of the exceptions to the post-AIA 35 U.S.C. § 102(a)(1) prior art. All of these exceptions are listed in post-AIA 35 U.S.C. § 102(b)(1). The following table provides an overview of these exceptions:

Relevant Statutory Language Governing the Exceptions to Prior Art Under Post-AIA 35 U.S.C. § 102(a)(1)	Conditions that Must be Met for the Exception to Apply
"(b) EXCEPTIONS.-- (1) DISCLO-SURES MADE 1 YEAR OR LESS BE-FORE THE EFFECTIVE FILING DATE OF THE CLAIMED INVENTION.-- A disclosure made 1 year or less before the effective filing date of a claimed invention shall not be prior art to the claimed invention under subsection (a)(1) if-- (A)the disclosure was made by the inventor or joint inventor . . . ." (Post-AIA 35 U.S.C. § 102(b)(1)(A))	A disclosure that would otherwise qualify as prior art under AIA 35 U.S.C. 102(a)(1) may be disqualified as prior art if the disclosure is made: 1) by the inventor or a joint inventor; and 2) no more than one year before the effective filing date of the claimed invention. (MPEP § 2153.01(a) (9th ed. 2014))

(Continued on next page)

[m]Note that at least one federal district court has already interpreted post-AIA 35 U.S.C. § 102(a)(1) as narrowing the body of sales and offers for sale that were available as prior art under pre-AIA 35 U.S.C. § 102(b) by requiring that such sales and offers for sale be "available to the public" in order to constitute prior art. *See Helsinn Healthcare S.A. v. Dr. Reddy's Laboratories Ltd.*, No. 11-3962 (MLC), 2016 U.S. Dist. LEXIS 27477, at *139, 141-42 (D.N.J. March 3, 2016) ("The Court, having considered the parties' arguments on the plain language meaning of *§ 102(a)(1)*, the USPTO's guidelines, the undisputed AIA Committee Report, and the public policy considerations underlying the passage of the AIA, concludes that *§ 102(a)(1)* requires a public sale or offer for sale of the claimed invention. The new requirement that the on-sale bar apply to public sales comports with the plain language meaning of the amended section, the USPTO's interpretation of the amendment, the AIA Committee Report, and Congress's overarching goal to modernize and streamline the United States patent system.") (emphasis in original); *Helsinn Healthcare S.A. v. Dr. Reddy's Laboratories, Ltd.*, No. 11-3962, 2015 U.S. Dist. LEXIS 167048, at *1, 6-7 (finding that the '219 patent was not invalid under the post-AIA 35 U.S.C. § 102(a)(1) on-sale bar because the contracts upon which the on-sale defense were premised were entirely confidential, and although the existence of one of the contracts was announced publicly, the claimed invention itself was not disclosed in the public announcement). The Federal Circuit, however, has yet to weigh in on this issue.

Relevant Statutory Language Governing the Exceptions to Prior Art Under Post-AIA 35 U.S.C. § 102(a)(1)	Conditions that Must be Met for the Exception to Apply
"(b) EXCEPTIONS.-- (1) DISCLOSURES MADE 1 YEAR OR LESS BEFORE THE EFFECTIVE FILING DATE OF THE CLAIMED INVENTION.-- A disclosure made 1 year or less before the effective filing date of a claimed invention shall not be prior art to the claimed invention under subsection (a)(1) if—(A) the disclosure was made ... by another who obtained the subject matter disclosed directly or indirectly from the inventor or a joint inventor ...." (Post-AIA 35 U.S.C. § 102(b)(1)(A))	A disclosure that would otherwise qualify as prior art under AIA 35 U.S.C. 102(a)(1) may be disqualified as prior art if the disclosure is made: 1) by a person who obtained the disclosed subject matter directly or indirectly from the inventor or a joint inventor; and 2) no more than one year before the effective filing date of the claimed invention. (MPEP § 2153.01(b) (9th ed. 2014))
"(b) EXCEPTIONS.—(1) DISCLOSURES MADE 1 YEAR OR LESS BEFORE THE EFFECTIVE FILING DATE OF THE CLAIMED INVENTION.-- A disclosure made 1 year or less before the effective filing date of a claimed invention shall not be prior art to the claimed invention under subsection (a)(1) if ... (B) the subject matter disclosed had, before such disclosure, been publicly disclosed by the inventor or a joint inventor or another who obtained the subject matter disclosed directly or indirectly from the inventor or a joint inventor." (Post-AIA 35 U.S.C. § 102(b)(1)(B))	A disclosure that would otherwise qualify as prior art under AIA 35 U.S.C. 102(a)(1) may be disqualified as prior art if: 1) the disclosure was made no more than one year before the effective filing date of the claimed invention; 2) the disclosed subject matter had been previously publicly disclosed by the inventor, a joint inventor, or another who obtained the subject matter directly or indirectly from the inventor or joint inventor; and 3) the previous public disclosure by the inventor, joint inventor, or other person who obtained the subject matter directly or indirectly from the inventor or joint inventor occurred no more than one year before the effective filing date of the claimed invention. (MPEP § 2153.02 (9th ed. 2014))

Upon an initial reading of post–AIA 35 U.S.C. § 102(b)(1), it may seem as though the exceptions provided therein do not apply to all types of prior art that are listed in post–AIA 35 U.S.C. § 102(a)(1). For instance, post–AIA 35 U.S.C. § 102(b)(1) only mentions that a "disclosure" meeting certain conditions is excluded from the prior art, while post–AIA 35 U.S.C. § 102(a)(1) explicitly lists specific activities such as "public uses" and "sales" of a claimed invention as prior art. Given the lack of parallelism in the wording of post–AIA 35 U.S.C. §§ 102(a)(1) and 102(b)(1), therefore, it may seem as though § 102(b)(1) only provides exceptions for some of the activities and references that may qualify as prior art under § 102(a)(1).

The legislative history of post–AIA 35 U.S.C. § 102 nevertheless makes it clear that any activity, reference, or document that could constitute prior art under post–AIA 35 U.S.C. § 102(a)(1) is subject to the exceptions of post–AIA 35 U.S.C. § 102(b)(1).[n] In fact, during the 2011 debates on the AIA, Senator Leahy explained:

> We intend that if an inventor's actions are such as to constitute prior art under subsection 102(a), then those actions necessarily trigger subsection 102(b)'s protections for the inventor and, what would otherwise have been section 102(a) prior art, would be excluded as prior art by the grace period provided by subsection 102(b). Indeed, as an example of this, subsection 102(b)(1)(A), as written, was deliberately couched in broader terms than subsection 102(a)(1). This means that any disclosure by the inventor whatsoever, whether or not in a form that resulted in the disclosure being available to the public, is wholly disregarded as prior art.
>
> **157 CONG. REC. S1496, 2011.**

A more detailed discussion of the exceptions provided in post–AIA 35 U.S.C. § 102(b)(1) appears in Section II.C.

ii. Prior Art Under Post-AIA 35 U.S.C. § 102(a)(2)

Post–AIA 35 U.S.C. § 102(a)(2) replaces pre–AIA 35 U.S.C. § 102(e) to define when disclosures in certain types of patent documents constitute prior art to a claimed invention. The subsequent subsections explain how this new statute expands the scope of the prior art that was available under pre–AIA 35 U.S.C. § 102(e) and discusses the exceptions to this category of prior art which are set forth in post–AIA 35 U.S.C. § 102(b)(2).

---

[n]Note that the legislative history is not always dispositive when determining how a statute will be interpreted by a court. *See, e.g., Hoechst Aktiengesellschaft v. Quigg*, 917 F.2d 522, 526 (Fed. Cir. 1990) ("It is well settled law that the plain and unambiguous meaning of the words used by Congress prevails in the absence of a clearly expressed legislative intent to the contrary...When faced with [ambiguity in the language of a statute] it is incumbent upon this court to examine the legislative history to discern Congress' intent.").

a. The Expansion of Pre-AIA 35 U.S.C. § 102(e) Prior Art by Post-AIA 35 U.S.C. § 102(a)(2)

Certain patent documents can qualify as prior art as of the date they are effectively filed as opposed to the date on which they are published or issued. This is significant because the U.S. PTO keeps patent applications confidential until they are published, and publication of a patent application does not occur until 18 months after earliest filing date for which a priority benefit is sought. 35 U.S.C. § 122(b)(1)(A). Thus, patent applications are generally not "available to the public" until approximately 18 months after they are filed. An overview of both the pre- and post–AIA statutes governing these types of patent documents is given in the following tables:

Relevant Statutory Language Governing When the Description of a Claimed Invention in a Patent Application Constitutes Prior Art Under Pre-AIA Law	Conditions that Must be Met for the Description of a Claimed Invention in a Patent or Patent Application to Constitute Prior Art Under Pre-AIA Statute
"A person shall be entitled to a patent unless … (e) the invention was described in – (1) an application for patent, published under section 122(b), by another filed in the United States before the invention by the applicant for patent …." (Pre-AIA 35 U.S.C. § 102(e)(1))	A patent application describing an invention° qualifies as prior art under pre–AIA 35 U.S.C. § 102(e)(1) if the following conditions are met: 1) the patent application **was published**ᵖ **under 35 U.S.C. § 122(b)**; 2) the patent application is **"by another"** (i.e., the inventive entity listed on the patent application differs from that listed on the pending application against which the patent application is being used as prior art); 3) the patent application was **filed in the U.S.**; and 4) the patent application was filed **before the date of invention** by the applicant for patent. (Pre-AIA 35 U.S.C. § 102(e)(1); MPEP §§ 2136.01, 2136.04, 2136.05 (9th ed. 2014))

(Continued on next page)

°Note that pre-AIA 35 U.S.C. § 102(e) does not address the situation where another U.S. patent application <u>claims</u> the same invention that is claimed in another pending U.S. patent application. Under the pre-AIA law, that situation would be addressed by 35 U.S.C. § 102(g). Under the post-AIA law, that situation would be addressed by the new provisions governing derivation proceedings, which are discussed in Chapter 2.

ᵖOne exception to the publication requirement is that Patent Examiners may provisionally reject claims in a patent application over an earlier-filed patent application that has not yet been published if that earlier-filed patent application shares at least one inventor, or has a common assignee, with the patent application against which it is being used as prior art. MPEP § 2136.01(II) (9th ed. 2014); *see also* MPEP §§ 2136.01(II) and 2136.05 (9th ed. 2015) (providing strategies for overcoming rejections, including provisional rejections, under pre-AIA 35 U.S.C. § 102(e)).

Relevant Statutory Language Governing When the Description of a Claimed Invention in a Patent Application Constitutes Prior Art Under Pre-AIA Law	Conditions that Must be Met for the Description of a Claimed Invention in a Patent or Patent Application to Constitute Prior Art Under Pre-AIA Statute
"A person shall be entitled to a patent unless . . . (e) the invention was described in . . . (2) a patent granted on an application for patent by another filed in the United States before the invention by the applicant for patent, except that an international application filed under the treaty defined in section 351(a)^q shall have the effects for the purposes of this subsection of an application filed in the United States only if the international application designated the United States and was published under Article 21(2) of such treaty in the English language . . . ." (Pre-AIA 35 U.S.C. § 102(e)(2))	The description of an invention in a patent qualifies as prior art under pre-AIA 35 U.S.C. § 102(e)(2) if the following conditions are met: 1) the patent application on which the patent was granted was filed "**by another**" (i.e., by an inventive entity that is different from that listed on the patent or patent application against which it is being used as prior art); 2) a **patent was granted**^{r} on the patent application that describes the invention claimed in the patent application at-issue; 3) the patent application on which the patent was granted was **filed in the U.S.** (except that PCT applications have the effect of having been filed in the U.S. if they were filed on or after November 29, 2000, designated the United States and were published in English under Article 21(2) of the PCT); and 4) the patent application on which the patent was granted was filed **before the date of invention** by the applicant for patent. (Pre-AIA 35 U.S.C. § 102(e)(2); MPEP § 2136.03 (9th ed. 2014))

^{q}The treaty under section 351(a) is the Patent Cooperation Treaty ("PCT"). The PCT allows patent applicants to preserve their patent rights in multiple countries that are signatories to the treaty by filing a single patent application.
^{r}Note that unlike the patents that constitute prior art under pre-AIA 35 U.S.C. § 102(a) and (b), the patents mentioned in pre-AIA 35 U.S.C. § 102(e)(2) do not have to be granted/issued before the date of invention of the patent or patent application against which they are being cited as prior art. Instead, the patent application from which the patent is granted need only have been <u>filed</u> before the date of invention under pre-AIA 35 U.S.C. § 102(e)(2).

Relevant Statutory Language Governing When the Description of a Claimed Invention in a Patent or Patent Application Constitutes Prior Art Under Post-AIA Law	Conditions that Must be Met for a Description of a Claimed Invention in a Patent or Patent Application to Constitute Prior Art Under the Post-AIA Statute	Descriptions of a Claimed Invention in a Patent or Patent Application that Do NOT Constitute Prior Art Under Post-AIA Law
"A person shall be entitled to a patent unless . . . (2) the claimed invention was described in a patent issued under section 151, or in an application for patent published or deemed published under section 122(b), in which the patent or application, as the case may be, names another inventor and was effectively filed before the effective filing date of the claimed invention." (Post-AIA 35 U.S.C. § 102(a)(2))	A description of an invention in a patent constitutes prior art under post-AIA 35 U.S.C. § 102(a)(2) if the following conditions are met: 1) the patent was **granted**; 2) the **inventive entity on the patent is different** from that listed on the patent or patent application against which the patent is being cited as prior art; and 3) the patent was effectively filed **before the effective filing date** of the claimed invention. A **patent application** (i.e., a U.S. or PCT application) constitutes prior art under post-AIA 35 U.S.C. § 102(a)(2) if the following conditions are met: 1) the patent application was **published**[a] **or was deemed published**[b] under 35 U.S.C. § 122(b); 2) the **inventive entity on the patent application is different** from that listed on the patent or patent application against which the patent application is being cited as prior art; and 3) the patent application was effectively filed **before the effective filing date** of the claimed invention. (MPEP §§ 2154.01, 2154.01(c) (9th ed. 2014))	1) Disclosure of an invention in a patent or patent application will not constitute prior art if the subject matter disclosed was obtained directly or indirectly from the inventor or a joint inventor (Post-AIA 35 U.S.C. § 102(b)(2)(A)); 2) disclosure of an invention in a patent or patent application will also not constitute prior art if the subject matter disclosed had, before such subject matter was effectively filed under post-AIA 35 U.S.C. § 102(a)(2), been publicly disclosed by the inventor or a joint inventor or another who obtained the subject matter disclosed directly or indirectly from the inventor or a joint inventor (Post-AIA 35 U.S.C. § 102(b)(2)(B)); and 3) disclosure of an invention in a patent or patent application will not constitute prior art under post-AIA 35 U.S.C. § 102(a)(2) if the subject matter disclosed and the claimed invention, not later than the effective filing date of the claimed invention, were owned by the same person or subject to an obligation of assignment to the same person (Post-AIA 35 U.S.C. § 102(b)(2)(C)).

[a] Note that unlike pre-AIA 35 U.S.C. § 102(e)(2), post-AIA 35 U.S.C. § 102(a)(2) does not require PCT applications to be published in English, or to be filed on or after November 29, 2000, in order to qualify as prior art. MPEP § 2154.01(a) (9th ed. 2014).

[b] WIPO publications of PCT applications that designate the United States are "deemed published" under 35 U.S.C. § 122(b). 35 U.S.C. § 374.

Relative to pre-AIA 35 U.S.C. § 102(e)(1) and (2), post-AIA 35 U.S.C. § 102(a)(2) both expands and narrows the scope of patents and patent applications that are available as prior art to a claimed invention. On the one hand, post-AIA 35 U.S.C. § 102(a)(2) expands the sphere of patents and patent applications that are available as prior art by: (1) enabling patents and patent applications that are entitled to a right of priority, or the benefit of, a foreign priority date to be used as prior art as of that foreign priority date, and (2) permitting patents and patent applications that are effectively filed before the **effective filing date** (and not just those that are effectively filed before the **date of invention**) to be used as prior art. On the other hand, post-AIA 35 U.S.C. § 102(a)(2) narrows the scope of patents and patent applications that are available as prior art under pre-AIA 35 U.S.C. § 102(e)(2) by excluding from the prior art those patents and patent applications that are owned by the same person, or subject to an obligation of assignment to the same person, as the claimed invention as of the effective filing date of the claimed invention.

The first way in which the AIA expands the pre-AIA scope of patent and patent application prior art is by allowing those patent documents that claim priority to a foreign application to be treated as prior art as of an earlier date. Indeed, under pre-AIA 35 U.S.C. § 102(e)(1) and (2), patents and patent applications are generally deemed to have been filed as of their U.S. filing dates for purposes of qualifying as prior art. MPEP § 2136.01(I) (9th ed. 2014). This means that if, for instance, a U.S. patent application was filed on September 1, 2002 and claims priority to a Brazilian patent application that was filed on September 5, 2001, the patent application could only be cited as prior art based on its U.S. filing date (not its Brazilian filing date).[u]

---

[u]This rule, which was known as the *Hilmer* doctrine, was rather frustrating for patent litigators seeking to invalidate patent claims because even though a patent application could only qualify as prior art under pre-AIA 35 U.S.C. § 102(e) on the basis of its U.S. filing date, patent applicants trying to obtain patent protection could rely on foreign priority dates to establish earlier dates of invention for purposes of obviating rejections over prior art. Take for example a situation where application A discloses invention X, was filed in the U.S. on September 3, 2004, and claims priority to a French application filed on September 4, 2003 while application B claims invention X, was filed in the U.S. on September 5, 2004, and claims priority to a German application filed on September 12, 2003. In this situation, application A is not likely to prevent the claim to invention X in application B from issuing in a patent under pre-AIA 35 U.S.C. § 102(e) because the applicant for application B could rely on the German priority date to antedate the U.S. filing date of application A in order to obtain a patent. The French priority date for application A, however, cannot be relied upon to antedate the German priority date of application B for prior art purposes.

MPEP § 2136.03(I). The main exception to this is that under pre-AIA law, international applications (i.e., PCT applications) meeting certain criteria could be cited as prior art based on their international filing dates. MPEP § 2136.03(II).

In contrast to the pre-AIA law, post-AIA 35 U.S.C. § 102(d)(2) provides that "[f]or purposes of determining whether a patent or application for patent is prior art to a claimed invention under [post-AIA 35 U.S.C. § 102(a)(2)], such patent or application shall be considered to have been effectively filed . . . if the patent or application for patent is entitled to claim a right of priority under section 119, 365(a), or 365(b), or to claim the benefit of an earlier filing date under section 120, 121, or 365(c), based upon 1 or more prior filed applications for patent, as of the filing date of the earliest such application that describes the subject matter." Sections 119 and 365(b) set forth the criteria for claiming a right of priority to foreign patent applications. Accordingly, the body of patent and patent application prior art is much greater under the post-AIA law than under the pre-AIA law because the effective prior art date of patents and patent applications is the earliest of their U.S. or foreign priority date. Thus, under the post-AIA law, the U.S. application mentioned previously that was filed on September 1, 2002 and claimed priority to a Brazilian patent application that was filed on September 5, 2001 could be cited as prior art based on its Brazilian filing date of September 5, 2001.

Relative to pre-AIA 35 U.S.C. § 102(e)(1) and (2), post-AIA 35 U.S.C. § 102(a)(2) also expands the scope of patent and patent application prior art by permitting patents and patent applications that are effectively filed before the **effective filing date** of a claimed invention to be used as prior art against that claimed invention. Under the pre-AIA law, a published patent application (or patent application upon which a patent was granted) qualifies as prior art under 35 U.S.C. § 102(e) only if it was filed before the **date of invention** of the claimed subject matter against which it is being used. As discussed previously, the **date of invention** is typically earlier than the **effective filing date** for claimed subject matter in a patent application. Accordingly, by only requiring that patents and patent applications be filed prior to the **effective filing date** of claimed subject matter, the AIA has expanded the scope of patent and patent application prior art that is available for use against any given claimed invention.

Finally, post-AIA 35 U.S.C. § 102(b)(2)(C) actually narrows the sphere of prior art that was available under pre-AIA 35 U.S.C. § 102(e) by excluding patents and patent applications from the prior art, wherein "the subject matter disclosed [in the patent or patent application that would otherwise qualify as prior art under post-AIA 35 U.S.C. § 102(a)(2)] and the claimed invention, not later than the effective filing date of the claimed invention, were owned by the same person or subject to an obligation of assignment to the same person."[v] This prior art exception is often referred to as the "common ownership" exception. Prior to the enactment of the AIA, the common ownership exception only applied to prior art that could be used to establish that a claimed invention was obvious under 35 U.S.C. § 103, not to prior art used for the purposes of demonstrating a lack of novelty under 35 U.S.C. § 102. MPEP § 2154.02(c) (9th ed. 2014). Under the post-AIA law, however, the common ownership exception will apply to prior art that could be used to demonstrate both lack of novelty and obviousness of a claimed invention.

Overall, post-AIA 35 U.S.C. § 102(a)(2) has the potential to both expand and narrow the scope of patents and patent applications that were available as prior art under pre-AIA 35 U.S.C. § 102(e)(1) and (2). More specifically, post-AIA 35 U.S.C. § 102(a)(2) has the potential to expand the scope of patents and patent applications that were available as prior art under pre-AIA 35 U.S.C. § 102(e)(1) and (2) by: (1) providing patents and patent applications that are entitled to a right of priority, or the benefit of, a foreign priority date with an earlier effective prior art date, and (2) permitting later-filed patents and patent applications to be used as prior art. Post-AIA 35 U.S.C. § 102(a)(2) will potentially narrow the scope of patents and patent applications that were available as prior art under pre-AIA 35 U.S.C. § 102(e)(2), however, by excluding from the prior art those patents and patent applications that are owned by the same person, or subject to an obligation of assignment to the same person, as the claimed invention as of the effective filing date of the claimed invention.

b. Exceptions to Post-AIA 35 U.S.C. § 102(a)(2) Prior Art

In certain instances, a patent or patent application that would otherwise qualify as prior art under post-AIA 35 U.S.C. § 102(a)(2) will be disqualified from doing so because it will meet one of the statutory exceptions to such prior art. All of these exceptions are set forth in post-AIA 35 U.S.C. § 102(b)(2). The following table provides an overview of these exceptions:

[v]Note, however, that even if a patent or published patent application is disqualified as prior art under post-AIA 35 U.S.C. § 102(b)(2)(C), the patent or patent application may still be cited against a claimed invention in a double patenting rejection or in an enablement rejection to establish the state of the art. MPEP § 2154.02(c) (9th ed. 2014).

Relevant Statutory Language Governing the Exceptions to Prior Art Under Post-AIA 35 U.S.C. 102(a)(2)	Conditions that Must be Met for the Exception to Apply
"(b) EXCEPTIONS . . . (2) DISCLOSURES APPEARING IN APPLICATIONS AND PATENTS.-- A disclosure shall not be prior art to a claimed invention under subsection (a)(2) if--(A) the subject matter disclosed was obtained directly or indirectly from the inventor or a joint inventor . . . ." (Post-AIA 35 U.S.C. § 102(b)(2)(A))	A disclosure that would otherwise qualify as prior art under post-AIA 35 U.S.C. § 102(a)(2) (i.e., a U.S. patent, U.S. patent application publication, or WIPO published application) may be disqualified as prior art if: 1) the disclosed subject matter was obtained directly or indirectly from the inventor or a joint inventor of the claimed invention. (MPEP § 2154.02(a) (9th ed. 2014))
"(b) EXCEPTIONS . . . (2) DISCLOSURES APPEARING IN APPLICATIONS AND PATENTS.-- A disclosure shall not be prior art to a claimed invention under subsection (a)(2) if . . . (B) the subject matter disclosed had, before such subject matter was effectively filed under subsection (a)(2), been publicly disclosed by the inventor or a joint inventor or another who obtained the subject matter disclosed directly or indirectly from the inventor or a joint inventor . . . ." (Post-AIA 35 U.S.C. § 102(b)(2)(B))	A disclosure that would otherwise qualify as prior art under post-AIA 35 U.S.C. § 102(a)(2) (i.e., a U.S. patent, U.S. patent application publication, or WIPO published application) may be disqualified as prior art if: 1) the disclosed subject matter had been previously publicly disclosed by the inventor, a joint inventor, or another person who obtained the disclosed subject matter directly or indirectly from the inventor or joint inventor; and 2) the previous public disclosure by the inventor, joint inventor, or other person who obtained the disclosed subject matter directly or indirectly from the inventor or joint inventor occurred no more than one year prior to the effective filing date of the claimed invention. (MPEP § 2154.02(b) (9th ed. 2014))
"(b) EXCEPTIONS . . . (2) DISCLOSURES APPEARING IN APPLICATIONS AND PATENTS.-- A disclosure shall not be prior art to a claimed invention under subsection (a)(2) if . . . (C) the subject matter disclosed and the claimed invention, not later than the effective filing date of the claimed invention, were owned by the same person or subject to an obligation of assignment to the same person." (Post-AIA 35 U.S.C. § 102(b)(2)(C))	A disclosure that would otherwise qualify as prior art under post-AIA 35 U.S.C. § 102(a)(2) (i.e., a U.S. patent, U.S. patent application publication, or WIPO published application) may be disqualified as prior art if: 1) the disclosed subject matter (i.e., the subject matter disclosed in the U.S. patent, U.S. patent application publication, or WIPO published application) and the claimed invention were owned by the same person, or subject to an obligation of assignment to the same person, as of a date that is not later than the effective filing date of the claimed invention. (MPEP § 2154.02(c) (9th ed. 2014))

The exceptions provided by post-AIA 35 U.S.C. §§ 102(b)(2)(A) and 102(b)(2)(B) are very similar to those that are seen in post-AIA 35 U.S.C. §§ 102(b)(1)(A) and 102(b)(1)(B). Unlike post-AIA 35 U.S.C. § 102(b)(1), however, post-AIA 35 U.S.C. § 102(b)(2) provides an exception to post-AIA 35 U.S.C. § 102(a)(2) prior art where a disclosure that would otherwise qualify as prior art under that statute, and the claimed invention, are owned by the same person, or subject to an obligation of assignment to the same person, as of a date that is not later than the effective filing date of the claimed invention (i.e., a "common ownership exception"). Post-AIA 35 U.S.C. § 102(c) delineates the requirements that must be met to demonstrate that disclosed subject matter and a claimed invention are owned by the same person or subject to an obligation of assignment to the same person.[w] As noted in Section II.B.ii.a., this "common ownership" exception to prior art formerly applied only to prior art that was being used to demonstrate a claimed invention's obviousness under pre-AIA 35 U.S.C. § 103. Pre-AIA 35 U.S.C. § 102; pre-AIA 35 U.S.C. § 103(c)(1). Now, however, the exception applies to prior art that is being used to demonstrate either lack of novelty under post-AIA 35 U.S.C. § 102(a)(2) or obviousness under post-AIA 35 U.S.C. § 103.

C. Abridgment of Inventor Grace Periods Under the AIA

U.S. patent law has always been somewhat unique in that it affords an inventor a relatively generous grace period during which he/she can file a new patent application if the inventor somehow discloses his/her invention to the public before filing a patent application. Other countries (e.g., Western European countries such as the United Kingdom and Germany) adhere to an absolute novelty standard that generally prohibits an inventor from obtaining patent protection for an invention if that invention has been made available to the public (e.g., through a written disclosure or sale of the invention) anywhere in the world before the inventor has filed a patent application. *See, e.g.,* European Patent Convention, Article 54(1) and (2) (explaining that an invention shall be considered to be novel if it does not form part of the state

---

[w]Notably, under the pre-AIA law, patent applicants were required to demonstrate that a joint research agreement was in effect before a claimed invention was made (i.e., before the date of invention) in order to qualify for the "common ownership" exception under pre-AIA 35 U.S.C. § 103(c)(1). Pre-AIA 35 U.S.C. § 103(c)(2)(A). Post-AIA 35 U.S.C. § 102(c), however, allows applicants to avail themselves of the "common ownership" exception by demonstrating that a joint research agreement was in effect on or before the effective filing date of the claimed invention. Post-AIA 35 U.S.C. § 102(c)(1). This change is significant because it provides parties with more time to enter into joint research agreements in order to disqualify disclosures that would otherwise constitute prior art under post-AIA 35 U.S.C. § 102(a)(2).

of the art, and the state of the art comprises everything made available to the public by means of a written or oral description, by use, or in any other way, before the date of filing of the European patent application).

Under pre-AIA 35 U.S.C. § 102, the inventor grace periods were set forth implicitly through the conditions that were required of prior art under the various subsections of the statute. For instance, by requiring that an invention be "described in a printed publication . . . more than one year prior to the date of the application for patent in the United States" in order for the printed publication to bar patentability of a claimed invention, pre-AIA 35 U.S.C. § 102(b) provided an inventor with a one-year grace period to file a patent application after he/she described his/her invention in a printed publication (e.g., a journal article). In contrast to the pre-AIA statute, post-AIA 35 U.S.C. § 102(b) explicitly lists exceptions to the prior art of post-AIA 35 U.S.C. § 102(a). In particular, post-AIA 35 U.S.C. § 102(b)(1) lists exceptions to the prior art of post-AIA 35 U.S.C. § 102(a)(1) and post-AIA 35 U.S.C. § 102(b)(2) lists exceptions to the prior art of post-AIA 35 U.S.C. § 102(a)(2). Importantly, the exceptions listed in post-AIA 35 U.S.C. § 102(b)(1) are what define an inventor's grace period under the new law.

Post-AIA 35 U.S.C. § 102(b)(1) specifically provides three types of inventor grace periods for references and activities that would otherwise qualify as prior art under post-AIA 35 U.S.C. § 102(a)(1): (1) a grace period for an inventor's own disclosures of his/her claimed invention (i.e., an "inventor's own disclosure" grace period), (2) a grace period for disclosures by someone other than the inventor, wherein that person obtained the disclosed subject matter directly or indirectly from the inventor (i.e., an "inventor-derived disclosure" grace period), and (3) a grace period for disclosures by someone other than an inventor that occur after the inventor, joint inventor, or another who obtained the disclosed subject matter directly or indirectly from the inventor or joint inventor, publicly discloses his/her invention (i.e., an "intervening disclosure" grace period).

Before discussing each of the three types of grace periods provided under post-AIA 35 U.S.C. § 102(b)(1), it is important to understand that all three grace periods differ from the grace periods provided under the pre-AIA law in that the starting date from which the grace periods are measured is not the same. In particular, the one-year grace period provided by pre-AIA 35 U.S.C. § 102(b) is measured from the filing date of the earliest application that is filed in the United States (i.e., an application that is either filed directly in the United States or an application that enters the national stage in the United States from a PCT application). The one-year grace period provided by post-AIA 35 U.S.C. § 102(b)(1), however, is measured from the filing date of the earliest U.S. or foreign patent application to which a proper benefit or priority claim has been asserted in the patent or application. MPEP § 2152.01 (9th ed. 2014). The practical effect of this change is that under the post-AIA law, an applicant may be able to overcome rejections over references and activities having an earlier effective prior art date than under the pre-AIA law where the applicant submits a proper claim to priority to a foreign patent application. Indeed, if a claimed invention is rejected over a reference or activity that pre-dates the U.S. filing date of the claimed invention, and qualifies as prior art under post-AIA 35 U.S.C. § 102(a)(1), a patent applicant can try to overcome the rejection by perfecting a priority or benefit claim to a foreign patent application and invoke an exception under post-AIA 35 U.S.C. § 102(b)(1). MPEP § 706.02(b)(1) (9th ed. 2014). Under pre-AIA 35 U.S.C. § 102(b), however, the filing date of a foreign patent application could not be used to overcome such prior art, even if the claimed invention was entitled to a priority or benefit claim to that foreign patent application. *See* MPEP § 706.02(b)(2) (9th ed. 2014) (explaining that a rejection under pre-AIA 35 U.S.C. § 102(b) can be overcome by perfecting benefit claims to U.S. applications under either 35 U.S.C. §§ 119(e) or 120 but not by perfecting a benefit claim to a foreign application under 35 U.S.C. § 119(a)-(d)). The following timeline demonstrates this difference:

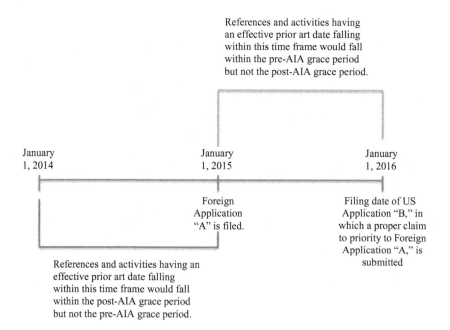

References and activities having an effective prior art date falling within this time frame would fall within the pre-AIA grace period but not the post-AIA grace period.

January 1, 2014   January 1, 2015   January 1, 2016

Foreign Application "A" is filed.

Filing date of US Application "B," in which a proper claim to priority to Foreign Application "A," is submitted

References and activities having an effective prior art date falling within this time frame would fall within the post-AIA grace period but not the pre-AIA grace period.

The three grace periods provided under post-AIA 35 U.S.C. § 102(b)(1) are discussed in more detail in sections II.C.i–iii.

i. Inventor's Own Disclosure Grace Period Under Post-AIA 35 U.S.C. § 102(b)(1)(A)

Other than the change in the date from which the grace period for certain patent applications is measured as discussed in Section II.C., the grace period that was provided to inventors who disclose their inventions before filing a patent application under pre-AIA 35 U.S.C. § 102 was left unchanged by post-AIA 35 U.S.C. § 102. In particular, both the pre-AIA and post-AIA law provide inventors with a one-year grace period to file a patent application[x] before that inventor's own public disclosure of his/her invention can be used as prior art against his/her patent application. *Compare* post-AIA 35 U.S.C. § 102(b)(1)(A) (explaining that a disclosure made 1 year or less before the effective filing date of a claimed invention shall not be prior art if the disclosure was made by the inventor or joint inventor) *with* pre-AIA 35 U.S.C. § 102(b) (barring patentability only where the invention was patented, described in a printed publication, in public use, or on sale *more than one year* prior to the date of the application for patent in the U.S.); pre-AIA 35 U.S.C. § 102(d) (barring patentability only where the patent applicant, or his legal representatives or assigns, filed an application for patent or inventor's certificate in a foreign country *more than twelve months before the filing of the application in the United States*).[y]

---

[x] Note that under the post-AIA law, the inventor is not the only person who may file a patent application. Indeed, the inventor, assignee, obligated assignee, or another party having sufficient interest in a claimed invention may file the patent application. 37 C.F.R. § 1.46(a); MPEP § 409.05 (9th ed. 2015).

[y] The grace period for an inventor's own disclosure applies only to references and activities that qualify as prior art under post-AIA 35 U.S.C. § 102(a)(1) (i.e., the only grace period for an inventor's own disclosures is set forth in post-AIA 35 U.S.C. § 102(b)(1)). There is no need for such a grace period for patent and published patent application prior art under post-AIA 35 U.S.C. § 102(a)(2) because in order for such patents and published patent applications to qualify as prior art under that statute, they must name someone other than the inventor(s) of the patent or patent application at-issue. *See* post-AIA 35 U.S.C. § 102(a)(2) (requiring that in order to qualify as prior art, a patent or patent application describing a claimed invention must name "another inventor" besides the inventor(s) of the claimed invention). Thus, if inventor A claims invention "x" in patent application #2 after disclosing invention "x" in a patent application #1 that was effectively filed, but not published, before the effective filing date of patent application #2, patent application #1 could only qualify as prior art to patent application #2 under post-AIA 35 U.S.C. § 102(a)(2) if patent application #1 lists as an inventor someone other than inventor A.

Moreover, the manner in which an inventor discloses his/her invention to the public is likely irrelevant to whether the grace period provided by post-AIA 35 U.S.C. § 102(b)(1)(A) applies. *See* 157 CONG. REC. S1496, 2011 (statement of Sen. Leahy) ("In particular, some in the small inventor community have been concerned that a disclosure by an inventor might qualify as patent-defeating prior art under [post-AIA 35 U.S.C. §] 102(a) because, for example, the inventor's public disclosure and by a 'public disclosure' I mean one that results in the claimed invention being 'described in a printed publication, or in public use, on sale, or otherwise available to the public'— might in some situation not be excluded as prior art under 35 U.S.C. § 102(b)'s grace period. There is absolutely no situation in which this could happen given the interplay between [post-AIA 35 U.S.C. §§] 102(a) and 102(b) as these subsections are drafted. We intend that if an inventor's actions are such as to constitute prior art under [post-AIA 35 U.S.C. §] 102(a), then those actions necessarily trigger [post-AIA 35 U.S.C. §] 102(b)'s protections for the inventor and, what would otherwise have been [post-AIA 35 U.S.C. §] 102(a) prior art, would be excluded as prior art by the grace period provided by [post-AIA 35 U.S.C. §] 102(b). Indeed, as an example of this, [post-AIA 35 U.S.C. §] 102(b)(1)(A), as written, was deliberately couched in broader terms than [post-AIA 35 U.S.C. §] 102(a)(1)."). If the disclosure occurs within the grace period that disclosure cannot be used as prior art against an inventor's claimed invention. Accordingly, if an inventor decides to disclose his/her invention in a journal article and forgets to contact his/her patent attorney before the article is published, he/she has one year from the effective prior art date of the journal article to file a patent application claiming his/her invention before that journal article can be cited as prior art against the patent application under either the pre- or post-AIA law.

ii. Inventor-Derived Disclosure Grace Period Under Post-AIA 35 U.S.C. § 102(b)(1)(A)

Like the grace period for an inventor's own disclosures, the grace period provided for inventor-derived disclosures is the same under both the pre- and post-AIA law except for the change in date from which the grace period for certain patent applications is measured, which is discussed in Section II.C. of this chapter.[z] In particular, under both the pre- and post-AIA law, an

[z]The inventor-derived grace period only applies if the inventor had not publicly disclosed the invention before the third party's public disclosure. Situations wherein the inventor first disclosed his/her invention to the public and then a third party subsequently makes a public disclosure of the invention are addressed by post-AIA 35 U.S.C. §§ 102(b)(1)(B) and 102(b) (2)(B) (discussed later).

inventor has one year to file a patent application claiming subject matter after the date on which a third party has publicly disclosed the subject matter so long as the inventor can demonstrate that the third party obtained the disclosed subject matter directly or indirectly from the inventor or joint inventor of the patent application. *Compare* post-AIA 35 U.S.C. § 102(b)(1) (A) (explaining that a disclosure that is made 1 year or less before the effective filing date of a claimed invention is not prior art if the person who disclosed the invention obtained the disclosed subject matter directly or indirectly from the inventor or a joint inventor) *with* MPEP § 2132.01(III) (9th ed. 2014) (allowing patent applicants to overcome pre-AIA novelty rejections over another's disclosure of their invention that is dated within a year of the patent application's filing date by filing an affidavit or declaration that establishes that the person disclosing the invention in the cited prior art derived the disclosed subject matter from the inventor of the patent application).

A common situation in which inventor-derived disclosures come into play arises when different inventive entities employed by the same company file patent applications that disclose overlapping subject matter. Imagine for instance that employees Bob and Sue of Unifever Corporation developed a new shampoo formulation incorporating aloe vera and claim that formulation in patent application #2, filed March 2, 2015. While Bob and Sue were developing their shampoo formulation, they also told Flo about an aloe vera-based liquid soap formulation who then disclosed that soap formulation in her own patent application #1, which is published February 5, 2015. If an Examiner relies on the aloe vera-based liquid soap formulation disclosure of Flo's patent application #1 to reject the claims in patent application #2, Bob and Sue could overcome the rejection by demonstrating through a declaration or affidavit that Flo derived the aloe vera–based liquid soap formulation from them.

iii. Intervening Disclosure Grace Period[aa] Under Post-AIA 35 U.S.C. § 102(b)(1)(B)

The most significant change to inventor grace periods can be seen in so-called "intervening disclosure" situations. "Intervening disclosure" situations occur when an inventor, joint inventor, or another who obtained subject

[aa]The grace period created by post-AIA 35 U.S.C. § 102(b)(1)(B) is often referred to as the "first-to-disclose" grace period because it is intended to protect inventors who have publicly disclosed their inventions from having their claimed inventions rendered unpatentable by any subsequent disclosures. 157 Cong. Rec. H4429, 2011 (statement of Rep. Smith).

matter from the inventor or joint inventor, publicly discloses a claimed invention[bb] and then a third party publicly discloses that claimed invention after the inventor, joint inventor, or another's public disclosure but before the inventor files a patent application claiming his/her invention. The following timeline demonstrates an example of such a scenario:

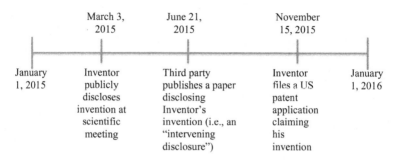

As compared to the pre-AIA law, post-AIA 35 U.S.C. § 102 can significantly restrict an inventor's right to a patent in intervening-disclosure situations because the new law only allows inventors to antedate so much of the intervening disclosure as was disclosed in their own public disclosure. MPEP § 2153.02 (9th ed. 2014). For example, imagine that the inventor in the aforementioned timeline publicly disclosed his newly invented pencil with an eraser and then the third party discloses both a pencil with an eraser and a pen having erasable ink with an eraser in his paper published on June 21, 2015. Then imagine that after carefully examining the inventor's claims, a Patent Examiner rejects the inventor's claims to the pencil with an eraser as lacking novelty over the third party's disclosure of the pencil with the eraser and as being obvious over the third party's disclosure of the pen with the eraser.

Under the pre-AIA law, the inventor could overcome both of these rejections by showing (through a declaration or affidavit under 37 C.F.R. § 1.131) that he/she invented the disclosed subject matter before the date of the third-party's disclosure. *See* MPEP § 706.02(b)(2) (9th ed. 2014) (explaining that a rejection under pre-AIA 35 U.S.C. § 102(a) can be overcome by filing an affidavit or declaration under 37 C.F.R. § 1.131(a) showing prior invention); MPEP § 715.03(I)(C) (9th ed. 2014) ("References

[bb]Note that for the "intervening disclosure" grace period to apply, the public disclosure by the inventor, joint inventor, or another who obtained the subject matter disclosed directly or indirectly from the inventor or joint inventor, must occur no earlier than one year prior to the effective filing date of the inventor's claimed invention. If the public disclosure occurs more than one year prior to the effective filing date of the claimed invention, then such public disclosure would render the claimed invention unpatentable for lack of novelty under post-AIA 35 U.S.C. § 102(a)(1).

or activities which disclose one or more embodiments of a single claimed invention, as opposed to species of a claimed genus, can be overcome by filing a 37 CFR § 1.131(a) affidavit showing prior completion of a single embodiment of the invention, whether it is the same or a different embodiment from that disclosed in the reference or activity."). The inventor could do this by providing evidence of his/her disclosure at the scientific meeting, which would establish that the inventor's date of invention for the claimed invention was prior to the third-party's intervening disclosure.

The date of invention for claimed subject matter is, however, irrelevant under post-AIA 35 U.S.C. § 102. Thus, the previously mentioned approach would not work under the new law. Instead, in order to overcome the Patent Examiner's rejections under the post-AIA law, the inventor would be required to demonstrate that the exception set forth in post-AIA 35 U.S.C. § 102(b)(1)(B)[cc] applies. That exception reads as follows:

> (b) EXCEPTIONS.-- (1) DISCLOSURES MADE 1 YEAR OR LESS BEFORE THE EFFECTIVE FILING DATE OF THE CLAIMED INVENTION.-- A disclosure made 1 year or less before the effective filing date of a claimed invention shall not be prior art to the claimed invention under subsection (a)(1) if ... (B) the subject matter disclosed had, before such disclosure, been publicly disclosed by the inventor or a joint inventor or another who obtained the subject matter disclosed directly or indirectly from the inventor or a joint inventor.

The U.S. PTO so far has taken the position that only the portions of the third-party's disclosure (i.e., the "intervening disclosure") that were also previously disclosed by the inventor, joint inventor, or another who obtained the disclosed subject matter directly or indirectly from the inventor or joint

---

[cc]Note that post-AIA 35 U.S.C. § 102(b)(2)(B) provides a parallel exception for intervening disclosures that appear in patent documents (i.e., a patent, U.S. patent application publication, or WIPO published application). While such intervening disclosures could qualify as prior art under post-AIA 35 U.S.C. § 102(a)(2), post-AIA 35 U.S.C. § 102(b)(2)(B) excludes such disclosures from the prior art where those disclosures occur after an inventor's own public disclosure. It is important to understand that while post-AIA 35 U.S.C. § 102(b)(2)(B) does not itself define a grace period or impose time restrictions for when disclosures mentioned therein must occur, the exception that is provided by this statute is constrained by the grace period set forth in post-AIA 35 U.S.C. § 102(b)(1)(A). For instance, if the inventor's own public disclosure (i.e., the disclosure that precedes the intervening disclosure) occurs more than one year prior to the effective filing date of the inventor's claimed invention, that public disclosure will constitute prior art under post-AIA 35 U.S.C. § 102(a)(1) and could not be disqualified under post-AIA 35 U.S.C. § 102(b)(1). Thus, while post-AIA 35 U.S.C. § 102(b)(2)(B) would serve to remove an intervening disclosure from the prior art, the inventor would be barred from obtaining a patent based on his/her own public disclosure, which could not be removed from the prior art.

inventor, will fall within the prior art exception outlined in post-AIA 35 U.S.C. § 102(b)(1)(B). MPEP § 2153.02 (9th ed. 2014). Accordingly, under the post-AIA law, the inventor in the previous example could overcome the novelty rejection over the third party's disclosure of the pencil with the eraser because that subject matter was disclosed in the inventor's own public disclosure dated March 3, 2015. Unlike under the pre-AIA law, however, the post-AIA law would not allow the inventor to overcome the obviousness rejection over the third party's disclosure of the pen with the eraser because the pen with the eraser was not included in the inventor's own public disclosure.

Importantly, the U.S. PTO has stated that for the prior art exception of post-AIA 35 U.S.C. § 102(b)(1)(B) to apply, there is no requirement that:

1) the mode of disclosure by the inventor or joint inventor or another who obtained the subject matter disclosed directly or indirectly from the inventor or a joint inventor (e.g., patenting, publication, sale) be the same as the mode of disclosure of the intervening grace period disclosure; or

2) the disclosure by the inventor or joint inventor or another who obtained the subject matter disclosed directly or indirectly from the inventor or a joint inventor be a verbatim disclosure of the intervening grace period disclosure.

MPEP § 2153.02 (9th ed. 2014). Unlike the exception for inventor-derived disclosures in post-AIA 35 U.S.C. § 102(b)(1)(A), there is also no requirement under post-AIA 35 U.S.C. § 102(b)(1)(B) that the third party must have obtained the disclosed subject matter from the inventor before providing the intervening disclosure. See 157 CONG. REC. S5320, 2011 (statement of Sen. Kyl) ("The U.S. inventor [who invokes § 102(b)(1)(B)] does not need to prove that the third party disclosures following his own disclosures are derived from him.").

In addition, if the subject matter of the third party's intervening disclosure is simply a more general description of the subject matter that was previously disclosed by the inventor, the intervening disclosure would fall within the post-AIA 35 U.S.C. § 102(b)(1)(B) exception and would therefore not be available as prior art under post-AIA 35 U.S.C. § 102(a)(1).[dd] MPEP

[dd]This is especially important for chemical inventions because it means that if an inventor discloses a chemical species and then a third party later discloses a genus encompassing that species before the inventor files a patent application claiming his/her species, the third-party's intervening disclosure will still fall within the exception to prior art of post-AIA 35 U.S.C. § 102(b)(1)(B) and would therefore not qualify as prior art under post-AIA 35 U.S.C. § 102(a)(1) (assuming that both the inventor's and third party's disclosures occur within one year of the effective filing date of the claimed species at-issue).

§ 2153.02 (9th ed. 2014). The converse is also true however. If an inventor, or joint inventor or another person who obtains subject matter from the inventor or joint inventor makes a public disclosure of a genus of compounds and then a third party discloses a species within that genus in an intervening disclosure such that the inventor's disclosure is simply a more general description of the subject matter that is described in the intervening disclosure, the intervening disclosure of the species would <u>not</u> fall within the exceptions of post-AIA 35 U.S.C. § 102(b)(1)(B) and <u>would be available</u> as prior art under post-AIA 35 U.S.C. § 102(a)(1). *Id.* Similarly, if an inventor, joint inventor, or another who obtains subject matter from the inventor or joint inventor makes a grace period public disclosure of a chemical species and then a third party discloses an alternate species in an intervening disclosure, the intervening disclosure of the alternate species <u>would be available</u> as prior art to the inventor's chemical species under post-AIA 35 U.S.C. § 102(a)(1). *Id.*

Yet another important distinction between the pre- and post-AIA grace periods in "intervening disclosure" situations is that under the post-AIA law, an inventor may only rely on his/her **public disclosure** to antedate an intervening disclosure by a third party. *See* post-AIA 35 U.S.C. § 102(b)(1)(B) (providing an exception to post-AIA 35 U.S.C. § 102(a)(1) prior art where "the subject matter disclosed had, before such disclosure, been *publicly* disclosed by the inventor or a joint inventor or another who obtained the subject matter disclosed directly or indirectly from the inventor or a joint inventor.") (emphasis added). Under the pre-AIA law, inventors could antedate such third party disclosures by relying on **non-public** information to establish that the claimed invention was invented on a date that is earlier than the date of the third party disclosure. *See* MPEP § 706.02(b)(2) (9th ed. 2014) (explaining that a rejection under pre-AIA 35 U.S.C. § 102(a) can be overcome by filing an affidavit or declaration under 37 C.F.R. § 1.131 showing prior invention); 37 C.F.R. § 1.131 (making no requirement that evidence used to establish prior invention be publicly available).

In sum, post-AIA 35 U.S.C. § 102(b)(1)(B) abridges the intervening disclosure grace period inventors enjoyed under pre-AIA 35 U.S.C. § 102 by: (1) allowing inventors to antedate only so much of an intervening disclosure as they themselves disclosed, and (2) requiring that an inventor's previous disclosure (i.e., the disclosure preceding an intervening disclosure) be public in order to be used to antedate an intervening disclosure.

D. Elimination of Certain Provisions of Pre-AIA 35 U.S.C. § 102

The last major change that the AIA made to pre-AIA 35 U.S.C. § 102 was that it eliminated several subsections known as the "loss of right" provisions from the statute. More specifically, subsections (c), (d), (f), and (g) of pre–AIA 35 U.S.C. § 102 were repealed. *Compare* Leahy-Smith America Invents Act of 2011, Pub. L. No. 112-29, § 3(b)(1), 125 Stat. 284, 285-87 *with* pre–AIA 35 U.S.C. § 102. The following table provides the language of the repealed subsections:

Subsections of Pre-AIA 35 U.S.C. 102 that were Repealed by the AIA
"A person shall be entitled to a patent unless . . . (c) he has abandoned the invention . . . ." (Pre–AIA 35 U.S.C. § 102(c))
"A person shall be entitled to a patent unless . . . (d) the invention was first patented or caused to be patented, or was the subject of an inventor's certificate, by the applicant or his legal representatives or assigns in a foreign country prior to the date of the application for patent in this country on an application for patent or inventor's certificate filed more than twelve months before the filing of the application in the United States . . . ." (Pre–AIA 35 U.S.C. § 102(d))
"A person shall be entitled to a patent unless . . . (f) he did not himself invent the subject matter sought to be patented . . . ." (Pre–AIA 35 U.S.C. § 102(f))
"A person shall be entitled to a patent unless . . . (g)(1) during the course of an interference conducted under section 135 or section 291, another inventor involved therein establishes, to the extent permitted in section 104, that before such person's invention thereof the invention was made by such other inventor and not abandoned, suppressed, or concealed, or (2) before such person's invention thereof, the invention was made in this country by another inventor who had not abandoned, suppressed, or concealed it. In determining priority of invention under this subsection, there shall be considered not only the respective dates of conception and reduction to practice of the invention, but also the reasonable diligence of one who was first to conceive and last to reduce to practice, from a time prior to conception by the other." (Pre–AIA 35 U.S.C. § 102(g))

Despite post–AIA 35 U.S.C. § 102's omission of the language of the "loss of right" subsections of pre–AIA 35 U.S.C. § 102, some of the activities in those subsections are still relevant to the patentability of an invention under the post–AIA law. For example, pre–AIA 35 U.S.C. § 102(f) prevents the patenting of an invention by an applicant who "did not himself invent the subject matter sought to be patented." Although this subsection of pre–AIA 35 U.S.C. § 102 was eliminated by post–AIA 35 U.S.C. § 102, the new law still recognizes that only the true inventor is entitled to a patent on a claimed invention. *See, e.g.,* 35 U.S.C. § 101 ("*Whoever* invents or discovers . . . may obtain a patent therefor, subject to the conditions and requirements of this title.") (emphasis added); post–AIA 35 U.S.C. § 135 (creating derivation proceedings to resolve inventorship disputes). Furthermore, activities that were mentioned in the "loss of right" subsections of pre-AIA 35 U.S.C. § 102 may produce prior art under post-AIA 35 U.S.C. § 102. For instance, if an inventor abandons his/her invention in a manner that results in that invention becoming publicly available, that abandonment may prevent the inventor from obtaining a patent under post-AIA 35 U.S.C. § 102(a)(1).

E. Putting It All Together

The following table gives a few examples that might help to show the similarities and differences between pre- and post-AIA 35 U.S.C. § 102:

Fact Pattern for Prior Art Rejection	Analysis of the Prior Art Rejection Under Pre-AIA Law	Analysis of the Prior Art Rejection Under Post-AIA Law
A patent application[ee] names Jan Smith as the sole inventor. Claim 1 of the patent application is rejected for lack of novelty under 35 U.S.C. § 102 over a journal article authored by Jan Smith and Fredrick Han. The journal article has an effective prior art date that is less than one year before the effective filing date of the invention claimed in claim 1.	The rejection here is under pre-AIA 35 U.S.C. § 102(a). The options for overcoming the rejection are: 1) Jan Smith could file a declaration or affidavit under 37 C.F.R. § 1.132 to establish that Frederick Han derived the subject matter of claim 1 that is disclosed in the journal article from her. By doing this, Jan Smith would establish that the invention disclosed in the journal article is not "by another."	The rejection here is under post-AIA 35 U.S.C. § 102(a)(1). The options for overcoming the rejection are: 1) Jan Smith could file an affidavit or declaration of attribution under 37 C.F.R. § 1.130(a) that establishes that the disclosure in the journal article was made by her pursuant to 35 U.S.C. § 102(b)(1)(A).

(Continued on next page)

[ee]In all of the examples in this section, it is assumed that the patent application filed by Jan Smith is not entitled to claim priority to any other patent application (either foreign or domestic). If Jan Smith's patent application were entitled to claim priority to a previous patent application that had an effective filing date that is prior to the effective date of the prior art cited against her, she could also perfect a claim to priority to either a foreign or domestic application to antedate the reference. *See* MPEP § 706.02(b)(1) and (2). This approach could be taken under either the pre-AIA or post-AIA law.

Analysis of the Prior Art Rejection Under Pre-AIA Law	Analysis of the Prior Art Rejection Under Post-AIA Law
2) Jan Smith could file a declaration or affidavit under 37 C.F.R. § 1.131 that establishes that she invented the subject matter of claim 1 on a date that is before the effective prior art date of the journal article that is cited against her patent application. 3) Jan Smith could amend claim 1 so that it claims subject matter that is patentably distinct from the subject matter disclosed in the journal article cited against her patent application. 4) Jan Smith could argue that the subject matter of claim 1 is patentably distinct from the subject matter disclosed in the journal article. 5) Assuming that Frederick Han is actually an inventor of the claimed subject matter of at least one claim in Jan Smith's patent application, Jan Smith could add Frederick Han as an inventor on her patent application. After doing this, the journal article would no longer be "by another" and could therefore not constitute prior art under pre-AIA 35 U.S.C. § 102(a). (MPEP §§ 706.02(b)(2), 2132.01(I) and (II) (9th ed. 2014))	2) Jan Smith could file an affidavit or declaration of attribution under 37 C.F.R. § 1.130(a) that establishes that Frederick Han directly or indirectly obtained the subject matter that is disclosed in the journal article, and claimed in claim 1 of the patent application, from her. *See* 35 U.S.C. § 102(b)(1)(A). 3) Jan Smith could amend claim 1 so that it claims subject matter that is patentably distinct from the subject matter disclosed in the journal article cited against her patent application. 4) Jan Smith could argue that the subject matter set forth in claim 1 is patentably distinct from the subject matter disclosed in the journal article. (MPEP §§ 706.02(b)(1), 2155 (9th ed. 2014))

Fact Pattern for Prior Art Rejection	Analysis of the Prior Art Rejection Under Pre-AIA Law	Analysis of the Prior Art Rejection Under Post-AIA Law
A patent application names Jan Smith as the sole inventor. Claim 1 of the patent application is rejected for lack of novelty under 35 U.S.C. § 102 over a journal article authored by Jan Smith and Fredrick Han. The journal article has an effective prior art date that is less than one year before the effective filing date of Jan Smith's patent application. Prior to the effective prior art date of the journal article, and within a year before Jan Smith filed her patent application, Jan Smith had also given a public presentation at a chemistry conference in which she disclosed the exact subject matter that was claimed in claim 1 of her patent application and that was disclosed in the journal article.	The rejection here is under pre-AIA 35 U.S.C. § 102(a). The options for overcoming the rejection are: 1) Jan Smith could file a declaration or affidavit under 37 C.F.R. § 1.132 to establish that Frederick Han derived the subject matter of claim 1 from her. By doing this, Jan Smith would establish that the invention disclosed in the journal article is not "by another." 2) Jan Smith could file a declaration or affidavit under 37 C.F.R. § 1.131 that establishes that she invented the subject matter of claim 1 on a date that is before the effective prior art date of the journal article that is cited against her patent application. Evidence of Jan Smith's presentation at the chemistry conference could be used to establish the earlier invention date.	The rejection here is under post-AIA 35 U.S.C. § 102(a)(1). The options for overcoming the rejection are: 1) Jan Smith could file an affidavit or declaration under 37 C.F.R. § 1.130(b) to establish that the subject matter disclosed in the journal article had, before such disclosure was made in the journal article, been publicly disclosed by Jan Smith. Jan Smith's affidavit or declaration would also need to include a description of her public presentation at the chemistry conference that has sufficient detail and particularity to determine what subject matter had been publicly disclosed. *See* post-AIA 35 U.S.C. § 102(b)(1)(B). 2) Jan Smith could amend claim 1 so that it claims subject matter that is patentably distinct from the subject matter disclosed in the journal article cited against her patent application.

(*Continued on next page*)

Analysis of the Prior Art Rejection Under Pre-AIA Law	Analysis of the Prior Art Rejection Under Post-AIA Law
3) Jan Smith could amend claim 1 so that it claims subject matter that is patentably distinct from the subject matter disclosed in the journal article cited against her patent application. 4) Jan Smith could argue that the subject matter of claim 1 is patentably distinct from the subject matter disclosed in the journal article. 5) Assuming that Frederick Han is actually an inventor of the claimed subject matter of at least one claim in Jan Smith's patent application, Jan Smith could add Frederick Han as an inventor on her patent application. By doing this, the journal article would no longer constitute prior art under pre-AIA 35 U.S.C. § 102(a) because it would not be authored "by another." (MPEP §§ 706.02(b)(2), 2132.01(I) and (II) (9th ed. 2014))	3) Jan Smith could argue that the subject matter of claim 1 is patentably distinct from the subject matter disclosed in the journal article. (MPEP §§ 706.02(b)(1), 2155 (9th ed. 2014))

Fact Pattern for Prior Art Rejection	Analysis of the Prior Art Rejection Under Pre-AIA Law	Analysis of the Prior Art Rejection Under Post-AIA Law
A patent application names Jan Smith as the sole inventor. Claim 1 of the patent application is rejected for lack of novelty under 35 U.S.C. § 102 over a pending published patent application that discloses, but does not claim, the subject matter claimed in claim 1 of Jan Smith's patent application (hereinafter referred to as the "prior art application"). The prior art application lists both Jan Smith and Fredrick Han as inventors. The prior art application has an effective prior art date only as of its filing date (i.e., it was filed before, but was published after, Jan Smith's patent application was filed), which is less than one year before the effective filing date of Jan Smith's patent application.	The rejection here is under pre-AIA 35 U.S.C. § 102(e)(1). The options for overcoming the rejection are: 1) Jan Smith could file a declaration or affidavit under 37 C.F.R. § 1.132 to establish that Frederick Han derived the subject matter of claim 1 from her. By doing this, Jan Smith would establish that the invention disclosed in the prior art application is not "by another" so that it would no longer qualify as prior art under pre-AIA 35 U.S.C. § 102(e)(1). 2) Jan Smith could file a declaration or affidavit under 37 C.F.R. § 1.131 that establishes that she invented the subject matter of claim 1 on a date that precedes the effective prior art date of the prior art application that is cited against her patent application. 3) Jan Smith could amend claim 1 so that it claims subject matter that is patentably distinct from the subject matter disclosed in the prior art application.	The rejection here is under post-AIA 35 U.S.C. § 102(a)(2). The options for overcoming the rejection are: 1) Jan Smith could file an affidavit or declaration of attribution under 37 C.F.R. § 1.130(a) that establishes that Frederick Han directly or indirectly obtained the subject matter that is disclosed in the prior art application, and claimed in claim 1 of the patent application, from her. See post-AIA 35 U.S.C. § 102(b)(2)(A). 2) Jan Smith could establish that the subject matter disclosed in the prior art application and the subject matter of claim 1, not later than the effective filing date of her patent application, were owned by the same person or subject to an obligation of assignment to the same person. See post-AIA 35 U.S.C. § 102(b)(2)(C).

(*Continued on next page*)

Analysis of the Prior Art Rejection Under Pre-AIA Law	Analysis of the Prior Art Rejection Under Post-AIA Law
4) Jan Smith could argue that the subject matter set forth in claim 1 is patentably distinct from the subject matter disclosed in the prior art application. 5) Assuming that Frederick Han is actually an inventor of the claimed subject matter of at least one claim in Jan Smith's patent application, Jan Smith could add Frederick Han as an inventor on her patent application. By doing this, Jan Smith would establish that the invention disclosed in the prior art application is not "by another" so that it would no longer qualify as prior art under pre-AIA 35 U.S.C. § 102(e)(1). (MPEP §§ 706.02(b)(2), 2132.01(I) and (II) (9th ed. 2014))	3) Jan Smith could amend claim 1 so that it claims subject matter that is patentably distinct from the subject matter disclosed in the prior art application cited against her patent application. 4) Jan Smith could argue that the subject matter set forth in claim 1 is patentably distinct from the subject matter disclosed in the prior art application. 5) Assuming that Frederick Han is actually an inventor of the claimed subject matter of at least one claim in Jan Smith's patent application, Jan Smith could add Frederick Han as an inventor on her patent application. *See* 37 C.F.R. § 1.48. By doing this, Jan Smith would establish that the prior art application does not "name another inventor" so that it would no longer qualify as prior art under post-AIA 35 U.S.C. § 102(a)(2). (MPEP §§ 706.02(b)(1), 2155 (9th ed. 2014))

III) NON-OBVIOUSNESS REQUIREMENT OF 35 U.S.C. § 103

Compared to the novelty requirement of pre-AIA 35 U.S.C. § 102, the non-obviousness requirement of pre-AIA 35 U.S.C. § 103 has undergone relatively minor changes under the AIA. A side-by-side comparison of the pre- and post-AIA statutes appears as follows:

Pre-AIA 35 U.S.C. § 103	Post-AIA 35 U.S.C. § 103
(a) A patent may not be obtained though the invention is not identically disclosed or described as set forth in section 102 of this title, if the differences between the subject matter sought to be patented and the prior art are such that the subject matter as a whole would have been obvious at the time the invention was made to a person having ordinary skill in the art to which said subject matter pertains. Patentability shall not be negatived by the manner in which the invention was made. (b) (1) Notwithstanding subsection (a), and upon timely election by the applicant for patent to proceed under this subsection, a biotechnological process using or resulting in a composition of matter that is novel under section 102 and nonobvious under subsection (a) of this section shall be considered nonobvious if– (A) claims to the process and the composition of matter are contained in either the same application for patent or in separate applications having the same effective filing date; and (B) the composition of matter, and the process at the time it was invented, were owned by the same person or subject to an obligation of assignment to the same person.	A patent for a claimed invention may not be obtained, notwithstanding that the claimed invention is not identically disclosed as set forth in section 102, if the differences between the claimed invention[ff] and the prior art are such that the claimed invention as a whole would have been obvious before the effective filing date of the claimed invention to a person having ordinary skill in the art to which the claimed invention pertains. Patentability shall not be negated by the manner in which the invention was made.

(Continued on next page)

[ff] Although post-AIA 35 U.S.C. § 103 recites "the differences between the claimed invention and the prior art" and pre-AIA 35 U.S.C. § 103(a) recites "the differences between the subject matter sought to be patented and the prior art," the U.S. PTO has indicated that the change in terminology in the new statute does not indicate the need for any difference in approach to the question of obviousness. (Fed. Reg. vol. 78, no. 31, Feb. 14, 2013, page 11082).

Pre-AIA 35 U.S.C. § 103	Post-AIA 35 U.S.C. § 103

(2) A patent issued on a process under paragraph (1)–
  (A) shall also contain the claims to the
    composition of matter used in or made
    by that process, or
  (B) shall, if such composition of matter is
    claimed in another patent, be set to expire
    on the same date as such other patent,
    notwithstanding section 154.
(3) For purposes of paragraph (1), the term
  "biotechnological process" means–
  (A) a process of genetically altering or
    otherwise inducing a single- or multi-
    celled organism to–
      (i) express an exogenous nucleotide
        sequence,
      (ii) inhibit, eliminate, augment, or
        alter expression of an endogenous
        nucleotide sequence, or
      (iii) express a specific physiological
        characteristic not naturally associated
        with said organism;
  (B) cell fusion procedures yielding a cell line
    that expresses a specific protein, such as a
    monoclonal antibody; and
  (C) a method of using a product produced by
    a process defined by subparagraph (A) or
    (B), or a combination of subparagraphs (A)
    and (B).
(c)
(1) Subject matter developed by another
  person, which qualifies as prior art only
  under one or more of subsections (e), (f),
  and (g) of section 102 of this title, shall not
  preclude patentability under this section
  where the subject matter and the claimed
  invention were, at the time the claimed
  invention was made, owned by the same
  person or subject to an obligation of
  assignment to the same person.

(*Continued on next page*)

Pre-AIA 35 U.S.C. § 103	Post-AIA 35 U.S.C. § 103
(2) For purposes of this subsection, subject matter developed by another person and a claimed invention shall be deemed to have been owned by the same person or subject to an obligation of assignment to the same person if – (A) the claimed invention was made by or on behalf of parties to a joint research agreement that was in effect on or before the date the claimed invention was made; (B) the claimed invention was made as a result of activities undertaken within the scope of the joint research agreement; and (C) the application for patent for the claimed invention discloses or is amended to disclose the names of the parties to the joint research agreement. (3) For purposes of paragraph (2), the term "joint research agreement" means a written contract, grant, or cooperative agreement entered into by two or more persons or entities for the performance of experimental, developmental, or research work in the field of the claimed invention.	

The most significant difference between pre- and post-AIA 35 U.S.C. § 103 is that under pre-AIA 35 U.S.C. § 103, whether claimed subject matter is obvious is determined as of the **date of its invention** whereas post-AIA 35 U.S.C. § 103 evaluates whether claimed subject matter is obvious as of its **effective filing date**.[gg] *Compare* pre-AIA 35 U.S.C. § 103(a) ("A patent may not be obtained . . . if the differences between the subject matter sought to be patented and the prior art are such that the subject matter as a whole would have been obvious *at the time the invention was made* to a person having ordinary skill in the art to which said subject matter pertains.") (emphasis added) *with* post-AIA 35 U.S.C. § 103 ("A patent for a claimed invention may not be obtained . . . if the differences between the claimed invention and the prior art are such that the claimed invention as a whole would have been obvious *before the effective filing date of the claimed invention*

[gg]Just as the scope of prior art that can be used to show obviousness of a claimed invention under pre-AIA 35 U.S.C. § 103 is defined in pre-AIA 35 U.S.C. § 102, the scope of prior art that can be used to show obviousness of a claimed invention under post-AIA 35 U.S.C. § 103 is defined in post-AIA 35 U.S.C. § 102. MPEP § 2141.01(I) (9th ed. 2014).

to a person having ordinary skill in the art to which the claimed invention pertains.") (emphasis added). The **effective filing date** for claimed subject matter is generally later in time than the **date of invention** for any given subject matter because, as discussed previously, it usually takes at least a few months to prepare and file a patent application after an inventor discloses an invention to his/her patent attorney or patent agent. Thus, the amount of prior art that is available to make obviousness rejections, or to assert invalidity defenses based on obviousness, will typically be greater for any given invention under the post-AIA law as compared to the pre-AIA law.

Another difference between pre- and post-AIA 35 U.S.C. § 103 is that post-AIA 35 U.S.C. § 103 no longer contains the provisions that were set forth in pre-AIA 35 U.S.C. § 103(b). Pre-AIA 35 U.S.C. § 103(b) specifically set forth a narrow exception to pre-AIA 35 U.S.C. § 103(a) for certain biotechnological inventions. In other words, biotechnological inventions that met the criteria of pre-AIA 35 U.S.C. § 103(b) would not be deemed obvious under pre-AIA 35 U.S.C. § 103(a). This exception is no longer available under the new law.

Lastly, it is worth pointing out that although it appears that post-AIA 35 U.S.C. § 103 differs from pre-AIA 35 U.S.C. § 103 in that it omits the common ownership exception to obviousness set forth in pre-AIA 35 U.S.C. § 103(c), the common ownership exception has actually been maintained through the revision of 35 U.S.C. § 102. More specifically, post-AIA 35 U.S.C. § 102(b)(2)(C) provides an exception to the prior art set forth in post-AIA 35 U.S.C. § 102(a)(2) in certain instances where the prior art and a claimed invention were owned by the same person or subject to an obligation of assignment to the same person. Only disclosures that qualify as prior art under post-AIA 35 U.S.C. § 102 may be used to make out an obviousness rejection or invalidity defense under post-AIA 35 U.S.C. § 103. MPEP § 2141.01 (9th ed. 2014). By providing a common ownership exception to the prior art of post-AIA 35 U.S.C. § 102(a)(2), therefore, the AIA has also maintained the common ownership exception for obviousness determinations under post-AIA 35 U.S.C. § 103.

IV) LEGISLATIVE GOALS ADDRESSED BY THE CHANGES TO 35 U.S.C. §§ 102 AND 103

Overall, the amendments to 35 U.S.C. §§ 102 and 103 were made to transition the U.S. patent law from a first-to-invent to a first-inventor-to-file system. The legislature's objectives in making this change to the U.S. patent system are discussed in Section V of Chapter 2.

V) EFFECTIVE DATE OF POST-AIA 35 U.S.C. §§ 102 AND 103

The amendments to 35 U.S.C. §§ 102 and 103 became effective 18 months after the AIA was passed (i.e., on March 16, 2013). Leahy-Smith America Invents Act of 2011, Pub. L. No. 112-29, § 3(n)(1), 125 Stat. 284, 293. New 35 U.S.C. §§ 102 and 103 apply to any claimed invention that has an effective filing date that is on or after March 16, 2013. MPEP § 2159.02 (9th ed. 2014). *Id.* The decision tree in Section IV of Chapter 2 can be used to determine whether the pre- or post-AIA versions of 35 U.S.C. §§ 102 and 103 apply to a given claim.

Importantly, the effective filing date of a claimed invention is determined on a claim-by-claim basis (i.e., different claims within the same patent or patent application may have different effective filing dates). Whether pre- or post-AIA 35 U.S.C. §§ 102 and 103 apply, however, is determined on an application-by-application basis. Thus, post-AIA 35 U.S.C. §§ 102 and 103 will apply to any patent application that contains, or contained at any time, a claim to an invention that has an effective filing date that is on or after March 16, 2013. MPEP § 2152.01 (9th ed. 2014).

VI) PRACTICAL IMPLICATIONS OF POST-AIA 35 U.S.C. §§ 102 AND 103

While most of the federal courts have yet to weigh in on the proper interpretations of the new statutory language of 35 U.S.C. § 102, the U.S. PTO's interpretation of the new law suggests that it will now be more important than ever that inventors file their patent applications as soon as possible. Indeed, given that the U.S. PTO has interpreted the intervening disclosure exception to prior art provided by post-AIA 35 U.S.C. §§ 102(b)(1)(B) and 102(b)(2)(B)[hh] very narrowly, the most prudent course of action for inventors is to file their patent applications before disclosing their inventions to the public in any manner to prevent a third-party's public disclosure of their invention from being used as prior art against them.

[hh]As discussed in Section II.C.iii., the intervening disclosure exception provided by post-AIA 35 U.S.C. §§ 102(b)(1)(B) and 102(b)(2)(B) pertains to the situation where an inventor publicly discloses his/her invention and then a third-party discloses the invention to the public after the inventor's disclosure but before the inventor files a patent application claiming the disclosed invention.

In instances where an inventor must disclose his/her invention to another party before filing a patent application (e.g., to seek funding from venture capitalists or to collaborate with others to further develop their technology), the inventor should make such disclosures only after a confidentiality or non-disclosure agreement has been executed by the party to whom he/she is making the disclosure. *See, e.g.,* 78 Fed. Reg. 11,075 (Feb. 14, 2013) ("The 'or otherwise available to the public' residual clause of AIA 35 U.S.C. 102(a)(1), however, indicates that AIA 35 U.S.C. 102(a)(1) does not cover secret sales or offers for sale. For example, an activity (such as a sale, offer for sale, or other commercial activity) is secret (non-public) if it is among individuals having an obligation of confidentiality to the inventor.").

CHAPTER FOUR

The AIA's Effects on Patentable Subject Matter and the Patentability Requirements Set Forth by 35 U.S.C. § 112

Contents

I) INTRODUCTION

The changes to the novelty requirement for patentability discussed in Chapter 3 were, by far, the most significant patent law reforms brought about by the AIA. Although less significant in comparison, the AIA also modified patent eligible subject matter and 35 U.S.C. § 112. More specifically, AIA § 14 eliminates tax strategies from the realm of subject matter that is eligible for patent protection, and AIA § 33 codifies the pre-AIA law that prohibited the patenting of human organisms. This chapter will examine these changes.

II) PATENT ELIGIBLE SUBJECT MATTER UNDER THE AIA

35 U.S.C. § 101 sets forth the types of subject matter that are eligible to be patented. For instance, while new and useful processes, machines, manufactured articles, and compositions of matter are patentable under 35 U.S.C. § 101, laws of nature, physical phenomena, and abstract ideas do not constitute patentable subject matter. MPEP § 2106(I) (9th ed. 2014) (citing *Bilski v. Kappos,* 561 U.S. 593, 601 (2010)). Although 35 U.S.C. § 101 was

America Invents Act Primer. http://dx.doi.org/10.1016/B978-0-12-812096-5.00004-1
Copyright © 2017 Sarah Hasford. Published by Elsevier Inc. All rights reserved.

not amended by the AIA, sections 14 and 33 of the new legislation provide further guidelines on the types of subject matter that are eligible for patent protection. These sections of the AIA are discussed in more detail in Sections II.A. and B.

A. AIA § 14—Tax Strategies No Longer Eligible for Patenting

Section 14(a) of the AIA reads as follows:

> In General — For purposes of evaluating an invention under section 102 or 103 of title 35, United States Code, any strategy for reducing, avoiding, or deferring tax liability,[a] whether known or unknown at the time of the invention or application for patent, shall be deemed insufficient to differentiate a claimed invention from the prior art.

In short, AIA § 14(a) provides that tax strategies cannot serve as the basis for finding that a claim is novel or non-obvious under 35 U.S.C. §§ 102 and 103, regardless of whether the tax strategy is actually in the prior art. By disallowing tax strategies to meet the novelty or non-obviousness requirements for patentability, AIA § 14(a) essentially renders tax strategies unpatentable as if 35 U.S.C. § 101 had been amended to exclude such inventions. *See* Gene Quinn, *AIA Oddities: Tax Strategy Patents and Human Organisms*, IPWatchdog, Sept. 12, 2013, http://www.ipwatchdog.com/2013/09/12/aia-oddities-tax-strategy-patents-and-human-organisms/id=45113/ ("In a bizarre circumstance Congress chose not to render tax strategy patents patent ineligible under 35 U.S.C. § 101. Rather they chose a far more convoluted route."); Joe Matal, *A Guide to the Legislative History of the America Invents Act: Part I of II*, 21 Fed. Cir. B.J. 435, 502 (2012) (citing Patent Reform Act of 2007, H.R. 1908, 110th Cong., sec. 10(a)(2)) (explaining that a provision that was similar to section 14 of the AIA was added to the 2007 House patent bill in the House Judiciary Committee and that the 2007 House patent bill provision would have amended 35 U.S.C. § 101 to ban tax strategy patents).

Notably, section 14(c) provides two exceptions to section 14(a) of the AIA where the invention at-issue is:

---

[a]AIA § 14(b) defines the term "tax liability" as "any liability for a tax under any Federal, State, or local law, or the law of any foreign jurisdiction, including any statute, rule, regulation, or ordinance that levies, imposes, or assesses such tax liability." Leahy-Smith America Invents Act of 2011, Pub. L. No. 112-29, 125 Stat. 284, 327.

(1) a method, apparatus, technology, computer program product, or system, that is used solely for preparing a tax or information return or other tax filing, including one that records, transmits, transfers, or organizes data related to such filing; or

(2) a method, apparatus, technology, computer program product, or system used solely for financial management, to the extent that it is severable from any tax strategy or does not limit the use of any tax strategy by any taxpayer or tax advisor.

Leahy-Smith America Invents Act of 2011, Pub. L. No. 112-29, 125 Stat. 284, 327-28. Although the inventions listed in AIA § 14(c) are presented as exclusions to the ban on the patenting of tax strategies, those inventions do not actually relate to tax strategies at all. In fact, paragraph (1) pertains to inventions that are used to prepare a tax return, information return, or other tax filing, and paragraph (2) relates to inventions that are used for financial management, wherein the inventions are completely separate from tax strategies. Under AIA § 14(c), therefore, a software program that is novel and non-obvious as software would be patentable even if it is used for tax purposes. *See California Inst. of Tech. v. Hughes Commc'ns Inc.*, 59 F. Supp. 3d 974, 985 (C.D. Cal. 2014) ("By excluding computer programs from subsection (a) [of AIA § 14], Congress contemplated that some computer programs were eligible for patent protection. Courts should not read *§ 101* to exclude software patents when Congress has contemplated their existence."). A software program that is not novel and non-obvious, however, could not meet the patentability requirements of 35 U.S.C. §§ 102 and 103 by implementing a novel and non-obvious tax strategy. H.R. Rep. No. 112-98, at 51-52, 79 (2011). Thus, although identified as "exclusions" to AIA § 14(a), the subparagraphs of AIA § 14(c) do not actually exclude any tax strategy inventions from being unpatentable under 35 U.S.C. §§ 102 and 103.

i. Legislative Goals Addressed by AIA § 14

The uncodified amendment set forth in AIA § 14(a) was enacted to counteract the rising trend of patenting tax strategies that was occurring during the time that the AIA was being debated. 157 Cong. Rec. S1198, 2011 (reprinting Aug. 1, 2007 article by the Tax Advisors). As explained in an August 1, 2007 Tax Advisors article that was submitted for the Congressional Record by Senator Grassley during his March 3, 2011 remarks, tax strategies that had been used routinely by tax practitioners for years were suddenly being patented so that anyone who wanted to reduce their tax liability using the same strategies would be forced to pay a royalty to a patent owner.[b] *Id.* And even though these tax strategies had been used for many years, and were therefore not novel or non-obvious, tax practitioners often could not challenge the validity of any patent claiming such strategies without violating client confidences. *Id.* at S1199. Absent viable defenses to infringement suits over tax strategy patents, these patents threatened to significantly undermine tax practitioners' practices by preventing them from best advising their clients on how to minimize their tax liability. *See id.* at S1200 (statement of Sen. Grassley) ("Tax strateg[y patents] are bad because they allow the tax law to be patented. A tax strategy patent makes taxpayers choose between paying more than legally required in taxes or providing a windfall to a tax strategy patentholder by paying a royalty to comply with the tax law.").

ii. Effective Date of AIA § 14

AIA § 14 took effect on September 16, 2011 and applies to any patent application that is pending on, or filed on or after that date, as well as to any patent that is issued on or after that date. Leahy-Smith America Invents Act of 2011, Pub. L. No. 112-29, § 14(e), 125 Stat. 284, 328. In other words, only patents that were issued before September 16, 2011 are exempt from AIA § 14's ban on patenting tax strategies.

iii. Practical Implications of AIA § 14

Practically speaking, AIA § 14 should prevent any patent application that claims a tax strategy and that is pending as of, or is filed on or after, September

---

[b]In 1998, *State Street Bank & Trust Co. v. Signature Fin. Group, Inc.*, 149 F.3d 1368, 1375-76 (Fed. Cir. 1998) paved the way for patents on tax strategies by confirming that business methods constitute patentable subject matter under 35 U.S.C. § 101. *See id.* at 1375 ("Since the 1952 Patent Act, business methods have been, and should have been, subject to the same legal requirements for patentability as applied to any other process or method."). Later on, tax practitioners became very concerned when a patent suit was brought in 2006 based on the so-called "SOGRAT" patent. 157 CONG. REC. S1199, 2011 (reprinting Aug. 1, 2007 article by the Tax Advisors). The SOGRAT patent described an estate planning technique that used grantor retained annuity trusts (GRATs) to transfer nonqualified stock options (NQSOs) to younger generations, with few or no gift tax consequences. *Id.* Many estate planners had routinely used GRATs to shift assets to younger generations. *Id.* Thus, these estate planners were quite surprised when an article touting the estate tax benefits of placing NQSOs into a GRAT noted that the technique had been patented by one of that article's authors. *Id.* Their surprise grew into concern when the patent holder instituted the aforementioned patent infringement suit against a taxpayer who implemented the technique without permission. *Id.* The patent case over the "SOGRAT" patent served as a major catalyst for the ban on tax strategy patents provided by AIA § 14.

16, 2011 from issuing as a patent. As such, tax practitioners should feel less restricted in advising their clients on how to minimize their tax liability. One question that remains is whether federal courts will refuse to enforce patent claims directed to a tax strategy that were issued before AIA § 14 took effect. *See* 157 CONG. REC. S1368, 2011 (statement of Sen. Levin) ("Although the bill does not apply on its face to the 130-plus tax [strategy] patents already granted, if someone tried to enforce one of those patents in court by demanding that a taxpayer provide a fee before using it to reduce their taxes, I hope a court will consider this bill's language and policy determination and refuse to enforce the patent as against public policy.").

B. AIA § 33—Prohibition on Patenting Human Organisms

Unlike section 14 of the AIA, which changed the status quo of the law on patenting tax strategies, section 33 of the AIA merely codifies the law that existed prior to the enactment of the legislation. In particular, section 33(a) of the AIA states that "[n]otwithstanding any other provision of law, no patent may issue on a claim directed to or encompassing a human organism." Leahy-Smith America Invents Act of 2011, Pub. L. No. 112-29, 125 Stat. 284, 340. Prior to the AIA, the U.S. PTO had a longstanding policy that prohibited the patenting of human organisms. *See* "Notice: Animals—Patentability," 1077 Off. Gaz. U.S. Pat. And Trademark Off. 8 (April 21, 1987) (explaining that human organisms are excluded from the scope of patentable subject matter under 35 U.S.C. § 101); MPEP § 2105 (8th ed. 2010) ("If the broadest reasonable interpretation of the claimed invention as a whole encompasses a human being, then a rejection under 35 U.S.C. 101 must be made indicating that the claimed invention is directed to nonstatutory subject matter."); *Media Press Release*, 98-06, U.S. PATENT AND TRADEMARK OFFICE (Apr. 1, 1998), *available at* http://www.uspto.gov/news/pr/1998/98-06.jsp (explaining that a human/non-human chimera may be ineligible for patent protection because of a failure to meet the moral utility requirement under § 101). The Supreme Court has also held that certain human organisms could not be patented. *See Ass'n for Molecular Pathology v. Myriad Genetics, Inc.*, 133 S. Ct. 2107, 2113, 2117 (2013) (holding that a claim to an isolated human gene did not constitute patentable subject matter under 35 U.S.C. § 101 where, other than being isolated from the genetic material surrounding it, the isolated gene had not undergone any other modifications).

Although the *Myriad* case provides some guidance for when a human organism may constitute patentable subject matter under 35 U.S.C. § 101, there is still a great deal of uncertainty as to how the federal courts will interpret the "directed to or encompassing a human organism" language of AIA § 33(a). *See* 149 CONG. REC. E2234, 2003 (statement of Rep. Dave Weldon) (mentioning arguments made by the Biotechnology Industry Organization (BIO) and the Coalition for the Advancement of Medical Research (CAMR) that § 33 of the AIA would "potentially prohibit[] patents on stem cell lines, procedures for creating human embryos, prosthetic devices, and in short almost any drug or product that might be used in or for human beings"); Ava Caffarini, *Directed to or Encompassing a Human Organism: How Section 33 of the America Invents Act may Threaten the Future of Biotechnology*, 12 J. MARSHALL. L. REV. INTELL. PROP. L., 768, 783 (2013) (arguing that a broad construction of AIA § 33 would have the destructive potential to invalidate patents directed to subject matter including, but not limited to, personalized medicine, pharmaceuticals, genes, prosthetics, artificial organs, research tools for embryonic stem cell research, medical devices, and human derivatives such as hormones and antibodies); *but see* 157 CONG. REC. E1183, 2011 (statement of Rep. Lamar Smith) (listing various patents, including patents claiming stem cells, stem cell derived tissues, stem cell lines, viable synthetic organs, a transgenic plant or animal, and animal models used for scientific research as patents that should not be affected by AIA § 33); 157 CONG. REC. E1178, 2011 (statement of Rep. Dave Weldon) (explaining that the term "human organism," as used in AIA § 33, should exclusively include human embryos, human fetuses, and human beings).

i. Legislative Goals Addressed by AIA § 33

Congress' main objective in enacting AIA § 33(a) is apparent from the plain language of the statute. Namely, Congress enacted the statute to prevent the U.S. PTO from issuing patents that are directed to, or encompass, a human organism. *See* 157 CONG. REC. E1177, 2011 (statement of Rep. Christopher H. Smith) (explaining that incorporating the Weldon Amendment into the AIA would "put the weight of law behind the USPTO [1987 P]olicy" by expressly prohibiting the patenting of human organisms). AIA § 33(a) codified the "Weldon Amendment,"[c] which was first adopted in 2004. *Id.* The Weldon Amendment states that "[n]one of the funds appropriated or otherwise made available under this [Appropriations] Act may be used to issue patents on claims directed to or encompassing a human organism." Consolidated Appropriations Act of 2004, Pub. L. No. 108-199, § 634, 118 Stat. 3, 101. The Weldon Amendment did not directly ban the issuance of

---

[c]For a more thorough discussion of the legislative history of the Weldon Amendment and AIA § 33, see: Yaniv Helev, *On Patenting Human Organisms or How the Abortion Wars Feed into the Ownership Fallacy*, 36 CARDOZO L. REV. 241, 252-257 (2014); Joe Matal, *A Guide to the Legislative History of the America Invents Act: Part I of II*, 21 FED. CIR. B.J. 435, 510-12 (2012).

patents on human organisms but instead tried to restrict the U.S. PTO from using its Congressionally appropriated funds to issue such patents. Yaniv Helev, *On Patenting Human Organisms or How the Abortion Wars Feed into the Ownership Fallacy*, 36 CARDOZO L. REV. 241, 253 (2014). There was concern, however, that the AIA might "authorize the USPTO to pay for the issuance of patents with 'user fees' instead of with Congressionally appropriated funds." 157 CONG. REC. E1184, 2011. Accordingly, section 33(a) was included in the AIA to ensure that the U.S. PTO would not issue patents on human organisms even if the issuance of patents was later funded by user fees instead of Congressional funds.

ii. Effective Date of AIA § 33

AIA § 33 applies to any patent application that is pending on, or filed on or after, September 16, 2011. Leahy-Smith America Invents Act of 2011, Pub. L. No. 112-29, § 33(b)(1), 125 Stat. 284, 340.

iii. Practical Implications of AIA § 33

Although AIA § 33 does not purport to change the status quo on the types of human organisms that are patentable under 35 U.S.C. § 101, the uncertainty as to the meaning of the "directed to or encompassing a human organism" language may cause a decline in investment and research for certain types of technologies. *See* Yaniv Helev, *On Patenting Human Organisms or How the Abortion Wars Feed into the Ownership Fallacy*, 36 CARDOZO L. REV. 241, 273 (2014) ("Such an increase in uncertainty as to patentability [of human organisms] is likely to raise doubts among potential investors about their ability to gain financially from research that they fund, which would dissuade them from investing in such research and development efforts.") (citation omitted).

III) AIA § 4—AMENDMENTS TO 35 U.S.C. § 112

Pre-AIA 35 U.S.C. § 112 underwent only very minor, non–substantive amendments[d] by the AIA.[e] A comparison of the pre- and post-AIA statutes appears in the following table:

---

[d]Note that while AIA § 4 did not substantively amend the text of pre-AIA 35 U.S.C. § 112, AIA § 15 eliminated the best mode defense that is based on the best mode requirement set forth in pre-AIA 35 U.S.C. § 112, ¶ 1 and post-AIA 35 U.S.C. § 112(a). AIA § 15's changes to the best mode defense are discussed in Section III of Chapter 7.

[e]The changes to pre-AIA 35 U.S.C. § 112 apply to any patent application that is filed on or after September 16, 2012. Leahy-Smith America Invents Act of 2011, Pub. L. No. 112-29, § 4(e), 125 Stat. 284, 297.

Pre-AIA 35 U.S.C. § 112	Post-AIA 35 U.S.C. § 112
The specification shall contain a written description of the invention, and of the manner and process of making and using it, in such full, clear, concise, and exact terms as to enable any person skilled in the art to which it pertains, or with which it is most nearly connected, to make and use the same, and shall set forth the best mode contemplated by the inventor of carrying out his invention.	(a) IN GENERAL.-- The specification shall contain a written description of the invention, and of the manner and process of making and using it, in such full, clear, concise, and exact terms as to enable any person skilled in the art to which it pertains, or with which it is most nearly connected, to make and use the same, and shall set forth the best mode contemplated by the inventor or joint inventor of carrying out the invention.
The specification shall conclude with one or more claims particularly pointing out and distinctly claiming the subject matter which the applicant regards as his invention.	(b) CONCLUSION.-- The specification shall conclude with one or more claims particularly pointing out and distinctly claiming the subject matter which the inventor or a joint inventor regards as the invention.
A claim may be written in independent or, if the nature of the case admits, in dependent or multiple dependent form.	(c) FORM.-- A claim may be written in independent or, if the nature of the case admits, in dependent or multiple dependent form.
Subject to the following paragraph, a claim in dependent form shall contain a reference to a claim previously set forth and then specify a further limitation of the subject matter claimed. A claim in dependent form shall be construed to incorporate by reference all the limitations of the claim to which it refers.	(d) REFERENCE IN DEPENDENT FORMS.-- Subject to subsection (e), a claim in dependent form shall contain a reference to a claim previously set forth and then specify a further limitation of the subject matter claimed. A claim in dependent form shall be construed to incorporate by reference all the limitations of the claim to which it refers.
A claim in multiple dependent form shall contain a reference, in the alternative only, to more than one claim previously set forth and then specify a further limitation of the subject matter claimed. A multiple dependent claim shall not serve as a basis for any other multiple dependent claim. A multiple dependent claim shall be construed to incorporate by reference all the limitations of the particular claim in relation to which it is being considered.	(e) REFERENCE IN MULTIPLE DEPENDENT FORM.-- A claim in multiple dependent form shall contain a reference, in the alternative only, to more than one claim previously set forth and then specify a further limitation of the subject matter claimed. A multiple dependent claim shall not serve as a basis for any other multiple dependent claim. A multiple dependent claim shall be construed to incorporate by reference all the limitations of the particular claim in relation to which it is being considered.

(Continued on next page)

Pre-AIA 35 U.S.C. § 112	Post-AIA 35 U.S.C. § 112
An element in a claim for a combination may be expressed as a means or step for performing a specified function without the recital of structure, material, or acts in support thereof, and such claim shall be construed to cover the corresponding structure, material, or acts described in the specification and equivalents thereof.	(f) ELEMENT IN CLAIM FOR A COMBINATION.-- An element in a claim for a combination may be expressed as a means or step for performing a specified function without the recital of structure, material, or acts in support thereof, and such claim shall be construed to cover the corresponding structure, material, or acts described in the specification and equivalents thereof.

The change that is most apparent when comparing the pre- and post-AIA statutes side-by-side is the addition of headings and lettering for each of the paragraphs of 35 U.S.C. § 112. Pre-AIA 35 U.S.C. § 112 does not provide any headings for each of the paragraphs, nor are the paragraphs identified using any numbers or letters. Nevertheless, patent practitioners commonly refer to the various paragraphs of pre-AIA 35 U.S.C. § 112 by the order in which they are listed. For example, the paragraph beginning with the phrase, "[t]he specification shall begin with a written description of the invention" is referred to as "112, first paragraph," while the paragraph beginning with the phrase "[a]n element in a claim for a combination may be expressed as a means or a step for performing a specified function" is referred to as "112, paragraph 6." Sections of post-AIA 35 U.S.C. § 112 will be referenced by the section letter to which they are assigned (e.g., "112(a)," "112(b)").

Besides changing the format of 35 U.S.C. § 112, the AIA amended the first and second paragraphs of the pre-AIA statute by referencing the "inventor or joint inventor" instead of only the "inventor" (as set forth in pre-AIA 35 U.S.C. § 112, first paragraph) or the "applicant" (as set forth in pre-AIA 35 U.S.C. § 112, second paragraph). The references to the "inventor or joint inventor" were likely included to add precision to the post-AIA statute and to bring that statute into conformity with post-AIA 35 U.S.C. § 100(f) which, for the first time in the history of U.S. patent law, provides a definition of the term "inventor."[f] More specifically, by referencing the "inventor or joint inventor," post-AIA 35 U.S.C. § 112(a) and (b) make clear

---

[f]Post-AIA 35 U.S.C. § 100(f) specifically defines the term "inventor" as "the individual or, if a joint invention, the individuals collectively who invented or discovered the subject matter of the invention."

that the requirements of those paragraphs must be met by *each and every* individual who is named as an inventor on a patent or patent application. *See* Robert A. Armitage, *Understanding the America Invents Act and its Implications for Patenting*, 40 AIPLA Q.J. 1, 18 (2012) ("[The definition of 'inventor'] provides a higher degree of readability and precision in the new law by allowing either a reference to 'the inventor' where the provision relates to the entity named in the patent or application for patent or 'the inventor or a joint inventor' where the intent is to reference *any or all* of the individuals so named in the patent or application for patent.") (emphasis in original).

CHAPTER FIVE

The AIA's Impact on Post-Grant Proceedings

Contents

I) INTRODUCTION

When patent practitioners use the term "post-grant proceeding," they are generally referring to two categories of U.S. PTO proceedings that occur after a patent already has been granted. The first category of proceedings involves third-party challenges to patents, wherein the third party seeks to have the U.S. PTO invalidate the patent on one or more grounds. The second category of proceedings allows a patent holder to strengthen his/her patent by having the U.S. PTO examine the patent further and/or by correcting one or more errors in the patent. In addition to these proceedings, a third post-grant procedure is the citation of prior art in a patent file, which is only occasionally used by third parties and used by patent holders even less often. This chapter provides an overview of these proceedings, as well as a discussion of how the AIA has impacted each one.

America Invents Act Primer. http://dx.doi.org/10.1016/B978-0-12-812096-5.00005-3

II) POST-GRANT PROCEEDINGS INVOLVING THIRD-PARTY CHALLENGES TO PATENTS

Before the AIA was enacted, the types of post-grant proceedings that were available to third parties to challenge the validity of a patent were somewhat limited. In fact, if a third party wanted the U.S. PTO to invalidate an issued patent, the only option was to file a request for either an inter partes reexamination or an ex parte reexamination. While both inter partes reexamination and ex parte reexamination had been established to reduce district court patent litigation by providing third parties with a mechanism for challenging patent validity in the U.S. PTO, the lengthy duration of the reexamination proceedings often prevented the proceedings from fulfilling this purpose. The slow pace of reexamination proceedings became particularly problematic in the years immediately preceding the enactment of the AIA, when many companies were being sued by patent trolls for allegedly infringing dubious patents. To address the shortcomings of reexamination practice, the AIA provided several new options for third parties to challenge patents in the U.S. PTO, namely, post grant review, inter partes review, and covered business method review. All of these new proceedings, which are discussed in further detail throughout this chapter, became available on September 16, 2012. Leahy-Smith America Invents Act of 2011, Pub. L. No. 112-29, §§ 6(c)(2)(A), 6(f)(2)(A), 18(a)(2), 125 Stat. 284, 304, 311, 330.

A. Post-Grant Proceedings Available Before the Enactment of the AIA

Prior to the AIA, there were two post-grant proceedings that were available to third parties who wished to challenge a patent's validity, namely ex parte reexamination and inter partes reexamination. Ex parte reexamination practice was established long before inter partes reexamination practice and differs from inter partes reexamination in that it: (1) allows for little participation by a third-party requester; (2) does not provide a third-party requester with a right to appeal to the Board of Patent Appeals and Interferences (BPAI); and (3) does not create any estoppel effect for parties involved in the proceeding. See 35 U.S.C. §§ 303-307 (omitting estoppel provisions from ex parte reexamination statutes); pre-AIA 35 U.S.C. § 314(b)(2) (explaining that during inter partes reexamination, "[e]ach time that the patent owner files a response to an action on the merits from the Patent and Trademark Office, the third-party requester shall have one opportunity to file written comments addressing issues raised by the action of the Office or the patent owner's response thereto . . . ."); MPEP § 2601 (8th ed. 2010) (explaining that inter partes reexamination differs from ex parte reexamination in that it provides third-party requesters with appeal rights to appeal to the Board of Patent Appeals and Interferences (Board) and with the right to participate in the patent owner's appeal to the Board, and that unlike ex parte reexamination, participation in inter partes reexamination is conditioned on the third-party requester accepting a statutory estoppel against subsequent review, either by the U.S. PTO or by a federal court, of the issues that were or could have been raised in the reexamination proceeding); MPEP § 2251 (9th ed. 2015) (explaining that a third-party requester may file a reply to the patent owner's statement in an ex parte reexamination, but is not permitted to file any further papers after such reply). The following tables provide further comparisons of ex parte and inter partes reexamination proceedings:

Type of Post-Grant Proceeding	Who Can Initiate the Proceeding	Proper Grounds for Initiating the Proceeding
Inter Partes Reexamination	Any person, other than the patent owner or the patent owner's privies, may file a request for inter partes reexamination. (37 C.F.R § 1.913(a); MPEP § 2609 (8th ed. 2010))	A third-party requester could only request inter partes reexamination on the basis that the claimed invention was: (1) anticipated under 35 U.S.C. § 102; and/or (2) obvious under 35 U.S.C. § 103. In addition, the anticipation and obviousness positions could only be based on prior patents or printed publications (i.e., prior art activities such as prior sales or public uses could not be used to establish the grounds for the reexamination request). (MPEP § 2617 (8th ed. 2010))
Ex Parte Reexamination	Any person, including the patent owner, may file a request for ex parte examination. (MPEP § 2209 (9th ed. 2014))	A requester could only request ex parte reexamination on the basis that the claimed invention was: (1) anticipated under 35 U.S.C. § 102; and/or (2) obvious under 35 U.S.C. § 103. In addition, the anticipation and obviousness positions could only be based on prior patents or printed publications (i.e., prior art activities such as prior sales or public uses could not be used to establish the grounds for the reexamination request). (MPEP § 2209 (9th ed. 2014))

Type of Post-Grant Proceeding	Burden on Requester for Having the Proceeding Initiated	Restrictions on Filing a Request for the Post-Grant Proceeding	Estoppel Effects
Inter Partes Reexamination	A third-party requester only needed to show that there was a substantial new question ("SNQ") of patentability for at least one claim of the patent. If such question was presented, reexamination would be ordered. (Pre-AIA 35 U.S.C. § 312(a); MPEP § 2609 (8th ed. 2010))	• Only patents that issued from an original application that was filed on or after November 29, 1999 were eligible for inter partes reexamination. (MPEP § 2609 (8th ed. 2010)) • Furthermore, requests for reexamination could only be filed during the period of enforceability for the patent at issue. (MPEP § 2609 (8th ed. 2010)) The period of enforceability is determined by adding 6 years to the date on which the patent expires. (MPEP § 2611 (8th ed. 2010)) • Inter partes reexamination practice was replaced with Inter Partes Review practice on September 16, 2012. Accordingly, the U.S. PTO has not, and will not, accept requests for inter partes reexamination filed on or after that date. (MPEP § 2610 (9th ed. 2014))	A third party who requests an inter partes reexamination is statutorily estopped from raising issues that were or could have been raised in the reexamination proceeding, in a subsequent proceeding, at either the PTO or in federal court. (MPEP § 2601 (8th ed. 2010))
Ex Parte Reexamination	A requester only needs to show that there is a substantial new question ("SNQ") of patentability for at least one claim of the patent. If such question is presented, reexamination will be ordered. (35 U.S.C. § 303(a); MPEP § 2209 (9th ed. 2014))	A request for ex parte reexamination must be filed during the period of enforceability of the patent at-issue. (MPEP § 2209 (9th ed. 2014))	Ex parte reexaminations do not create any estoppel effects. (See 35 U.S.C. § 304–307 (omitting estoppel provisions from the ex parte reexamination statutes))

One of the main reasons inter partes reexamination and ex parte reexamination proceedings were established was to curb the number of patent litigations brought in district courts by providing third parties with the option to challenge the validity of patents in the U.S. PTO. *See* 145 CONG. REC. E1789-90, 1999 ("Title V is intended to reduce expensive patent litigation in U.S. district courts by giving third-party requesters, in addition to the existing ex parte reexamination in Chapter 30 of title 35, the option of inter partes reexamination proceedings in the PTO. Congress enacted legislation to authorize ex parte reexamination of patents in the PTO in 1980, but such reexamination has been used infrequently since a third party who requests reexamination cannot participate at all after initiating the proceedings. Numerous witnesses have suggested that the volume of lawsuits in district courts will be reduced if third parties can be encouraged to use reexamination by giving them an opportunity to argue their case for patent invalidity in the PTO.").

The reexamination proceedings, however, proved to be extremely lengthy and thus did not curb district court patent litigation to the extent desired. *See* 157 CONG. REC. S5326, 2011 (statement of Sen. Leahy) ("The current inter partes reexamination process has been criticized for being too easy to initiate and used to harass legitimate patent owners, while being too lengthy and unwieldy to actually serve as an alternative to litigation when users are confronted with patents of dubious validity."). In fact, as of November 22, 2013, the U.S. PTO posted statistics indicating that the average pendency of inter partes reexaminations was 36 months[a] and the average pendency of ex parte reexaminations was about 28 months.[b] In some of the most popular jurisdictions for patent suits (e.g., the Eastern District of Texas) cases can progress to trial in as little as one year's time. As a result, reexamination proceedings have generally failed to provide a meaningful alternative to many district court litigations because those proceedings often drag out beyond the completion of district court litigations.[c]

[a] *See* U.S. PTO, http://www.uspto.gov/patents/stats/inter_parte_historical_stats_roll_up_EOY2013.pdf (last visited Aug. 2, 2016).
[b] *See* U.S. PTO, http://www.uspto.gov/patents/stats/ex_parte_historical_stats_roll_up_EOY2013.pdf (last visited Aug. 2, 2016).
[c] During the years immediately preceding enactment of the AIA, it was common for patent trolls to file suit in jurisdictions that are known for progressing to trial very quickly (e.g., the Eastern District of Texas). By doing this, the patent trolls could extract a desirable settlement from defendants because defendants knew that if they litigated the case all the way through trial, they would likely spend several million dollars on attorneys' fees and discovery costs. And with reexaminations often taking two to three years to complete, ▶

B. Post-Grant Proceedings Available After the Enactment of AIA §§ 6 and 18

To provide better alternatives to district court patent litigation, the AIA introduced several new post-grant proceedings through which third parties may challenge the validity of a patent. These new proceedings are: (1) post grant review ("PGR"), (2) inter partes review ("IPR"), and (3) a transitional program for covered business method ("CBM") review. PGRs[d] and IPRs replace inter partes reexamination practice, which was phased out by the AIA as of September 16, 2012, while the transitional program for CBM review was established to curtail patent litigations involving patent claims to business methods. MPEP § 2601 (9th ed. 2014); *see also* 157 CONG. REC. H4421, 2011 ("The America Invents Act reduces frivolous litigation over weak or overbroad patents by establishing a pilot program to review a limited group of business method patents that never should have been awarded in the first place."). Although Congress eliminated inter partes reexamination proceedings, ex parte reexamination proceedings were maintained and were left largely unchanged by the AIA. The only substantive change to ex parte reexamination practice was that the AIA amended pre-AIA 35 U.S.C. § 306 to clarify that patent owners who are dissatisfied with a decision in an ex parte reexamination may only appeal to the Federal Circuit—not to the Eastern District of Virginia.[e] *See* Leahy-Smith America Invents Act of 2011, Pub. L. No. 112-29, § 6(h)(2)(A), 125 Stat. 284, 312 (amending pre-AIA 35 U.S.C.

◄ defendants in patent troll cases really had only one option for invalidating the patent— litigate the case through trial and obtain a judgment of invalidity. Accordingly, to avoid the high cost of litigation, defendants often settled early with patent trolls for some amount of money that was less than what they expected to spend to litigate the case. As a result, patent trolls asserting dubious patents were able to continue asserting those patents against other defendants with little chance that any one defendant would take them all the way to trial and invalidate their patents.

[d]PGRs also replace public use proceedings that were formerly brought under 37 C.F.R. § 1.292. *See* Final Rule on Changes to Implement the Preissuance Submissions by Third Parties Provision of the Leahy-Smith America Invents Act, 77 Fed. Reg. 42,150 (July 17, 2012) ("This final rule also eliminates the public use proceeding provisions of 37 CFR 1.292. Because section 6 of the AIA makes available a post grant review proceeding in which prior public use may be raised, the pre-grant public use proceeding previously set forth in 37 CFR 1.292 is no longer necessary."). Public use proceedings were extremely rare but could be brought by individuals who had sufficient evidence to make a prima facie showing that the invention claimed in a patent application believed to be on file had been in public use or on sale more than one year before the filing of the patent application. Pre-AIA 37 C.F.R. § 1.292(a).

[e]The amendment to pre-AIA 35 U.S.C. § 306 took effect on September 16, 2011 and applies to any appeal of a reexamination that was pending on, or brought on or after, that date. Leahy-Smith America Invents Act of 2011, Pub. L. No. 112-29, § 6(h)(2)(B), 125 Stat. 284, 312.

§ 306 to remove the reference to 35 U.S.C. § 145, which provided a party who was dissatisfied with a decision of the Board of Patent Appeals and Interferences with the right to bring a civil action against the Director of the U.S. PTO in the United States District Court for the Eastern District of Virginia).

Like reexamination proceedings, PGRs, IPRs, and CBM Reviews allow petitioners to prove invalidity of one or more patent claims using a lower burden of proof than the burden that is required in district court litigation. *Compare* 35 U.S.C. §§ 316(e), 326(e) and Leahy–Smith America Invents Act of 2011, Pub. L. No. 112-29, § 18(a)(1), 125 Stat. 284, 329 (explaining that petitioners in PGRs, IPRs, and CBM Reviews must prove a proposition of unpatentability by a preponderance of the evidence); *Dome Patent L.P. v. Lee,* 799 F.3d 1372, 1379 (Fed. Cir. 2015) (explaining that the district court did not err by requiring the U.S. PTO to show by a preponderance of the evidence that a reexamined claim was obvious during an ex parte reexamination proceeding); MPEP § 2686.04(IV) (8th ed. 2010) (explaining that non–patentability must be shown by a preponderance of the evidence in inter partes reexamination proceedings) *with Microsoft Corp. v. i4i Ltd. P'ship, et al.,* 131 S. Ct. 2238, 2242 (2011) (holding that 35 U.S.C. § 282 requires an invalidity defense to be proved by clear and convincing evidence). PGRs, IPRs, and CBM Reviews all differ from reexamination proceedings, however, in that they must be completed within a certain time period that is set forth by statute. *See* 35 U.S.C. §§ 316(a)(11), 326(a)(11) and Leahy–Smith America Invents Act of 2011, Pub. L. No. 112-29, § 18(a)(1), 125 Stat. 284, 329 (explaining that a final determination in any PGR, IPR or CBM Review will usually be issued not later than one year after the date on which the Director notices the institution of the proceeding); pre–AIA 35 U.S.C. § 305 (explaining that after the time periods for filing statements under pre–AIA 35 U.S.C. § 304 have expired, ex parte reexamination will be conducted according to the procedures established for initial examination but providing no deadline by which the proceeding must be completed); pre–AIA 35 U.S.C. § 314(a) (explaining that, subject to certain exceptions, inter partes reexamination will be conducted according to the procedures established for initial examination, and that such reexamination will be conducted with special dispatch, but providing no deadline by which the proceeding must be completed). Accordingly, the new post-grant proceedings present a much better alternative to district court litigation because parties who petition for such proceedings are better able to predict when such proceeding will conclude, and the proceedings often progress at a much faster pace than district court litigation. A comparison of the new post-grant proceedings is provided in the following tables:

Type of Post-Grant Proceeding	Who Can Bring the Post-Grant Proceeding?	On What Grounds Can the Patent be Challenged?	Burden of Proof for Establishing Invalidity
Post Grant Review (PGR)	Anyone who is not the patent owner (post–AIA 35 U.S.C. § 321(a)) so long as that person (or a real party–in–interest of that person) has not filed a civil action (i.e., a Declaratory Judgment action) challenging the validity of a claim of the patent before filing the PGR petition. (Post–AIA 35 U.S.C. § 325(a)(1); see also post-AIA 35 U.S.C. § 325(a)(3) (explaining that a counterclaim does not constitute a civil action for purposes of post-AIA 35 U.S.C. § 325(a))	The grounds for seeking PGR include any ground that could be raised under 35 U.S.C. § 282(b)(2) or (3). These include grounds that can be raised under 35 U.S.C. § 102 or § 103 as well as 35 U.S.C. § 101 and § 112 (with the exception of compliance with the best mode requirement). (Post-AIA 35 U.S.C. § 321(b); Inter Partes, Post Grant, and Covered Business Method Review Final Rules, 77 Fed. Reg. 48,684 (Aug. 14, 2012))	A petitioner in a PGR has the burden of proving a proposition of unpatentability by a preponderance of the evidence. (Post-AIA 35 U.S.C. § 326(e))
Inter Partes Review (IPR)	Anyone who is not the patent owner (post–AIA 35 U.S.C. § 311(a)) so long as that person has not: (1) filed a civil action (i.e., a Declaratory Judgment action) challenging the valid-ity of a claim of the patent before the date on which the IPR petition is filed (post-AIA 35 U.S.C. § 315(a)(1)); OR (2) been served with (or had their privy or real party–in–interest served with) a complaint alleging infringement of the patent more than one year before filing the petition for the IPR. (Post-AIA 35 U.S.C. § 315(b))	A petitioner may challenge one or more claims of a patent only on a ground that could be raised under 35 U.S.C. § 102 (anticipation) or 35 U.S.C. § 103 (obviousness) and only on the basis of prior art consisting of patents or printed publications. (Post-AIA 35 U.S.C. § 311(b))	A petitioner in an IPR has the burden of proving a proposition of unpatentability by a preponderance of the evidence. (Post-AIA 35 U.S.C. § 316(e))

| Covered Business Method (CBM) Review | A person may not file a petition for a CBM review unless the person or person's real party-in-interest or privy has been sued for infringement of the patent or has been charged with infringement under that patent. (Leahy–Smith America Invents Act of 2011, Pub. L. No. 112–29, § 18(a)(1)(B), 125 Stat. 284, 330; 37 C.F.R. § 42.302(a)) | Same grounds as for post grant reviews, except that when a petitioner challenges the validity of one or more claims in a covered business method patent on a ground raised under 35 U.S.C. §§ 102 or 103, as in effect on the day before the effective date set forth in AIA § 3(n)(1) (i.e., brings a challenge under pre-AIA 35 U.S.C. §§ 102 or 103), the prior art or that can be used is limited by AIA § 18(a)(1)(C). Unfortunately, AIA § 18(a)(1)(C) is fairly ambiguous so it is not clear what types of prior art are included under that provision. (Leahy–Smith America Invents Act of 2011, Pub. L. No. 112–29, §§ 18(a)(1) and 18(a)(1)(C), 125 Stat. 284, 329–30; *see also* Inter Partes, Post Grant, and Covered Business Method Review Final Rules, 77 Fed. Reg. 48,680 (Aug. 14, 2012) (explaining that AIA § 18(a)(1)(C) limits the prior art that may be used to challenge business method patents granted under the first-to-invent system but offering no explanation as to the meaning of that provision); Donald S. Chisum, *America Invents Act of 2011: Analysis and Cross-References*, Dec. 5, 2011, http://www.chisum.com/wp-content/uploads/AIAOverview.pdf) (explaining that the awkward wording of subpart (ii) of AIA § 18(a)(1)(C) is a puzzle)). | A petitioner in a CBM review has the burden of proving a proposition of unpatentability by a preponderance of the evidence. (Post-AIA 35 U.S.C. § 326(e); Leahy–Smith America Invents Act of 2011, Pub. L. No. 112–29, § 18(a)(1), 125 Stat. 284, 329) |

Type of Post-Grant Proceeding	What is the Standard for Instituting the Proceeding?	When Can a Petition for the Post-Grant Proceeding be Filed?
Post Grant Review (PGR)	• The Director may not authorize institution of a PGR unless the Director determines that the information presented in the petition, if not rebutted, would demonstrate that it is more likely than not that at least one of the challenged claims is unpatentable. (Post-AIA 35 U.S.C. § 324(a)). • A PGR may also be instituted if the petition raises a novel or unsettled legal question that is important to other patents or patent applications. (Post-AIA 35 U.S.C. § 324(b))	• A petition for PGR must be filed no later than the date that is nine months after a patent subject to the first-inventor-to-file system has been granted or the reissue has been issued. (Post-AIA 35 U.S.C. § 321(c))
Inter Partes Review (IPR)	• In order for an IPR to be instituted, the information presented in a petition for IPR, and any response and/or amendment to the petition, must show that there is a reasonable likelihood that the petitioner will prevail with respect to at least one of the claims challenged in the petition. (Post-AIA 35 U.S.C. § 314(a))	• For patents subject to the first-inventor-to-file system, an IPR petition may be filed after the later of: (i) 9 months after the patent is granted (or reissue is granted, if applicable); or (ii) the date of termination of any post grant review of the patent. (Post-AIA 35 U.S.C. § 311(c) (1) and (2)) • For patents subject to the first-to-invent system, an IPR petition may be filed upon issuance of the patent or reissued patent. (Leahy-Smith America Invents Technical Corrections Act of 2013, Pub. L. No. 112-274, § 1(d)(1), 126 Stat. 2456; Inter Partes Review Technical Correction Final Rule, 78 Fed. Reg. 17,871 (Mar. 25, 2013))

| Covered Business Method (CBM) Review | • The Director may not authorize institution of a CBM review unless the Director determines that the information presented in the petition, if not rebutted, would demonstrate that it is more likely than not that at least one of the challenged claims is unpatentable. (Post-AIA 35 U.S.C. § 324(a); Leahy-Smith America Invents Act of 2011, Pub. L. No. 112–29, § 18(a)(1), 125 Stat. 284, 330)

• A CBM review may also be instituted if the petition raises a novel or unsettled legal question that is important to other patents or patent applications. (Post-AIA 35 U.S.C. § 324(b); Leahy-Smith America Invents Act of 2011, Pub. L. No. 112–29, § 18(a)(1), 125 Stat. 284, 330). | • For patents subject to the first-inventor-to-file system, a petition for CBM review may be filed at any time except during the period in which a petition for a post grant review of the patent may be filed (i.e., the petition may not be filed during the first nine months after the patent has been granted). (Leahy-Smith America Invents Act of 2011, Pub. L. No. 112–29, § 18(a)(2), 125 Stat. 284, 330; 37 C.F.R. § 42.303).

• For patents subject to the first-to-invent system, a petition for CBM review may be filed at any time, including within the first nine months after the patent has been granted. (*See* Inter Partes, Post Grant, and Covered Business Method Review Final Rules, 77 Fed. Reg. 48,709 (Aug. 14, 2012) ("The transitional review program is available for non-first-to-file patents, even within the first nine months of the grant of such patents. The rule is consistent with the limitation set forth in section 18(a)(2) of the AIA, and therefore no change was made.")).

• Furthermore, the AIA provision governing CBM reviews is repealed as of September 16, 2020. (Leahy-Smith America Invents Act of 2011, Pub. L. No. 112–29, § 18(a)(3)(A), 125 Stat. 284, 330). Accordingly, the U.S. PTO will not consider petitions for Covered Business Method Reviews that are filed on or after that date. (37 C.F.R. § 42.300(d)). |

Type of Post-Grant Proceeding	Does the Post-Grant Proceeding Have any Estoppel Effects with Respect to Other Proceedings?	What Types of Patents May be Challenged in the Post-Grant Proceeding?	Length of Post-Grant Proceeding
Post Grant Review (PGR)	• Yes, a petitioner (as well as any real-parties-in-interest and privies of the petitioner) involved in a post grant review of a claim that results in a final written decision under post-AIA 35 U.S.C. § 328(a), is estopped from subsequently challenging the patent claim(s) at the U.S. PTO, in federal court or at the ITC, on any ground that the petitioner raised or reasonably could have raised during the post grant review. (Post-AIA 35 U.S.C. § 325(e)(1) and (2))	• Only patents that are subject to the first-inventor-to-file system are subject to post grant review. (Inter Partes, Post Grant, and Covered Business Method Review Final Rules, 77 Fed. Reg. 48,687 (Aug. 14, 2012))	• A final determination in any post grant review will be issued not later than one year after the date on which the Director notices the institution of the proceeding, except that the one-year period may be extended up to six months upon a showing of good cause by the Director and the one-year period may be adjusted in the case of joinder. (Post-AIA 35 U.S.C. § 326(a)(11)).
Inter Partes Review (IPR)	• Yes, a petitioner (as well as any real parties-in-interest and privies of the petitioner) involved in an inter partes review of a claim that results in a final written decision under post-AIA 35 U.S.C. § 318(a), is estopped from subsequently challenging the patent claim(s) at the U.S. PTO, in federal court or at the ITC, on any ground that the petitioner raised or reasonably could have raised during the *inter partes review.* (Post-AIA 35 U.S.C. § 315(e)(1) and (2))	• Patents subject to the first-to-invent system; and • Patents subject to the first-inventor-to-file system (*See Inter Partes,* Post Grant, and Covered Business Method Review Final Rules, 77 Fed. Reg. 48,680 (Aug. 14, 2012) (indicating that the changes for inter partes review proceedings apply to any patent issued before, on, or after September 16, 2012))	• A final determination in an inter partes review will be issued not later than one year after the date on which the review is instituted, except that the one-year period may be extended up to six months upon a showing of good cause by the Director and the one-year period may be adjusted in the case of joinder. (Post-AIA 35 U.S.C. § 316(a)(11))

Covered Business Method (CBM) Review	• Yes, a petitioner in a transitional covered business method proceeding that results in a final written decision under post-AIA 35 U.S.C. § 328(a), or the petitioner's real party-in-interest, may not assert, either in a civil action or in a proceeding before the ITC, that the claim is invalid on any ground that the petitioner raised during the covered business method proceeding. (Leahy-Smith America Invents Act of 2011, Pub. L. No. 112-29, § 18(a)(1)(D), 125 Stat. 284, 330). • In addition, petitioners in transitional covered business method proceedings are subject to the same estoppel before the U.S. PTO that PGR petitioners are. (Post-AIA 35 U.S.C. § 325(e)(1); Leahy-Smith America Invents Act of 2011, Pub. L. No. 112-29, § 18(a)(1), 125 Stat. 284, 330).	• Covered business method patents subject to the first-to-invent system; and • Covered business method patents subject to the first-inventor-to-file system. (See Inter Partes, Post Grant, and Covered Business Method Review Final Rules, 77 Fed. Reg. 48,687 (Aug. 14, 2012) ("The AIA provides that the transitional program for the review of covered business method patents will take effect on September 16, 2012, one year after the date of enactment, and applies to any covered business method patent issued before, on, or after September 16, 2012.")).	• A final determination in any covered business method review will be issued not later than one year after the date on which the Director notices the institution of the proceeding, except that the one-year period may be extended up to six months upon a showing of good cause by the Director and the one-year period may be adjusted in the case of joinder. (Post-AIA 35 U.S.C. § 326(a)(11); Leahy-Smith America Invents Act of 2011, Pub. L. No. 112-29, § 18(a)(1), 125 Stat. 284, 330).

Of all of the post-grant proceedings that were created by the AIA, IPRs have been utilized the most. In fact, as of June 30, 2016, the U.S. PTO posted statistics showing that a total of 4,704 IPR petitions had been filed since the post-grant proceeding became available on September 16, 2012, whereas only 469 CBM review petitions and 29 PGR petitions had been filed during that same time period.[f] The fact that far more petitions for IPR have been filed than for either PGR or CBM review should come as no surprise because: (1) far more patents are eligible to be challenged through IPR than are eligible for either PGR or CBM review, and (2) the potential estoppel effects against petitioners are much more limited for IPR than for either PGR or CBM review.

First, far more patents are eligible to be challenged through IPR proceedings than through PGR proceedings because an IPR may be used to challenge patents that are subject to both the first-to-invent and first-inventor-to-file systems. Accordingly, patents issued before, on, or after September 16, 2012 are eligible to be challenged through IPR proceedings. Inter Partes, Post Grant, and Covered Business Method Review Final Rules, 77 Fed. Reg. 48,680 (Aug. 14, 2012). In contrast, only patents that are subject to the first-inventor-to-file system are eligible for PGR challenges. Id. at 48,687. The first-inventor-to-file system took effect on March 16, 2013 and, as discussed in Section IV of Chapter 2, only claims filed on or after that date have the potential to be subject to the first-inventor-to-file system. Given that it often takes 2–3 years for the U.S. PTO to grant a patent after a patent application has been filed, relatively few patents subject to the first-inventor-to-file system are currently in existence.

Furthermore, while it is possible to challenge patents that are subject to both the first-to-invent and first-inventor-to-file systems through CBM reviews, the pool of patents that is eligible to be challenged through CBM reviews is still far more limited than the pool that is eligible to be challenged via IPRs. In fact, only "covered business method patents"[g] are eligible to be challenged through the AIA's transitional CBM review program, whereas patents pertaining to any subject matter may be challenged through an IPR. Leahy-Smith America Invents Act of 2011, Pub. L. No. 112-29, § 18(a)(1), 125 Stat. 284,

---

[f] See U.S. PTO Patent Trial and Appeal Board Statistics, http://www.uspto.gov/sites/default/files/documents/2016-6-30%20PTAB.pdf (last visited Aug. 2, 2016).

[g] AIA § 18(d)(1) defines a "covered business method patent" as "a patent that claims a method or corresponding apparatus for performing data processing or other operations used in the practice, administration, or management of a financial product or service, except that the term does not include patents for technological inventions." Leahy-Smith America Invents Act of 2011, Pub. L. No. 112-29, 125 Stat. 284, 331.

329; *see also* post–AIA 35 U.S.C. §§ 311–319 (omitting any requirement that a patent claim certain subject matter in order to be eligible to be challenged through an IPR proceeding). Additionally, a petition for a CBM review cannot be filed unless a person, or a person's real party-in-interest or privy, has been sued for infringement of the patent or has been charged with infringement under that patent. Leahy-Smith America Invents Act of 2011, Pub. L. No. 112-29, § 18(a)(1)(B), 125 Stat. 284, 330; 37 C.F.R. § 42.302(a). No such requirement exists for IPR challenges. *See* post–AIA 35 U.S.C. §§ 311 and 315 (omitting any requirement that a person have been sued for infringement of a patent or have been charged with infringement under a patent in order for that person to file an IPR petition to challenge the validity of the patent).

The differences in estoppel effects also explain why IPRs are far more popular than either PGRs or CBM reviews. Indeed, even though petitioners (and their real parties-in-interest and privies) are estopped from subsequently challenging patent claims at the U.S. PTO, in federal court or the ITC on any ground the petitioner "raised or reasonably could have raised" in an IPR and PGR, the estoppel effect for PGRs can be far broader than for IPRs because a PGR may be sought on far more invalidity grounds. Post-AIA 35 U.S.C. § 315(e)(1) and (2); post-AIA 35 U.S.C. § 325(e)(1) and (2). Indeed, as indicated in the previous tables, the grounds for seeking PGR include any ground that can be raised under 35 U.S.C. § 282(b)(2) or (3) (e.g., defenses under 35 U.S.C. §§ 101, 102, 103 and 112) while IPRs may only be sought on grounds that can be raised under 35 U.S.C. § 102 or § 103. Post-AIA 35 U.S.C. § 311(b); post-AIA 35 U.S.C. § 321(b). Thus, the number of defenses that were "raised or reasonably could have been raised" in a PGR proceeding can far exceed those that were "raised or reasonably could have been raised" in an IPR proceeding, thereby creating a broader estoppel effect.

Although CBM reviews are a type of PGR, the estoppel effect created by CBM reviews differs slightly from the estoppel effect created by PGRs. Specifically, petitioners (and their real parties-in-interest and privies) in CBM review proceedings are only estopped from subsequently challenging patents in federal court or in the ITC on grounds they <u>actually raised</u> instead of on grounds they raised or "reasonably could have raised."[h] Leahy-Smith America Invents Act of 2011, Pub. L. No. 112-29, § 18(a)(1)(D), 125 Stat. 284, 330. Even with this difference, the estoppel effect on subsequent federal court and/or ITC proceedings that is created by CBM reviews has

---

[h]Note that the estoppel effect for subsequent challenges in the U.S. PTO is the same for both PGRs and CBM reviews. Post-AIA 35 U.S.C. § 325(e)(1); Leahy-Smith America Invents Act of 2011, Pub. L. No. 112-29, § 18(a)(1), 125 Stat. 284, 330. Only the estoppel effect for subsequent challenges in federal courts or the ITC differs.

the potential to be greater than that created by IPRs where, for instance, a CBM petitioner raised challenges other than those that can be raised under 35 U.S.C. § 102 or § 103. *See* Leahy-Smith America Invents Act of 2011, Pub. L. No. 112-29, § 18(a)(1)(D), 125 Stat. 284, 330 (explaining that a petitioner in a CBM proceeding that results in a final written decision under post-AIA 35 U.S.C. § 328(a), or the petitioner's real party-in-interest, may not assert, either in a civil action or in a proceeding before the ITC, that the claim is invalid on any ground that the petitioner raised during the covered business method proceeding); post-AIA 35 U.S.C. § 315(e)(1) and (2) (explaining that a petitioner (as well as any real parties-in-interest and privies of the petitioner) involved in an inter partes review of a claim that results in a final written decision under post-AIA 35 U.S.C. § 318(a), is estopped from subsequently challenging the patent claim(s) at the U.S. PTO, in federal court or at the ITC, on any ground that the petitioner raised or reasonably could have raised during the inter partes review).

Overall, the new post-grant proceedings created by the AIA have been game changers for patent troll cases[i] because a defendant who has been sued on a dubious patent can now file a request for an IPR, CBM review and/or PGR, and there is a good chance that such proceeding will conclude before the patent litigation reaches trial, or at least before the district court enters a final judgment. In some jurisdictions (e.g., the District of Delaware and the Northern District of California), there is even a good chance that the district court litigation will be stayed pending the outcome of the post-grant proceeding. *See* Docket Report by Docket Navigator, *Motions to Stay District Court Cases Pending Post-Grant Proceedings*, Aug. 24, 2015, http://docketreport.blogspot.com/2015/08/motions-to-stay-district-court-cases.html ("The Northern District of California (CAND), the Eastern District of Texas (TXED), and the District of Delaware (DED) all decided a similar number of motions to stay. CAND decided the most overall, and had the highest grant rate, 57%. TXED is the opposite, with only 25% granted and over 67% denied. Delaware granted 55%, denied 38%, and partially granted 7%."). Post-grant proceedings therefore give alleged infringers an incredible amount of leverage against a patent owner because if the U.S. PTO finds that a patent is invalid before a final judgment is reached in a district court

---

[i]Much to the chagrin of many branded pharmaceutical companies, IPRs have also been used to challenge pharmaceutical patents involved in Hatch-Waxman litigation. *See, e.g., Senju Pharm. Co., Ltd. et al. v. Metrics, Inc. et al.*, No. 14-3962, 2015 U.S. Dist. LEXIS 41504, at *39 (D.N.J. Mar. 31, 2015) (explaining that the statutory framework of the Hatch-Waxman Act would not be undermined by allowing inter partes review before the PTAB to proceed in parallel with a district court litigation).

litigation, the patent troll no longer has a patent to assert against that defendant (or any other defendant) in the district court litigation and the case is over. *See Fresenius USA, Inc. v. Baxter Int'l Inc.*, 721 F.3d 1330, 1340 (Fed. Cir. 2013) (explaining that if a claim is cancelled or amended to cure invalidity in a post-grant proceeding while an infringement suit over the claim is still pending in district court, "the patentee's cause of action [in the district court] is extinguished and the suit fails.").

III) POST-GRANT PROCEEDINGS FOR PATENT OWNERS DESIRING TO STRENGTHEN THEIR PATENT RIGHTS

Before the AIA was enacted, more post-grant proceedings were available to patent owners desiring to strengthen their patents than for third parties wishing to challenge a patent in the U.S. PTO. None of the available proceedings, however, was specifically designed to insulate a patent from being held unenforceable for inequitable conduct. The lack of ability to redress inequitable conduct issues was problematic in the years preceding enactment of the AIA because inequitable conduct was pleaded as a defense in the majority of patent infringement cases, regardless of whether there was any evidence suggesting that a patentee had engaged in fraudulent conduct when obtaining his/her patent. To address this issue, the AIA provided supplemental examination, a brand new proceeding that specifically enables patent owners to insulate their patents from inequitable conduct allegations. In addition, the AIA modified some of the existing post-grant proceedings to enable patent owners to more easily correct errors in their patents and strengthen their patent rights.

A. Post-Grant Proceedings Available to Patent Owners Before the Enactment of the AIA

It is not uncommon for a patent holder to discover a mistake in his/her patent or, worse yet, additional prior art that could impact the scope of his/her patent claims after the patent has already been issued. Patent holders already had several options for remedying these issues before the AIA was enacted. In fact, depending on the type of issue a patent holder was trying to address, the patent holder could: (1) file a reissue application, (2) file a request for an ex parte reexamination, or (3) request a certificate of correction,[j] including

[j]Certificates of correction were also available to third parties who believed a patent contains certain types of errors. *See* 37 C.F.R. § 1.322(a)(1)(iii) (explaining that upon receiving information about a mistake from a third party, the Director may issue a certificate of correction pursuant to 35 U.S.C. § 254 to correct a mistake in a patent, incurred through the fault of the Office, where such mistake is clearly disclosed in the records of the Office).

a request to correct inventorship. An overview of these options is provided in the following tables:

Type of Post-Grant Procedure	Types of Errors that Can be Corrected Through the Post-Grant Procedure
Reissue	(1) A defective specification or drawing; and/or (2) A claim that is either too narrow or too broad, so long as the error in the specification, drawing or claim scope renders the patent wholly or partly inoperative or invalid. (Pre-AIA 35 U.S.C. § 251(a))
Ex Parte Reexamination	(1) Patent claims that are too broad under 35 U.S.C. § 102 or § 103 in light of patent or printed publication prior art; and/or (2) Failure to cite certain prior art during the prosecution of a patent. (MPEP § 2209 (9th ed. 2014))
Certificate of Correction	(1) Virtually any type of mistake made by the U.S. PTO can be corrected via a certificate of correction so long as it is clearly disclosed by the U.S. PTO's records. (Pre-AIA 35 U.S.C. § 254). (2) Only minor mistakes (i.e., those mistakes that do not materially affect the scope or meaning of the patent) made by patent applicants in good faith may be corrected using a certificate of correction. (Pre-AIA 35 U.S.C. § 255; MPEP § 1481 (9th ed. 2014)) Such mistakes include clerical errors and typographical errors. (3) An applicant's failure to perfect a claim for a foreign priority benefit in a patented application may be remedied using a certificate of correction if the requirements of 35 U.S.C. § 119(a)-(d) or (f) had been satisfied in the patented application or in a parent application prior to issuance of the patent and the requirements of 37 CFR § 1.55 are met. (MPEP § 216.01 (9th ed. 2014)) (4) An applicant's failure to perfect a claim for a domestic priority benefit by either (i) failing to make reference to a prior co-pending application pursuant to 37 C.F.R. § 1.78(a)(2); or (ii) making an incorrect reference to a prior co-pending application pursuant to 37 C.F.R. § 1.78(a)(2). (MPEP § 1481.03 (9th ed. 2014))
Request for Correction of Inventorship	(1) Naming someone as an inventor on a patent who is not actually an inventor; or (2) Failing to name someone who is an inventor on the patent. (Pre-AIA 35 U.S.C. § 256)

Type of Post-Grant Procedure	Examples of Types of Errors that Cannot be Corrected Through the Post-Grant Procedure
Reissue	(1) Spelling errors, grammatical errors, or typographical errors that would not render the patent wholly or partly invalid or inoperative; (2) Editorial or clerical errors that would not render the patent wholly or partly invalid or inoperative; and (3) Any other error that would not render the patent wholly or partly invalid or inoperative. (MPEP § 1402 (9th ed. 2014))
Ex Parte Reexamination	(1) Claims that are too narrow (i.e., claims cannot be broadened during an ex parte reexamination); and (2) Minor errors such as spelling, typographical, or clerical errors. (MPEP § 2209 (9th ed. 2014))
Certificate of Correction	(1) Failure of the patent applicant to cite certain prior art during the prosecution of a patent; (2) Defects in the specification that would materially affect the scope or meaning of the patent where such defects were caused by the patent applicant; and (3) Defects in drawings that would materially affect the scope or meaning of the patent where such defects were caused by the patent applicant. (Pre-AIA 35 U.S.C. § 255; MPEP § 1481 (9th ed. 2014))
Request for Correction of Inventorship	Any error that does not involve the inventorship entity listed on the patent cannot be corrected by a request for correction of inventorship. (Pre-AIA 35 U.S.C. § 256; MPEP § 1481.02 (9th ed. 2014))

Type of Post-Grant Procedure	Do Intervening Rights Apply?	Deadlines for Post-Grant Procedure
Reissue	Yes, alleged infringers of reissued patents may be able to defend against claims of infringement based on two types of intervening rights: (1) "Absolute" intervening rights are available for a party that "prior to the grant of a reissue, made, purchased, offered to sell, or used within the United States, or imported into the United States, anything patented by the reissued patent"; and	(1) Any reissue that is filed to broaden one or more patent claims must be filed on or before the date that is two years from the date that the patent was issued. (Pre-AIA 35 U.S.C. § 251(d))

(*Continued on next page*)

Type of Post-Grant Procedure	Do Intervening Rights Apply?	Deadlines for Post-Grant Procedure
	(2) "Equitable" intervening rights may be provided where an alleged infringer of the reissue patent made "substantial preparation" to manufacture, use, offer to sell, sell, purchase, or import anything patented in the reissue before the reissue issued. (Pre–AIA 35 U.S.C. § 252)	(2) Reissue applications must be filed during the pendency of the patent. Once a patent has expired, the U.S. PTO is divested of any authority to reissue the patent. (MPEP § 1415.01 (9th ed. 2014))
Ex parte Reexamination	Yes, intervening rights apply to ex parte reexamination in the same manner as for reissue. (Pre–AIA 35 U.S.C. § 307(b); MPEP § 2293 (9th ed. 2014))	(1) Requests for ex parte reexaminations may be filed at any time during the period of enforceability for a patent (i.e., generally up to six years after the expiration of the patent). (MPEP § 2211 (9th ed. 2014))
Certificate of Correction	No. (*See Superior Fireplace Co. v. Majestic Prods. Co.*, 270 F.3d 1358, 1373 (Fed. Cir. 2001) (distinguishing certificates of correction from reissues on the basis that intervening rights apply to the latter but not to the former))	Requests for certificates of correction can be filed at any time but must be filed before a patent infringement suit is brought if the patentee wishes to rely on the certificate of correction. (*E.I. Du Pont de Nemours and Co. v. Macdermid Printing Solutions, L.L.C.*, 525 F.3d 1353, 1362 (Fed. Cir. 2008); *Southwest Software, Inc. v. Harlequin Inc.*, 226 F.3d 1280, 1295 (Fed. Cir. 2000)).
Request for Correction of Inventorship	Not applicable because claim scope cannot be changed using a request for correction of inventorship. (Pre–AIA 35 U.S.C. § 256; MPEP § 1481.02 (9th ed. 2014))	Requests for Correction of Inventorship under pre-AIA 35 U.S.C. § 256 must be filed after the patent has issued. A separate statute (pre-AIA 35 U.S.C. § 116) governs the correction of inventorship for pending non-provisional applications.

Type of Post-Grant Procedure	Limitations of Post-Grant Procedure
Reissue	(1) If claims are being changed in a reissue, the claims must be for the same invention that was disclosed as being the invention in the original patent (MPEP § 1412.01 (9th ed. 2014)) (2) Claims cannot be broadened to recapture subject matter that was claimed in the original application but that was canceled in order to obtain an allowance of the application (MPEP § 1412.02 (9th ed. 2014)) (3) No new matter can be added to the patent during reissue. (37 C.F.R. § 1.173(a))
Ex parte Reexamination	(1) Claims cannot be broadened during an ex parte reexamination. (MPEP § 2209 (9th ed. 2014))
Certificate of Correction	Certificates of correction that are filed to correct a patent applicant's mistake(s) cannot: (1) introduce new matter; or (2) make a correction that would require the patent to be reexamined. (MPEP § 1481 (9th ed. 2014))
Request for Correction of Inventorship	Requests to correct inventorship must include a statement from each person who is being added as an inventor and each person who is currently named as an inventor either agreeing to the change of inventorship or stating that he or she has no disagreement with regard to the requested change. (37 C.F.R. § 1.324(b)(1)). If such statement cannot be obtained, then the proper method for correcting inventorship is via a reissue. (MPEP § 1412.04 (9th ed. 2014))

All of the post-grant proceedings highlighted in the previous tables are still available today, but reissue and correction of inventorship proceedings underwent some changes through the AIA. These changes are highlighted in the discussions provided in the following sections.

i. Reissue

Patentees have long relied on the reissue process to correct a myriad of errors that would render their patent partly or wholly inoperative or invalid. For instance, the process can be used to fix claims that are too broad or too narrow, correct inaccuracies in the specification, correct a claim to foreign priority (or include such a claim if it was omitted), and/or correct erroneous references to prior co-pending applications used for domestic priority claims. MPEP § 1402 (9th ed. 2014). More specific examples of the types of errors that can and cannot be fixed through reissue are listed in the following table:

Errors that Can be Corrected via Reissue	Errors that Cannot be Corrected via Reissue
A failure to present narrower claims in the original patent. *See In re Tanaka*, 640 F.3d 1246, 1251 (Fed. Cir. 2011) ("[T]he omission of a narrower claim from a patent can render a patent partly inoperative by failing to protect the disclosed invention to the full extent allowed by law.").	Failure to timely file a divisional application to claim non-elected subject matter following a restriction requirement. (MPEP § 1402(I) (9th ed. 2014))
Misjoinder of inventors (i.e., either an error of a person being incorrectly named in an issued patent as the inventor, or through error of an inventor incorrectly not named in an issued patent). (MPEP §§ 1402(II); 1412.04(II) (9th ed. 2014))	Non-substantive drawing errors (e.g., a reference numeral correction or addition, the addition of shading, or even the addition of an additional figure merely to "clarify" the disclosure). (MPEP § 1402(V) (9th ed. 2014))
Failure to perfect a priority claim to a foreign patent application (e.g., by failing to file a certified copy of the original foreign application before the patent was granted). (MPEP § 1402(III) (9th ed. 2014))	Applicants cannot use reissue to delete an earlier-obtained claim to priority in order to try to extend the term of their patent. (MPEP § 1405 (9th ed. 2014))
Failure to claim priority to another domestic application (i.e., either a provisional application or non-provisional patent application). (MPEP § 1402(IV) (9th ed. 2014))	
Errors in priority claims (e.g., where a patent claims priority to the wrong domestic or foreign patent application). (MPEP § 1402(III) and (IV) (9th ed. 2014))	
Substantive drawing errors. (MPEP § 1402(V) (9th ed. 2014))	
Misrepresentations made during patent prosecution that would render the patent wholly or partly invalid or inoperable. (Post-AIA 35 U.S.C. § 251)	

One of the major limitations to using reissue practice before the AIA was passed, however, was that the oath or declaration that was required for filing a reissue application had to state that all errors being corrected in the reissue application arose "without any deceptive intention" on the part of the applicant. Pre-AIA 35 U.S.C. § 251; pre-AIA 37 C.F.R. § 1.175(a)(2). This meant that if a patentee had made a misrepresentation to the U.S. PTO, or intentionally failed to cite a relevant prior art reference of which it was aware during the original prosecution of a patent, the patentee could not file a reissue application to correct such error (e.g., by citing the reference to the U.S. PTO and amending his/her claims). Misrepresentations and intentional failure to cite prior art are grounds upon which a patent can be held unenforceable for inequitable conduct. Thus, prior to the AIA, patentees could not use reissue to cure inequitable conduct problems.

As shown in the following table, AIA § 20(d)(1)(B) amended pre-AIA 35 U.S.C. § 251[k] so that reissue applicants are no longer required to assert that the error(s) being corrected arose "without any deceptive intention" in their reissue oath or declaration.[l] Leahy-Smith America Invents Act of 2011, Pub. L. No. 112-29, 125 Stat. 284, 333-34.

[k]The amendments to pre-AIA 35 U.S.C. § 251 became effective on September 16, 2012 and apply to reissues commenced on or after that date. Leahy-Smith America Invents Act of 2011, Pub. L. No. 112-29, § 20(l), 125 Stat. 284, 335.

[l]AIA § 20 also removed "without deceptive intent" language from pre-AIA 35 U.S.C. §§ 116, 184, 185, 253, 256, and 288. Leahy-Smith America Invents Act of 2011, Pub. L. No. 112-29, §§ 20(a), (b), (c), (e), (f) and (h), 125 Stat. 284, 333-34. All of the statutory amendments made by AIA § 20 took effect on September 16, 2012 and apply to proceedings commenced on or after that date. *Id.* at § 20(l). Pre-AIA 35 U.S.C. §§ 116 and 256 permit the correction of certain inventorship errors in patents and patent applications. AIA §§ 20(a) and (f) therefore amend the pre-AIA statutes to permit correction of such inventorship errors even where the errors are made with deceptive intent. *Compare* pre-AIA 35 U.S.C. § 116 ("Whenever through error a person is named in an application for patent as the inventor, or through an error an inventor is not named in an application, and such error arose *without any deceptive intention* on his part, the Director may permit the application to be amended accordingly, under such terms as he prescribes.") (emphasis added); pre-AIA 35 U.S.C. § 256 ("Whenever through error a person is named in an issued patent as the inventor, or through error an inventor is not named in an issued patent and such error arose *without any deceptive intention* on his part, the Director may, on application of all the parties and assignees, with proof of the facts and such other requirements as may be imposed, issue a certificate correcting such error.") (emphasis added) *with* post-AIA 35 U.S.C. § 116(c) ("Whenever through error a person is named in an application for patent as the inventor, or through an error an inventor is not named in an application, the Director may permit the application to be amended accordingly, under such terms as he prescribes."); post-AIA 35 U.S.C. § 256(a) ("Whenever through error a person is named in an issued patent as the inventor, or through error an inventor is not named in an issued patent, the Director may, on application of all the parties and assignees, with proof of the facts and such other requirements as may be imposed, issue a certificate correcting such error.").

Pre- and post-AIA 35 U.S.C. §§ 184 and 185 require that patent applicants obtain a license before filing a patent application in a foreign country after a corresponding patent application has been filed in the U.S. and provide that a retroactive license may be granted in instances where an application has been filed abroad through error. AIA § 20(b) amends pre-AIA 35 U.S.C. § 184 to make clear that a license may be granted retroactively even where the erroneous foreign filing was made with deceptive intent. *Compare* pre-AIA 35 U.S.C. § 184 ("The [foreign filing] license may be granted retroactively where an application has been filed abroad through error and *without deceptive intent* and the application does not disclose an invention within the scope of section 181 of this title.") (emphasis added) *with* post-AIA 35 U.S.C. ▶

Pre-AIA 35 U.S.C. § 251	Post-AIA 35 U.S.C. § 251
Whenever any patent is, through error *without any deceptive intention,* deemed wholly or partly inoperative or invalid, by reason of a defective specification or drawing, or by reason of the patentee claiming more or less than he had a right to claim in the patent, the Director shall, on the surrender of such patent and the payment of the fee required by law, reissue the patent for the invention disclosed in the original patent, and in accordance with a new and amended application, for the unexpired part of the term of the original patent. No new matter shall be introduced into the application for reissue.	(a) IN GENERAL.–Whenever any patent is, through error, deemed wholly or partly inoperative or invalid, by reason of a defective specification or drawing, or by reason of the patentee claiming more or less than he had a right to claim in the patent, the Director shall, on the surrender of such patent and the payment of the fee required by law, reissue the patent for the invention disclosed in the original patent, and in accordance with a new and amended application, for the unexpired part of the term of the original patent. No new matter shall be introduced into the application for reissue.

(Continued on next page)

◄ § 184(a) ("The [foreign filing] license may be granted retroactively where an application has been filed abroad through error and the application does not disclose an invention within the scope of section 181."). Furthermore, AIA § 20(c) amends pre-AIA 35 U.S.C. § 185 to remove the requirement that an error in failing to procure a foreign filing license occur without deceptive intent in order for a U.S. patent to remain valid after such error. *Compare* pre-AIA 35 U.S.C. § 185 ("A United States patent issued to such person, his successors, assigns, or legal representatives shall be invalid, unless the failure to procure such [foreign filing] license was through error *and without deceptive intent,* and the patent does not disclose subject matter within the scope of section 181 of this title.") (emphasis added) *with* post-AIA 35 U.S.C. § 185 ("A United States patent issued to such person, his successors, assigns, or legal representatives shall be invalid, unless the failure to procure such [foreign filing] license was through error, and the patent does not disclose subject matter within the scope of section 181.").

Pre- and post-AIA 35 U.S.C. § 253 provide that if one claim of a patent is invalid, the remaining claims are not automatically rendered invalid. AIA § 20(e) amends pre-AIA 35 U.S.C. § 253 to remove the requirement that the invalidity of one claim be "without any deceptive intention" in order to maintain validity of the other claims. *Compare* pre-AIA 35 U.S.C. § 253 ("Whenever, *without any deceptive intention,* a claim of a patent is invalid the remaining claims shall not thereby be rendered invalid.") (emphasis added) *with* post-AIA 35 U.S.C. § 253 ("Whenever a claim of a patent is invalid the remaining claims shall not thereby be rendered invalid."). Similarly, pre- and post-AIA 35 U.S.C. § 288 allow litigants to maintain a patent infringement suit over patent claims that may be valid even if other claims in the patent have been invalidated. AIA § 20(h) therefore amends the pre-AIA statute to remove the requirement that the invalidity of one or more claims be "without deceptive intention" in order to maintain such patent suits over claims that have not yet been invalidated. *Compare* pre-AIA 35 U.S.C. § 288 ("Whenever, *without deceptive intention,* a claim of a patent is invalid, an action may be maintained for the infringement of a claim of the patent which may be valid.") (emphasis added) *with* post-AIA 35 U.S.C. § 288 ("Whenever a claim of a patent is invalid, an action may be maintained for the infringement of a claim of the patent which may be valid.").

Pre-AIA 35 U.S.C. § 251	Post-AIA 35 U.S.C. § 251
The Director may issue several reissued patents for distinct and separate parts of the thing patented, upon demand of the applicant, and upon payment of the required fee for a reissue for each of such reissued patents. The provisions of this title relating to applications for patent shall be applicable to applications for reissue of a patent, except that application for reissue may be made and sworn to by the assignee of the entire interest if the application does not seek to enlarge the scope of the claims of the original patent. No reissued patent shall be granted enlarging the scope of the claims of the original patent unless applied for within two years from the grant of the original patent. (Emphasis added.)	(b) MULTIPLE REISSUED PATENTS.– The Director may issue several reissued patents for distinct and separate parts of the thing patented, upon demand of the applicant, and upon payment of the required fee for a reissue for each of such reissued patents. (c) APPLICABILITY OF THIS TITLE.– The provisions of this title relating to applications for patent shall be applicable to applications for reissue of a patent, except that application for reissue may be made and sworn to by the assignee of the entire interest if the application does not seek to enlarge the scope of the claims of the original patent or the application for the original patent was filed by the assignee of the entire interest. (d) REISSUE PATENT ENLARGING SCOPE OF CLAIMS.– No reissued patent shall be granted enlarging the scope of the claims of the original patent unless applied for within two years from the grant of the original patent.

Proponents of the amendment to pre-AIA 35 U.S.C. § 251 believed that by allowing patent holders to correct errors that could render a patent unenforceable for inequitable conduct, patent issues could be adjudicated more efficiently. *See* 157 CONG. REC. S1378, 2011 (statement of Sen. Kyl) ("Eliminating the various deceptive-intent requirements moves the U.S. patent system away from the 19th century model that focused on the patent owner's subjective intent, and towards a more objective-evidence-based system that will be much cheaper to litigate and more efficient to administer.").

The AIA also changed the reissue statute to allow certain assignees to seek a reissue in order to broaden the claims of their patent. As shown previously, pre-AIA 35 U.S.C. § 251 expressly prohibited assignees from seeking broadening reissues. *See id.* ("[E]xcept that application for reissue may be made and sworn to by the assignee of the entire interest if the application does not seek to enlarge the scope of the claims of the original patent.").

Post-AIA 35 U.S.C. § 251(c), however, allows an assignee who holds the entire interest in a reissue application to seek a broadening reissue if that assignee had filed the patent application that issued as the original patent.[m] *See also* 37 C.F.R. § 1.175(c) (explaining that a reissue oath or declaration may be signed by the assignee of the entire interest if (a) the reissue application does not seek to enlarge the scope of the claims of the original patent), or (b) the application for the original patent was filed under 37 C.F.R. § 1.46 by the assignee of the entire interest). The main purpose of amending pre-AIA 35 U.S.C. § 251 to allow certain assignees to apply for a broadening reissue was to bring the reissue statute into conformity with other statutes that were amended by the AIA to enable assignees to file patent applications without seeking an oath or declaration from the inventor(s). *See* 157 CONG. REC. S1373, 2011 (statement of Sen. Kyl) ("Expanding an assignee's right to seek broadening reissue is consistent with the bill's changes to [pre-AIA 35 U.S.C. §§] 115 and 118, which expand assignees' rights by allowing assignees to apply for a patent against the inventor's wishes. If an assignee exercises his right to apply for a patent against the inventor's wishes, there is no reason not to allow the same assignee to also seek a broadening reissue within the [post-AIA 35 U.S.C. §] 251 time limits.").

ii. Ex Parte Reexamination

While reissue is used to correct certain types of errors in patents, ex parte reexamination is generally utilized to test the validity of a patent's claims against certain prior art references that may or may not have been considered during the original prosecution of a patent. Indeed, as shown previously, a request for ex parte reexamination can only be filed on the basis that one or more prior art patents and/or publications raise(s) a substantial new question of patentability for at least one claim of a patent under 35 U.S.C. § 102 and/or 35 U.S.C. § 103. MPEP § 2209 (9th ed. 2014). If the U.S. PTO finds that a substantial new question of patentability is raised, and that the patent's claims are too broad in view of the prior art that is being considered during reexamination, a patent owner can narrow the claims through claim amendments,[n] thereby shielding the claims from

---

[m]If the patent application that issued as the original patent was instead filed by an inventor or another assignee from whom the assignee of the reissue application acquired its rights, the assignee of the reissue application cannot seek a broadening reissue under post-AIA 35 U.S.C. § 251. 37 C.F.R. § 1.175(c)(2).

[n]Note that unlike in reissue applications, claims undergoing ex parte reexamination cannot be broadened. MPEP § 2209 (9th ed. 2014).

future invalidity attacks. *See* MPEP § 2234 (9th ed. 2015) (setting forth the guidelines for amending claims in patents undergoing ex parte reexamination). If, on the other hand, the U.S. PTO finds that the newly cited prior art does not raise a substantial new question of patentability for the claims at issue, the U.S. PTO will decline to reexamine the claims and future alleged infringers likely will have a more difficult time establishing invalidity based on such prior art. *See* MPEP § 2209 (9th ed. 2015) (explaining that in order for ex parte reexamination to be ordered, a substantial new question of patentability must be found); *see also Am. Hoist & Derrick Co. v. Sowa & Sons, Inc.,* 725 F.2d 1350, 1360 (Fed. Cir. 1984) ("When new evidence touching validity of the patent not considered by the PTO is relied on [in patent litigation], the tribunal considering it is not faced with having to *disagree* with the PTO or with *deferring* to its judgment or with taking its expertise into account. The evidence may, therefore, carry more weight and go further toward sustaining the attacker's unchanging burden.") (emphasis in original). Finally, even if the U.S. PTO does institute an ex parte reexamination based on newly cited prior art, the U.S. PTO may ultimately decide not to reject the claims as being anticipated or obvious over that art. In that case, future alleged infringers will also have a more difficult time establishing invalidity based on that prior art.

Other than increasing the fee for filing a request for an ex parte reexamination proceeding, and clarifying a jurisdictional issue for appeals of decisions in ex parte reexamination proceedings,[o] the AIA left ex parte reexamination practice unchanged. *See* Changes to Implement the Supplemental Examination Provisions of the Leahy-Smith America Invents Act and to Revise Reexamination Fees, 77 Fed. Reg. 48,828 (Aug. 14, 2012) (raising the fee for ex parte reexamination requests from $2,520 to $17,750). Ex parte reexamination was utilized by patent owners relatively frequently before the AIA was enacted because it was the only proceeding through which a patentee could have the U.S. PTO evaluate the validity of a patent by considering prior art that was not presented to the U.S. PTO during the original prosecution of a patent without admitting that his/her patent claims were too broad in view of that prior art. *See* 37 C.F.R. § 1.510(b)(2) (explaining that even where a patent owner submits prior art

---

[o]As mentioned in Section II.B., the AIA amended pre-AIA 35 U.S.C. § 306 to clarify that patent owners who are dissatisfied with a decision in an ex parte reexamination may only appeal to the Federal Circuit—not to the Eastern District of Virginia. Leahy-Smith America Invents Act of 2011, Pub. L. No. 112-29, § 6(h)(2)(A), 125 Stat. 284, 312.

to be considered by the U.S. PTO in a reexamination, he/she can include an explanation as to why the patent claims are distinguishable over that prior art). Reissue practice could not necessarily be used for this purpose because in order to file a reissue application, a patent owner must state that he/she believes that the original patent is wholly or partly invalid or inoperative and specifically identify an error on which he/she bases the reissue application. 37 C.F.R. § 1.175(a); MPEP § 1414 (9th ed. 2014).

The number of total requests for ex parte reexaminations has decreased significantly since the AIA provisions pertaining to post-grant proceedings went into effect on September 16, 2012. See http://www.uspto.gov/sites/default/files/documents/ex_parte_historical_stats_roll_up_EOY2014.pdf (showing that requests for ex parte reexaminations decreased from 787 in fiscal year 2012 to 305 and 343 in fiscal years 2013 and 2014, respectively). While the U.S. PTO's statistics do not indicate whether the total number of requests for ex parte reexamination filed by patent owners (as opposed to third parties) has decreased, it is likely such requests have declined because supplemental examination (discussed in Section III.B.) now provides a far more attractive option for patent owners seeking to bolster the validity of their patents.

iii. Certificates of Correction, Including Certificates Requesting Correction of Inventorship

Certificates of correction are generally used to correct very minor errors in a patent. There are three types of certificates of correction: (1) certificates that correct mistakes made by the U.S. PTO, (2) certificates that correct mistakes made by an applicant, and (3) certificates that correct inventorship errors.

Virtually any type of error made by the U.S. PTO can be corrected using a certificate of correction so long as the error is incurred through the fault of the U.S. PTO and the error is apparent from the patent's file history. See 35 U.S.C. § 254 ("Whenever a mistake in a patent, incurred through the fault of the Patent and Trademark Office, is clearly disclosed by the records of the Office, the Director may issue a certificate of correction stating the fact and nature of such mistake, under seal, without charge, to be recorded in the records of patents."). A common example of a U.S. PTO error that can be corrected using a certificate of correction occurs when an applicant makes claim amendments, and the Patent Examiner enters those amendments and allows the patent application, but then the

U.S. PTO erroneously prints the unamended claims in the issued patent. In this case, remedying the error through a certificate of correction filed pursuant to 35 U.S.C. § 254 is typically appropriate because the file history for the patent will clearly disclose the U.S. PTO's mistakes.

The types of errors that can be corrected using a certificate of correction are far more limited when the error is the patent applicant's fault. Indeed, certificates of correction can only redress minor mistakes (e.g., typographical or clerical errors) that are made by an applicant. *See* 35 U.S.C. § 255 ("Whenever a mistake of a clerical or typographical nature, or of minor character, which was not the fault of the Patent and Trademark Office, appears in a patent and a showing has been made that such mistake occurred in good faith, the Director may, upon payment of the required fee, issue a certificate of correction, if the correction does not involve such changes in the patent as would constitute new matter or would require reexamination."); *Japanese Found. for Cancer Research v. Lee*, 773 F.3d 1300, 1306-07 (Fed. Cir. 2014) (holding that a certificate of correction filed under 35 U.S.C. § 255 could not be used to withdraw a terminal disclaimer that was erroneously filed where the alleged error was the filing of the terminal disclaimer itself and not an error in the patent number or application that was apparent on its face, such as a transposed number or the number of a related patent).

The MPEP explicitly defines "minor mistakes" as those errors that do not materially affect the scope or meaning of the patent. MPEP § 1481 (9th ed. 2014). The Federal Circuit has nevertheless held that a certificate of correction filed under 35 U.S.C. § 255 can be used to broaden claims so long as the mistake that led to the erroneously narrow claims "is clearly evident from the specification, drawings, and prosecution history . . . ." *Superior Fireplace Co. v. Majestic Prods. Co.*, 270 F.3d 1358, 1373 (Fed. Cir. 2001); *see also Central Admixture Pharmacy Servs., Inc., et al. v. Advanced Cardiac Solutions, P.C., et al.*, 482 F.3d 1347, 1350, 1355 (Fed. Cir. 2007) (holding that a certificate of correction filed under 35 U.S.C. § 255 to replace all instances of the word "osmolarity" with the word "osmolality" was invalid because the error being corrected was not clearly evident to one of skill in the art and the result of its correction was to broaden the claims).

One type of error that might seem significant, but that can be remedied using a certificate of correction, is an error in the inventorship entity listed on the patent. *See* 35 U.S.C. § 256(a) (permitting the U.S. PTO to issue a certificate of correction if a person who should have been listed as an inventor was omitted from the inventorship entity listed on a patent or a person who was not an inventor was erroneously included in the inventorship entity listed on a patent). Notably, the request to correct inventorship must include a statement from each person who is being added as an inventor, and each person who is currently named as an inventor, either agreeing to the change of inventorship or stating that he or she has no disagreement with regard to the requested change. 37 C.F.R. § 1.324(b)(1). If the inventors do not agree to the change in inventorship, then the inventorship must be corrected via a reissue application instead of a request for correction of inventorship. MPEP § 1412.04 (9th ed. 2014).

The AIA did not change the certificate of correction practices under either 35 U.S.C. §§ 254 or 255, but the legislation did make it easier for patentees to correct inventorship for an issued patent using a certificate of correction filed under 35 U.S.C. § 256. Indeed, under pre-AIA 35 U.S.C. § 256, errors in inventorship must have occurred "without any deceptive intention" on the part of the relevant inventor(s) in order for the inventorship error to be corrected using a certificate of correction. MPEP § 1481.02 (9th ed. 2014). Consistent with the changes made to the reissue statute, however, AIA § 20(f) eliminated the "without any deceptive intention" language from 35 U.S.C. § 256. Leahy-Smith America Invents Act of 2011, Pub. L. No. 112-29, 125 Stat. 284, 334. Thus, for any requests to correct inventorship errors that are filed on or after September 16, 2012, there is no requirement that the errors arose without deceptive intent.

B. AIA § 12—Supplemental Examination

As discussed previously, most of the post-grant proceedings used by patent owners to strengthen their patents were left largely unchanged by the AIA. AIA § 12 does, however, provide supplemental examination, which is a new post-grant procedure that patent owners can use to strengthen their patent rights. Supplemental examination, which became available on September 16, 2012, and can be used for any patent issued before, on, or after September 16, 2012, allows a patent owner to request further examination of a patent to consider, reconsider, or correct information believed to be relevant to the patent. Leahy-Smith America Invents Act of 2011, Pub. L. No. 112-29, § 12(c), 125 Stat. 284, 327; post-AIA 35 U.S.C. § 257(a); MPEP § 2801 (9th ed. 2014). The following tables provide an overview of supplemental examination:

Who can Initiate a Supplemental Examination Proceeding?	Burden Patent Owner Must Meet in Order for Reexamination to be Ordered Pursuant to a Request for Supplemental Examination	Examples of Types of Errors that Can be Corrected Through Supplemental Examination	Examples of Types of Errors that Cannot be Corrected Through Supplemental Examination
Only a patent owner may file a request for supplemental examination of a patent. Post-AIA 35 U.S.C. § 257(a) does not authorize the U.S. PTO to accept a request for supplemental examination from a party who is not the patent owner. For example, an exclusive licensee or other person with sufficient proprietary interest under 35 U.S.C. § 118 is not eligible to file a request for supplemental examination. (Post-AIA 35 U.S.C. § 257(a); 37 C.F.R. § 1.601(a); MPEP § 2803 (9th ed. 2015))	The criteria for making the determination on the request for supplemental examination is whether any of the items of information submitted with or as part of the request raise a substantial new question of patentability affecting at least one claim of the patent. (Post-AIA 35 U.S.C. § 257(a)) If the certificate issued under post-AIA 35 U.S.C. § 257(a) indicates that a substantial new question of patentability is raised by 1 or more items of information in the request, the Director will order reexamination of the patent. (Post-AIA 35 U.S.C. § 257(b))	(1) Patent claims that are too broad in view of the prior art under 35 U.S.C. § 102 or 35 U.S.C. § 103. (MPEP § 2802 (9th ed. 2014)) (2) Patent claims that may be invalid under 35 U.S.C. § 101 or 35 U.S.C. § 112, for double patenting, for a prior use or prior sale, etc. (MPEP § 2802 (9th ed. 2014)) (3) Failure to cite certain prior art during the prosecution of a patent. (35 U.S.C. § 257(a) and (c)(1))	(1) Errors in factual information in the patent, wherein such errors do not raise a substantial new question of patentability (e.g., errors in foreign or domestic priority or benefit claims where no intervening prior art has been cited, or missing or erroneous information pertaining to the common ownership of the claimed invention); (2) Errors in inventorship; and (3) Clerical, typographical, or spelling errors. (MPEP § 2809.01 (9th ed. 2014))

Do Intervening Rights Apply to Patents Corrected via Supplemental Examination?	Deadlines for Supplemental Examination	Limitations of Supplemental Examination
If a substantial new question of patentability is found such that ex parte reexamination is ordered, intervening rights should apply in the same way they apply to ex parte reexaminations (See MPEP § 2823 (9th ed. 2014) (explaining that reexaminations ordered pursuant to a supplemental examination will be conducted in the same manner as those ordered pursuant to 35 U.S.C. § 302 with certain exceptions, none of which pertain to intervening rights)). If no substantial new question of patentability is found, no intervening rights would apply because the patent would not undergo reexamination and no changes to the claims could be made. (*See* 37 C.F.R. § 1.625(c) (explaining that ex parte reexamination will not be ordered where no substantial new question is found)).	(1) A request for supplemental examination may be filed any time during the period of enforceability for a patent. (37 C.F.R. § 1.601(c); MPEP § 2808)	(1) A request for supplemental examination must be filed by the owner(s) of the entire right, title and interest in the patent. (37 C.F.R. § 1.601(a); MPEP § 2803) Thus, exclusive licensees cannot file such request. (2) Third parties cannot participate in supplemental examinations. (37 C.F.R. § 1.601(b)) (3) Each request for supplemental examination may include no more than twelve items of information believed to be relevant to the patent. (37 C.F.R. § 1.605(a); MPEP § 2809)

Supplemental examination offers several advantages over the post-grant proceedings that were available to patent owners prior to the AIA. The most significant advantage of supplemental examination is that it provides patent owners with the opportunity to insulate their patents from inequitable conduct allegations before they are involved in litigation.[P] In particular,

[P]In the years leading up to the AIA's enactment, inequitable conduct defenses were overpleaded and were often referred to as a "plague" on the patent system. *See e.g., Hoffman-La Roche, Inc. v. Promega Corp.*, 323 F.3d 1354, 1372 (Fed. Cir. 2003) (Newman, J., dissenting) ▶

post-AIA 35 U.S.C. § 257(c)(1) points out that "[a] patent shall not be held unenforceable on the basis of conduct relating to information that had not been considered, was inadequately considered, or was incorrect in a prior examination of the patent if the information was considered, reconsidered, or corrected during a supplemental examination of the patent." This means that if, for example, a patent owner discovers that he/she did not submit a key piece of prior art during the original examination of a patent, he/she may still submit that prior art in a request for supplemental examination and the patent cannot be held unenforceable for inequitable conduct on the basis that the prior art submitted during the supplemental examination was withheld during the original prosecution of the patent.

There are, however, two major exceptions to a patent owner's ability to insulate his/her patent from inequitable conduct defenses using supplemental examination. First, if a defendant has alleged that inequitable conduct has occurred in a pleading filed in a civil action, or in a notice letter received by the patent owner in an ANDA case, before the date on which the patent owner files a request for supplemental examination to try to cleanse his/her patent of the alleged inequitable conduct, the provision of post-AIA 35 U.S.C. § 257(c)(1) does not apply and the patent may be held unenforceable for inequitable conduct. Post-AIA 35 U.S.C. § 257(c)(2)(A). Second, in order for the provision of post-AIA 35 U.S.C. § 257(c)(1) to apply in an ITC case, the supplemental examination, and any reexamination ordered pursuant to the supplemental examination request, must be concluded (not merely filed) before the date on which the ITC action is brought. Post-AIA 35 U.S.C. § 257(c)(2)(B). Put more simply, the protection of post-AIA 35 U.S.C. § 257(c)(1) only applies where patent owners proactively use supplemental examination to address any possible inequitable conduct issues before they are formally accused of having committed inequitable conduct by another party involved in a civil suit or ITC proceeding.

Another significant advantage of supplemental examination is that when a patent owner submits a complete request for supplemental examination,

◀ ("Litigation-induced assaults on the conduct of science and scientists, by aggressive advocates intent on destruction of reputation and property for private gain, produced the past 'plague' of charges of 'inequitable conduct.'"). Although the Federal Circuit greatly curtailed the use of inequitable conduct defenses in *Therasense, Inc. v. Becton, Dickinson & Co.*, 649 F.3d 1276 (Fed. Cir. 2011) (en banc), Congress also felt an obligation to curb use of the defense. *See* 153 CONG. REC. S4691, 2007 (statement of Sen. Hatch) ("Attorneys well know that the inequitable conduct defense has been overpleaded and has become a drag on the litigation process.")

the U.S. PTO has only three months to conduct the supplemental examination and conclude the examination by issuing a certificate that indicates whether the information presented in the request raises a substantial new question of patentability.[q] Post-AIA 35 U.S.C. § 257(a). If the supplemental examination certificate indicates that a substantial new question of patentability is raised by one or more items of information in the request, the U.S. PTO will order reexamination of the patent to consider the items of information that raise a substantial new question of patentability. Post-AIA 35 U.S.C. § 257(b). If not, the certificate will indicate that no substantial new question of patentability exists, and no reexamination will be ordered. 37 C.F.R. § 1.625(c). The ability to conclude a supplemental examination within three months[r] could be invaluable to a patent owner who is about to file suit on his/her patent, particularly where the U.S. PTO finds that no substantial new question of patentability exists.

Supplemental examination also offers an advantage over reissues in that not all claims of a patent will necessarily be reexamined if reexamination is ordered pursuant to a request for supplemental examination. 37 C.F.R. § 1.176(a); MPEP §§ 1440, 2816.01 (9th ed. 2014); *see also* 37 C.F.R. § 1.610(b)(4) (requiring requests for supplemental examinations to identify each claim of the patent for which supplemental examination is requested).[s] Furthermore, unlike in reissues where the Patent Examiner will consider all of the prior art cited during the original prosecution of the patent, Patent Examiners will not necessarily consider all of the prior art that was cited during the original prosecution for reexaminations that are instituted from a supplemental examination proceeding. *Compare* 37 C.F.R. § 1.625(d)(2) (indicating that any ex parte reexamination ordered pursuant to a request for supplemental examination is not limited

---

[q]An item of information raises a substantial question of patentability where there is a substantial likelihood that a reasonable examiner would consider the item important in deciding whether or not the claim is patentable. MPEP § 2242 (9th ed. 2014).

[r]Note that the U.S. PTO must also decide whether to order reexamination within three months of a filing for a request for ex parte reexamination. *See* MPEP § 2241 (9th ed. 2015) ("The determination of whether or not to reexamine must be made within 3 months following the filing date of a request.").

[s]Because Examiners must reexamine all claims of a patent during reissue, patentees who engage in this process risk having all of their claims rejected even if they only seek reissue on the basis that a subset of their claims are inoperative or invalid. *See* MPEP § 1445 (9th ed. 2015) ("As stated in 37 C.F.R. 1.176, a reissue application, including all the claims therein, is subject to 'be examined in the same manner as a non-reissue, non-provisional application.'").

to the items of information provided during supplemental examination but making no requirement that Patent Examiners consider all prior art cited during the original prosecution) *with* MPEP § 1406 (9th ed. 2014) (explaining that Examiners should consider all references that have been cited during the original prosecution of the patent when examining a reissue application).

A further advantage of supplemental examination is that a much broader range of materials and issues can be considered than can be assessed via ex parte reexamination. Indeed, while only anticipation and obviousness issues that are based on patent and printed publication prior art may be considered in ex parte reexamination proceedings, any information that a patent owner believes to be relevant to the patent may be considered in a supplemental examination. MPEP § 2209 (9th ed. 2014); MPEP § 2801 (9th ed. 2014). For instance, if a patent owner is concerned that one or more claims of his/her patent is invalid based on a prior sale or use, he/she can submit evidence of the sale or use to the U.S. PTO for consideration. MPEP § 2801 (9th ed. 2014). Patent owners cannot submit such information in an ex parte reexamination. MPEP § 2209 (9th ed. 2014). Additionally, while invalidity issues arising under only 35 U.S.C. § 102 and § 103 can be considered in an ex parte reexamination, any ground of patentability, such as, for example, patent eligible subject matter, written description, enablement, indefiniteness, and double-patenting can be considered during a supplemental examination. MPEP § 2801 (9th ed. 2014).

In sum, supplemental examination promises to be a powerful tool for those patent owners who wish to strengthen their patents and who utilize the proceeding in a timely manner. While inequitable conduct defenses are not pleaded as often as they used to be, the ability to protect one's patent from such charges could be invaluable. The process could also be extremely effective where, for example, a patent owner wants to undermine an alleged infringer's ability to advance a particular defense by obtaining a certificate from the U.S. PTO indicating that certain information does not raise any substantial new questions of patentability for his/her patent.

IV) AIA § 6(g)—CITATION OF PRIOR ART IN A PATENT FILE UNDER 35 U.S.C. § 301

Pre- and post-AIA 35 U.S.C. § 301 allow anyone (i.e., a patent owner or a third party) to cite to the U.S. PTO prior art that they believe has a bearing on the patentability of any claim of a particular patent. If such citation of prior art meets the requirements of pre- and post-35 U.S.C. § 301, the U.S. PTO includes the citation, as well as the prior art that is cited, as part of the official file of the patent. It is important to note that the U.S. PTO does not reexamine the claims of the patent, or make any other determinations regarding the validity of a patent's claim, even if a citation of prior art is included as part of the official file of the patent. *See* post-AIA 35 U.S.C. § 301(d) ("A written statement submitted pursuant to subsection (a)(2), and additional information submitted pursuant to subsection (c), shall not be considered by the Office for any purpose other than to determine the proper meaning of a patent claim in a proceeding that is ordered or instituted pursuant to [35 U.S.C. §§ 304, 314, or 324]."); MPEP § 2202 (8th ed. 2010) ("The basic purpose for citing prior art in patent files is to inform the patent owner and the public in general that such patents or printed publications are in existence and should be considered when evaluating the validity of the patent claims. Placement of citations in the patent file along with copies of the cited prior art will also ensure consideration thereof during any subsequent reissue or reexamination proceeding."). Thus, although citation of prior art under pre- and post-AIA 35 U.S.C. § 301 is technically a post-grant proceeding, the proceeding does not fit neatly within either of the two other categories of post-grant proceedings that were discussed earlier. Indeed, while third parties may cite prior art to the U.S. PTO, a third party who cites prior art, and takes no further action, is not seeking to have the U.S. PTO invalidate the patent on the basis of the prior art. *Id.* Additionally, even though a patent owner can use this procedure to cite prior art in his/her own patent file, doing so would likely weaken, not strengthen, his/her patent.[t] An overview of the post-AIA citation of prior art process appears in the following tables:

[t] If a patent owner submits references through the citation of prior art process, the submission would likely only serve to draw attention to the prior art and would potentially serve as an admission that such prior art is pertinent to the claims of the issued patent. Such an admission would undoubtedly be used against the patent owner if he/she ever brought a patent infringement action. If a patent owner is concerned that newly discovered prior art could impact his/her claims, he/she would generally be better served by filing a request for ex parte reexamination, an application for reissue, or a request for supplemental examination so that the U.S. PTO can consider the scope of the patent claims in light of that prior art.

Who Can Submit Prior Art Under the Citation of Prior Art Procedure?	Requirements for Citations of Prior Art Under Post-AIA 35 U.S.C. § 301
Any person (i.e., a patent owner or a third party) may submit prior art under the citation of prior art procedure. (Post-AIA 35 U.S.C. § 301(a)).	A citation of prior art must include: (1) prior art consisting of patents or printed publications which that person believes to have a bearing on the patentability of any claim of a particular patent; OR (2) statements of the patent owner filed in a proceeding before a Federal court or the Office in which the patent owner took a position on the scope of any claim of a particular patent. In addition, all citations of prior art must include: (1) an explanation of how the prior art or written statement is pertinent and applicable to the patent; (2) an explanation of how the claims of the patent differ from the submitted prior art or written statement; (3) copies of all the submitted prior art patents, printed publications, or written statements and any necessary English translation(s); (4) proof of service (if the citation is made by someone other than the patent owner); AND (5) a cover sheet identifying the patent to which the citation pertains. (Post-AIA 35 U.S.C. § 301(a)(1) and (2); MPEP § 2205 (9th ed. 2015))

Types of Errors that Can be Corrected Through a Citation of Prior Art	Do Intervening Rights Apply to Patents in Which a Citation of Prior Art Has Been Made?	Notable Deadlines for Citations of Prior Art	Limitations of the Citation of Prior Art Procedure
No errors can be corrected through the citation of prior art procedure. Instead, the procedure allows certain prior art, which someone believes to have a bearing on the patentability of any claim of the patent, to become part of the official file of the patent. In addition, the procedure allows patent owner statements filed in a proceeding before a federal court or the U.S. PTO, in which the patent owner took a position on the scope of any claim of a particular patent, to become part of the official file of the patent. (Post-AIA 35 U.S.C. § 301(a) and (b))	No, because claim scope cannot be changed via citation of prior art. (Post-AIA 35 U.S.C. § 301(d))	Any person may cite prior art pursuant to 35 U.S.C. § 301 at any time during the period of enforceability of a patent. (37 C.F.R. § 1.501(a) and MPEP § 2204 (9th ed. 2014)). The period of enforceability typically extends six years beyond the expiration of a patent. (MPEP § 2204 (9th ed. 2014))	(1) Patent owner statements regarding the scope of claims in one patent may not be submitted to the U.S. PTO to be included in the official file of another patent (even if the patent is related). (Patent Owner Claim Scope Statement Final Rule, 77 Fed. Reg. 46,619 (Aug. 6, 2012)) (2) If a third party submits prior art or a patent owner statement, the patent owner cannot respond to the third-party submission. (Patent Owner Claim Scope Statement Final Rule, 77 Fed. Reg. 46,619–20 (Aug. 6, 2012)) A patent owner may, however, file the prior art or statement submitted by the third party in their own submission and then explain how the claims differ from the prior art and/or the patent owner statement. (37 C.F.R. § 1.501(b)(2)) (3) Patent owner statements made to the ITC may not be submitted to the U.S. PTO under 35 U.S.C. § 301 because the ITC is not a federal court, it is a federal agency. (Patent Owner Claim Scope Statement Final Rule, 77 Fed. Reg. 46,621 (Aug. 6, 2012)) (4) When material is cited under 35 U.S.C. § 301, the person submitting the prior art or written statement must include an explanation of how the prior art or written statement is pertinent and applicable to the patent. (Post-AIA 35 U.S.C. § 301(b))

The AIA materially altered the citation of prior art practice by expanding the scope of information that can be submitted to the U.S. PTO for an issued patent. In particular, pre-AIA 35 U.S.C. § 301 only permitted the citation of prior art consisting of patents or printed publications. Once AIA § 6(g)(1) became effective on September 16, 2012,[11] however, patents, printed publications, and statements made by the patent owner filed in any proceeding before a federal court or the U.S. PTO, in which the patent owner took a position on the scope of any claim of the patent, could be filed. Leahy-Smith America Invents Act of 2011, Pub. L. No. 112-29, 125 Stat. 284, 311-12. The purpose of the AIA's amendment to 35 U.S.C. § 301 was to limit a patent owner's ability to put forward different positions with respect to the prior art in different proceedings regarding the same patent. *See* H.R. REP. No. 112-98, at 46 (2011) ("This addition will counteract the ability of patent owners to offer differing interpretations of prior art in different proceedings."). A comparison of pre- and post-AIA 35 U.S.C. § 301 is shown in the following table:

[11]Post-AIA 35 U.S.C. § 301 applies to citations to prior art that are submitted on or after September 16, 2012, and such prior art citations can be made for any patent that issued before, on, or after September 16, 2012. Leahy-Smith America Invents Act of 2011, Pub. L. No. 112-29, § 6(g)(3), 125 Stat. 284, 312.

Pre-AIA 35 U.S.C. § 301	Post-AIA 35 U.S.C. § 301
Any person at any time may cite to the Office in writing prior art consisting of patents or printed publications which that person believes to have a bearing on the patentability of any claim of a particular patent. If the person explains in writing the pertinency and manner of applying such prior art to at least one claim of the patent, the citation of such prior art and the explanation thereof will become a part of the official file of the patent. At the written request of the person citing the prior art, his or her identity will be excluded from the patent file and kept confidential.	(a) IN GENERAL.– Any person at any time may cite to the Office in writing- (1) prior art consisting of patents or printed publications which that person believes to have a bearing on the patentability of any claim of a particular patent; or (2) statements of the patent owner filed in a proceeding before a Federal court or the Office in which the patent owner took a position on the scope of any claim of a particular patent. (b) OFFICIAL FILE.– If the person citing prior art or written statements pursuant to subsection (a) explains in writing the pertinence and manner of applying the prior art or written statements to at least 1 claim of the patent, the citation of the prior art or written statements and the explanation thereof shall become a part of the official file of the patent. (c) ADDITIONAL INFORMATION.– A party that submits a written statement pursuant to subsection (a)(2) shall include any other documents, pleadings, or evidence from the proceeding in which the statement was filed that addresses the written statement. (d) LIMITATIONS.– A written statement submitted pursuant to subsection (a)(2), and additional information submitted pursuant to subsection (c), shall not be considered by the Office for any purpose other than to determine the proper meaning of a patent claim in a proceeding that is ordered or instituted pursuant to section 304, 314, or 324. If any such written statement or additional information is subject to an applicable protective order, such statement or information shall be redacted to exclude information that is subject to that order. (e) CONFIDENTIALITY.– Upon the written request of the person citing prior art or written statements pursuant to subsection (a), that person's identity shall be excluded from the patent file and kept confidential.

Although not often used, the citation of prior art process provided by post-AIA 35 U.S.C. § 301 may be an attractive option for third parties who want to bring attention to certain prior art references and/or ensure that a patent owner does not present inconsistent arguments regarding its claim scope to different tribunals (i.e., federal courts and the U.S. PTO). Third parties can make submissions under post-AIA 35 U.S.C. § 301 anonymously to avoid retaliation by patent owners. Post-AIA 35 U.S.C. § 301(e). Also, unlike inter partes review, covered business method review, and post grant review proceedings, no estoppel effect flows from citing materials under post-AIA 35 U.S.C. § 301. Post-AIA 35 U.S.C. §§ 301, 315(e)(1) and (2), 325(e)(1) and (2); Leahy-Smith America Invents Act of 2011, Pub. L. No. 112-29, § 18(a)(1), 125 Stat. 284, 330. Such submissions might therefore be attractive to third parties who do not wish to bring a post-grant challenge to a patent themselves but would like to make certain prior art or patent owner statements part of the official file for a patent so that others may bring such challenges.

CHAPTER SIX

The AIA's Effects on Claims That May Be Brought in Patent Infringement Suits

Contents

I) INTRODUCTION

AIA §§ 16, 17, and 19 all set forth provisions that impact the claims that can be brought in patent infringement suits. For instance, AIA § 16 reforms the patent marking statutes and impacts both the damages patentees can seek in patent suits as well as the types of claims that can be brought in such litigations. Additionally, AIA § 17 enacted a new patent statute that impacts the type of evidence that may be used to prove willful and induced infringement claims. Lastly, AIA § 19(d) enacted a new patent statute that affects

America Invents Act Primer. http://dx.doi.org/10.1016/B978-0-12-812096-5.00006-5

how defendants may be joined in a patent infringement suit. This chapter provides a discussion of these provisions.

II) AIA § 16: AMENDMENTS TO THE PATENT MARKING STATUTES

35 U.S.C. §§ 287 and 292 are the two patent statutes that govern the marking of patented articles with relevant patent information. 35 U.S.C. § 287(a) encourages patentees to notify the public of the existence of their patents by limiting the damages they can obtain for infringement of their patents in instances where they choose not to mark their patented articles with applicable patent information.[a] More specifically, 35 U.S.C. § 287(a) provides that in the event that a patentee fails to properly mark its patented articles, "no damages shall be recovered by the patentee in any action for infringement, except on proof that the infringer was notified of the infringement and continued to infringe thereafter, in which event damages may be

---

[a]There are several exceptions to 35 U.S.C. § 287(a)'s limitation on damages. For instance, the statute does not apply to patents that contain only method or process claims. Thus, a patentee's damages will not be limited where there has been a failure to mark an article with the number of a patent that contains only method or process claims, even if the article is made using a method or process claimed in the patent or the article can be used to carry out such method or process. See State Contracting & Eng'g v. Condotte Am., Inc., 346 F.3d 1057, 1073 (Fed. Cir. 2003) (quoting Bandag, Inc. v. Gerrard Tire Co., 704 F.2d 1578, 1581 (Fed. Cir. 1983)) (explaining that it is "settled in the case law that the notice requirement of [pre-AIA 35 U.S.C. § 287(a)] does not apply where the patent is directed to a process or method."). But see Am. Med. Sys., Inc. v. Med. Eng'g Corp., 6 F.3d 1523, 1538-39 (Fed. Cir. 1993) ("Where the patent contains both apparatus and method claims, however, to the extent that there is a tangible item to mark by which notice of the asserted method claims can be given, a party is obliged to do so if it intends to avail itself of the constructive notice provisions of section 287(a)."). 35 U.S.C. § 287(a) also does not apply to "a patent owner who neither sells nor authorizes others to sell articles covered by the patent." Loral Fairchild Corp. v. Victor Co. of Japan, Ltd., 906 F. Supp. 813, 816 (E.D.N.Y. 1995). As a result, it is more difficult for a patentee who manufactures a product covered by his patent to recover damages than it is for a non-practicing patentee or the owner of a process patent to do so. Joe Matal, A Guide to the Legislative History of the America Invents Act: Part II of II, 21 FED. CIR. B.J. 585 n.303 (2012). Some courts have also recognized a de minimis exception to 35 U.S.C. § 287(a)'s limitation on damages so that if a patentee produces thousands of patented articles but mistakenly sells a few hundred articles that are unmarked with its patent number, the damages to which the patentee is entitled will not be limited by 35 U.S.C. § 287(a). Joel Voelzke, Patent Marking Under 35 U.S.C. § 287(a): Products, Processes, and The Deception of the Public, INTELL. PROP. L. REV. 27 (1997) (citing Hazeltine Corp. v. Radio Corp. of Am., 20 F. Supp. 668, 671-72 (S.D.N.Y. 1937)).

recovered only for infringement occurring after such notice." 35 U.S.C. § 287(a) serves the three related purposes of:

1) helping the public to avoid innocent infringement,
2) encouraging patentees to give notice to the public that an article is patented, and
3) aiding the public to identify whether an article is patented.

Nike, Inc. v. Wal-Mart Stores, Inc., 138 F.3d 1437, 1443 (Fed. Cir. 1998) (citations omitted).

35 U.S.C. § 292(a) expands upon 35 U.S.C. § 287(a) by imposing liability on parties who mark their articles with inaccurate patent information, i.e., on parties who are found to have falsely marked their articles. More specifically, 35 U.S.C. § 292(a) imposes a fine of $500 per false marking offense on:

> *Whoever marks upon, or affixes to, or uses in advertising in connection with any unpatented article the word "patent" or any word or number importing the same is patented, for the purpose of deceiving the public; or [w]hoever marks upon, or affixes to, or uses in advertising in connection with any article the words "patent applied for," "patent pending," or any word importing that an application for patent has been made, when no application for patent has been made, or if made, is not pending, for the purpose of deceiving the public. . . .*

Under pre-AIA 35 U.S.C. § 292(a), a party could incur liability for falsely marking a product with patent information if that party:

1) marks an article with a number of an expired patent (*see, e.g., Pequignot v. Solo Cup Co.*, 608 F.3d 1356, 1362 (Fed. Cir. 2010) ("In sum, we agree with Pequignot and the district court that articles marked with expired patent numbers are falsely marked."));
2) marks an article with several patent numbers, wherein less than all of the listed patents cover the article (*see, e.g., Astec Am., Inc. v. Power-One, Inc.*, No. 6:07-cv-464, 2008 U.S. Dist. LEXIS 30365, at *27-30 (E.D. Tex. Apr. 11, 2008) (explaining that marking an article as "protected by one or more of the following US patents" when not all of the patents cover the article constitutes false marking "when the patentee practices this form of marking with an intent to deceive the public."));
3) marks an article with the phrase "Patent Pending" or "Pat. Pending" when in fact no patent application claiming the article has been filed (*see, e.g., Sadler-Cisar, Inc. v. Commercial Sales Network, Inc.*, 786 F. Supp. 1287, 1296 (N.D. Ohio 1991) (finding that defendants were liable for false marking under 35 U.S.C. § 292 where they "marked [their] product with the designation 'patent pending' and offered it for sale

when no patent application was in fact pending" and the mismarking was not "solely the result of inadvertence."));

4) marks an article with the number of a patent that covers a method of making the article but that does not cover the article itself (*see, e.g., Clontech Laboratories, Inc. v. Invitrogen Corp.*, 406 F.3d 1347, 1357 (Fed. Cir. 2005) (suggesting that a patentee may not have been liable for false marking had it indicated that its cDNA library products were made by patented methods instead of marking its products with the statement "[t]his product is the subject of U.S. Patent No. [number of the method patent]."));

5) marks an article with the number of a patent that does not cover the article (*see, e.g., Brose v. Sears, Roebuck & Co.*, 455 F.2d 763, 766 (5th Cir. 1972) ("The mismarking comes not from the fact that the kit bears the true legend 'licensed under U.S. Patent [Kraly] 3,095,342.' Rather, it is because [Sears] although licensed by Patentee Kraly under 3,095,342, does not in the kit make a device which [is covered by] the claims of the cited patent.")); or

6) marks an article with the number of a patent that has been adjudged to be invalid and/or unenforceable (*see* Donald W. Rupert, *Trolling for Dollars: A New Threat to Patent Owners*, 21 Intell. Prop. & Tech. L.J. 2 (2009)).

A party does not incur liability under pre-AIA 35 U.S.C. § 292(a) merely by committing one or more of the aforementioned offenses. Indeed, in order for a party to be held liable under that statute, one must demonstrate that the party falsely marked an article "for the purpose of deceiving the public." Pre- and post-AIA 35 U.S.C. § 292(a); *see also Pequignot*, 608 F.3d at 1361-62 (pointing out that in order to be liable for false marking, a party must prove by a preponderance of the evidence that a patentee: (1) marked an "unpatented article," and (2) acted "for the purpose of deceiving the public."). Moreover, just because a party knew that an article was marked with false patent information does not mean that that party acted "for the purpose of deceiving the public." *Id.* at 1363.

Both of the patent marking statutes were amended by the AIA. Specifically, AIA § 16(a) amended pre-AIA 35 U.S.C. § 287(a), and AIA § 16(b) amended pre-AIA 35 U.S.C. § 292. A detailed discussion of AIA § 16's amendments to the patent marking statutes is provided in Sections II.A. and B.

A. AIA § 16(a)(1): Amendments to Pre-AIA 35 U.S.C. § 287(a)—Virtual Marking

AIA § 16(a)(1) amended pre-AIA 35 U.S.C. § 287(a) to allow patentees to more easily mark their patented articles with applicable patent information. More specifically, the new legislation added language to the pre-AIA statute to enable patentees to "virtually mark" their patented articles, thereby alleviating them from the burden of modifying and updating patent information on the articles themselves.

i. Brief History of Pre- and Post-AIA 35 U.S.C. § 287(a)

While patent marking statutes have existed for many years, they have not always burdened patentees with the task of notifying the public of their patents. In fact, prior to the Patent Act of 1842, it was the public who carried the burden of determining whether something was patented. *See Boyden v. Burke*, 55 U.S. 575, 582 (1853) (explaining that because patents were public records, all persons were "bound to take notice of their contents."). The Patent Act of 1842 removed this burden from the general public, however, by requiring "all patentees and assignees of patents . . . to stamp . . . on each article vended, or offered for sale, the date of the patent." *Nike, Inc.*, 138 F.3d at 1443 (citing The Patent Act of 1842, 5 Stat. 543, 544). If a patentee failed to mark each patented article, he/she incurred a fine of "not less than one hundred dollars." *Id.*

The patent marking statute did not resemble pre- and post-AIA 35 U.S.C. § 287(a) until 1861. Indeed, during that year, the statutory penalty set forth in the marking statute of the Patent Act of 1842 was replaced with a limitation on the patentee's right to recover for infringement, thereby allowing patentees to choose whether to mark their patented articles. *Id.* (citing the Patent Act of 1861, 12 Stat. 246, 249); *see also* Elizabeth I. Winston, *The Flawed Nature of the False Marking Statute*, 77 Tenn. L. Rev. 120 (2009) ("[The limitation on damages set forth in pre-AIA 35 U.S.C. § 287(a)] allows the marking party to perform a cost-benefit analysis of marking the product based on the likelihood of infringement, the costs of marking, and the economic recovery possible from an infringement suit before marking a product."). Similar to pre- and post-AIA 35 U.S.C. § 287(a), the Patent Act of 1861 specifically provided that "no damage shall be recovered by the plaintiff" unless that plaintiff marked the article as patented or the infringer received actual notice of the patent. *Nike, Inc.*, 138 F.3d at 1443 (citing the Patent Act of 1861, 12 Stat. 246, 249). After 1861, the marking statute remained substantially unchanged until the AIA was enacted.

ii. AIA § 16(a)(1)'s Amendments to Pre-AIA 35 U.S.C. § 287(a)

A side-by-side comparison of pre- and post-AIA 35 U.S.C. § 287(a) is provided in the following table:

Pre-AIA 35 U.S.C. § 287(a)	Post-AIA 35 U.S.C. § 287(a)
Patentees, and persons making, offering for sale, or selling within the United States any patented article for or under them, or importing any patented article into the United States, may give notice to the public that the same is patented, either by fixing thereon the word "patent" or the abbreviation "pat.", together with the number of the patent, or when, from the character of the article, this cannot be done, by fixing to it, or to the package wherein one or more of them is contained, a label containing a like notice. In the event of failure so to mark, no damages shall be recovered by the patentee in any action for infringement, except on proof that the infringer was notified of the infringement and continued to infringe thereafter, in which event damages may be recovered only for infringement occurring after such notice. Filing of an action for infringement shall constitute such notice.	Patentees, and persons making, offering for sale, or selling within the United States any patented article for or under them, or importing any patented article into the United States, may give notice to the public that the same is patented, either by fixing thereon the word "patent" or the abbreviation "pat.", together with the number of the patent, or by fixing thereon the word 'patent' or the abbreviation 'pat.' together with an address of a posting on the Internet, accessible to the public without charge for accessing the address, that associates the patented article with the number of the patent, or when, from the character of the article, this cannot be done, by fixing to it, or to the package wherein one or more of them is contained, a label containing a like notice. In the event of failure so to mark, no damages shall be recovered by the patentee in any action for infringement, except on proof that the infringer was notified of the infringement and continued to infringe thereafter, in which event damages may be recovered only for infringement occurring after such notice. Filing of an action for infringement shall constitute such notice. (Emphasis added.)

The only difference between pre- and post-AIA 35 U.S.C. § 287(a) is that the latter statute contains the added language "or by fixing thereon the word 'patent' or the abbreviation 'pat.' together with an address of a posting on the Internet, accessible to the public without charge for accessing the address, that associates the patented article with the number of the patent."

This language is commonly referred to as the "virtual marking" provision because it permits patentees to mark their patented articles with a web address. Patentees can therefore comply with post-AIA 35 U.S.C. § 287(a) by marking their patented articles using any of the following methods:

1) Patentees may fix the word "patent" or the abbreviation "pat." together with the number of the patent on the patented article;

2) If it is not practical to fix the aforementioned text to the patented article itself, patentees may fix the text on a label that is fixed to either: (a) the patented article or (b) the package wherein one or more of the patented articles is contained;

3) Patentees may fix on the patented article the word "patent" or the abbreviation "pat." together with an address of a posting on the Internet, accessible to the public without charge for accessing the address, that associates the patented article with the number of the patent; or

4) If it is not practical to fix the aforementioned text in #3 to the patented article itself, patentees may fix the text on a label that is fixed to either: (a) the patented article or (b) the package wherein one or more of the patented articles is contained.

Pre-AIA 35 U.S.C. § 287(a) only permitted patentees to mark their patented articles using the aforementioned methods 1 and 2. Thus, post-AIA 35 U.S.C. § 287(a) provides patentees with a greater number of options when deciding how to mark their patented products with applicable patent information.

iii. Goals Addressed by the Amendment to Pre-AIA 35 U.S.C. § 287(a) and Practical Implications of the Same

Congress' main objective in amending pre-AIA 35 U.S.C. § 287(a) was to make it easier for patentees to mark their patented products with relevant patent information. Indeed, the final Senate Judiciary Committee Report on the provision that was enacted as AIA § 16(a) explains:

> The Act permits patent holders to "virtually mark" a product by providing the address of a publicly available website that associates the patented article with the number of the patent. The burden will remain on the patent holder to demonstrate that the marking was effective. This amendment will save costs for producers of products that include technology on which a patent issues after the product is on the market, and will facilitate effective marking on smaller products.
>
> **Joe Matal, *A Guide to the Legislative History of the America Invents Act: Part II of II*, 21 Fed. Cir. B.J. 539, 585 (2012) (citing H.R. Rep. No. 112-98, at 52-53 (2011)).**

Before the enactment of AIA § 16(a), the task of updating or modifying patent markings on patented articles could be extremely onerous where, for instance, a patentee produces a large volume of patented articles or where a patentee produces its patented articles using a mold or other apparatus that is costly or difficult to replace. *See, e.g., Pequignot*, 608 F.3d at 1359 (explaining that Solo Cup Co. had not updated its patent markings on its lids for disposable cups because the lids were manufactured using mold cavities and wholesale replacement of the mold cavities would be costly and burdensome). As referenced in the Senate Judiciary Committee Report mentioned previously, marking challenges also arose when a patentee sold his/her articles before the patent claiming such articles issued. Indeed, in those instances, patentees were required to mark their articles with "patent pending" text before the patent issues but then had to update the marking with the appropriate patent number after issuance of the associated patent to avoid the damages limitation of pre-AIA 35 U.S.C. § 287(a). *See also* pre- and post-AIA 35 U.S.C. § 292(a) (imposing liability on parties who mark their articles with inaccurate patent or patent application information). By enabling patentees to mark their patented articles, or the labels that are attached to their patented articles, with the address of a website that lists all of the patents that cover the article, post-AIA 35 U.S.C. § 287(a) should make patent markings easier. Indeed, patentees that elect to mark their patented articles in this manner can simply update or modify any applicable patent information on their websites, thereby avoiding the burdensome task of modifying patent markings on the patented articles themselves.

iv. Effective Date of AIA § 16(a)(1)'s Amendment to Pre-AIA 35 U.S.C. § 287(a)

AIA § 16(a)(2) provides that the amendment to pre-AIA 35 U.S.C. § 287(a) "shall apply to any case that is pending on, or commenced on or after, the date of the enactment of this Act." Leahy-Smith America Invents Act of 2011, Pub. L. No. 112-29, 125 Stat. 284, 328. The AIA was enacted on September 16, 2011. *Id.* at 284. Thus, patentees who have complied with the "virtual marking" provision of post-AIA 35 U.S.C. § 287(a) will not incur any limitations on damages under that statute so long as their patent infringement suit was pending on September 16, 2011, or was commenced on or after that date.

B. AIA § 16(b): Changes to False Marking *Qui Tam* Actions

As discussed in Section II, pre- and post-AIA 35 U.S.C. § 292(a) impose liability on parties who mark their articles with inaccurate patent information, i.e., on parties who are found to have falsely marked their articles. In the years immediately preceding the enactment of the AIA, parties mostly incurred liability for false marking offenses through false marking *qui tam* suits. As explained in Section II.b.ii., AIA § 16(b) amended pre-AIA 35 U.S.C. § 292(a) to put an end to these suits, which had become a nuisance to both patentees and the judiciary alike.

i. Background on False Marking Qui Tam Actions Under Pre-AIA 35 U.S.C. § 292(a)

Qui tam lawsuits are different than most other lawsuits in that the plaintiff who brings a *qui tam* suit is not necessarily the person who has been harmed by the defendant's alleged misconduct.[b] Indeed, *qui tam* statutes authorize private citizens to bring an action on behalf of the government and himself/herself for an injury that the government alone suffered. *See Stauffer v. Brooks Brothers, Inc.*, 619 F.3d 1321, 1325 (Fed. Cir. 2010) (citing *Vt. Agency of Natural Res.*, 529 U.S. at 773) ("[A] *qui tam* plaintiff, or relator, can establish standing based on the United States' implicit partial assignment of its damages claim . . . to 'any person' . . . In other words, even though a relator may suffer no injury himself, a *qui tam* provision operates as a statutory assignment of the United States' rights, and 'the assignee of a claim has standing to assert the injury in fact suffered by the assignor.'"). Pre-AIA 35 U.S.C. § 292 is an example of a *qui tam* statute. That statute authorizes any person to bring a lawsuit on behalf of the U.S. government against a party who falsely marks an article with a patent number or who falsely marks an article with text indicating that a patent is pending when no patent application directed to the article has actually been filed. Pre-AIA 35 U.S.C. § 292(a) and (b). Moreover, those who succeed in proving their false marking claims in a *qui tam* suit are entitled to recover one half of the damages that are provided by pre-AIA 35 U.S.C. § 292(a), i.e., $500 per false marking offense. Pre-AIA 35 U.S.C. § 292(b).

---

[b]The term *"qui tam"* is short for the Latin phrase *"qui tam pro domino rege quam pro se ipso in hac parte sequitur,"* which means "who pursues this action on our Lord the King's behalf as well as his own." *Vt. Agency of Natural Res. v. U.S. ex rel. Stevens,* 529 U.S. 765, 768 n.1 (2000). *Qui tam* actions promote the efficient use of government resources by allowing those who sue on behalf of the government to recover at least part of the damages to which the government is entitled. Elizabeth I. Winston, *The Flawed Nature of the False Marking Statute,* 77 TENN. L. REV. 117 (2009). Indeed, because of its limited resources, the U.S. government cannot investigate every legal offense that is committed but must instead focus only on those offenses that are likely to cause the greatest public harm and/or those offenses that are easiest to detect and investigate. *Id.* *Qui tam* actions, therefore, promote the efficient use of government resources by incentivizing private parties to step into the role of the government to investigate offenses that are not likely to greatly harm the public and/or for which the private party is in a better position than the government to investigate. *Id.*

Even though statutes that provide for false marking claims have existed since 1842, very few false marking suits were brought before 2007, and the few that were brought were not *qui tam* actions because they were filed by patentees' competitors who alleged they had been harmed by a patentees' false marking offenses. Patent Act of 1842 ch. 263 § 5, 5 Stat. 543, 544 (1842) (codified as amended at 35 U.S.C. § 292 (2006)); *see also Clontech Laboratories, Inc.*, 406 F.3d at 1351-52 (citing *Arcadia Mach. & Tool, Inc. v. Sturm, Ruger & Co.,* 786 F.2d 1124, 1125 (Fed. Cir. 1986)) ("The case law of this circuit on [pre-AIA 35 U.S.C. § 292(a)] is sparse. In fact, only one precedent has substantively addressed the statute, and in that case, we affirmed, without discussion of the text of the statute, the trial court's holding that no violation of the statute had occurred because the plaintiff failed 'to produce any evidence of intent to deceive the public.'"); *Icon Health & Fitness, Inc. v. The Nautilus Group, Inc.*, No. 1:02-cv-109, 2006 U.S. Dist. LEXIS 24153, at *8 (D. Utah Mar. 23, 2006) ("Case law addressing what exactly constitutes a false-marking 'offense' is scant."); Donald W. Rupert, *Trolling for Dollars: A New Threat to Patent Owners*, 21 INTELL. PROP. & TECH. L.J. 3 (2009) ("Prior to 2007, cases involving the false marking statute were typically between competitors; the patent owner would sue for infringement and the defendant-competitor would counterclaim for, among other things, the false marking penalty.").

Beginning in 2007, however, individuals who were not competitors of patent owners began filing false marking *qui tam* suits in an attempt to reap enormous windfalls in damages. Donald W. Rupert, *Trolling for Dollars: A New Threat to Patent Owners*, 21 INTELL. PROP. & TECH. L.J. 3 (2009).[c] These individuals looked for high-volume products that were marked with patent numbers, investigated whether the products were covered by the identified patents, and then filed lawsuits to try to recover huge amounts of damages.[d]

---

[c]One of the most infamous false marking suits of this type was *Pequignot v. Solo Cup Co.,* 608 F.3d 1356 (Fed. Cir. 2010). In that case, Mathew A. Pequignot, who was an attorney registered to practice before the U.S. Patent and Trademark Office, sued Solo Cup Co. for falsely marking 21,757,893,672 of its drinking cup lids with the numbers of expired patents and sought the statutory damages of $500 per falsely marked article. *Id.* at 1359. As noted by the Federal Circuit, the $5.4 trillion in damages sought by Pequignot would have been sufficient to pay back 42% of the national debt. *Id.* at n.1.

[d]Note that while pre-AIA 35 U.S.C. § 292(a) provides statutory damages of $500 for every false marking offense, pre-AIA 35 U.S.C. § 292(b) provides that only half of the statutory damages may go to the person who brings the false marking suit while the other half goes to the U.S. government.

Id. The Federal Circuit had opened the door to the possibility of exorbitant damages awards in false marking cases when it ruled that the "not more than $500 for every such offense" language of pre-AIA 35 U.S.C. § 292(a) imposed fines of up to $500 for every article that was found to be falsely marked, thereby rejecting a patentee's argument that the statutory language merely imposed a fine of $500 for every decision a patentee made to falsely mark its products. *See Forest Group, Inc. v. Bon Tool Co.*, 590 F.3d 1295, 1303 (Fed. Cir. 2009) (finding that the clear language of 35 U.S.C. § 292 imposes a $500 fine for every article that is falsely marked despite Forest's argument that such interpretation of the statute would encourage "a new cottage industry" of false marking litigation by plaintiffs who have not suffered any direct harm). Not surprisingly, a "'cottage industry' of false marking litigation by plaintiffs who had not suffered any harm" emerged shortly after the *Forest Group* case was decided. Indeed, at least 600 *qui tam* false marking suits were filed in 2010 alone. Posting of Justin E. Gray to http://www.grayonclaims.com/home/2011/1/28/2010-false-marking-year-in-review-looking-forward.html (Jan. 28, 2011); *see also* Joe Matal, *A Guide to the Legislative History of the America Invents Act: Part II of II*, 21 FED. CIR. B.J. 539, 586 n.315 (2012) ("In March 2011, during a House Intellectual Property Subcommittee hearing on the AIA, a witness noted that 'over 800 [false marking] *qui tam* actions have been filed since the [*Forest Group Inc.*] decision was handed down.'") (citation omitted).

The surge in filings of *qui tam* false marking suits provoked Congress to amend pre-AIA 35 U.S.C. § 292. *See, e.g.,* H.R. REP. NO. 112-98, at 53 (2011) ("To address the recent surge in litigation, the bill replaces the qui tam remedy for false marking with a new action that allows a party that has suffered a competitive injury as a result of such marking to seek compensatory damages. The United States would be allowed to seek the $500-per-article fine, and competitors may recover in relation to actual injuries that they have suffered as a result of false marking, but the bill would eliminate litigation brought by unrelated, private third parties."). The amendments to the statute are discussed in Section II.B.ii.

ii. AIA § 16(b)'s Changes to Pre-AIA 35 U.S.C. § 292

The AIA made two major changes to pre-AIA 35 U.S.C. § 292, which can be seen in the side-by-side comparison of pre- and post-AIA 35 U.S.C. § 292 shown in the following table:

Pre-AIA 35 U.S.C. § 292	Post-AIA 35 U.S.C. § 292
(a) Whoever, without the consent of the patentee, marks upon, or affixes to, or uses in advertising in connection with anything made, used, offered for sale, or sold by such person within the United States, or imported by the person into the United States, the name or any imitation of the name of the patentee, the patent number, or the words "patent," "patentee," or the like, with the intent of counterfeiting or imitating the mark of the patentee, or of deceiving the public and inducing them to believe that the thing was made, offered for sale, sold, or imported into the United States by or with the consent of the patentee; or Whoever marks upon, or affixes to, or uses in advertising in connection with any unpatented article the word "patent" or any word or number importing the same is patented, for the purpose of deceiving the public; or Whoever marks upon, or affixes to, or uses in advertising in connection with any article the words "patent applied for," "patent pending," or any word importing that an application for patent has been made, when no application for patent has been made, or if made, is not pending, for the purpose of deceiving the public – Shall be fined not more than $500 for every such offense. (b) ~~Any person may sue for the penalty, in which event one-half shall go to the person suing and the other to the use of the United States.~~ (Emphasis added.)	(a) Whoever, without the consent of the patentee, marks upon, or affixes to, or uses in advertising in connection with anything made, used, offered for sale, or sold by such person within the United States, or imported by the person into the United States, the name or any imitation of the name of the patentee, the patent number, or the words "patent," "patentee," or the like, with the intent of counterfeiting or imitating the mark of the patentee, or of deceiving the and inducing them to believe that the thing was made, offered for sale, sold, or imported into the United States by or with the consent of the patentee; or Whoever marks upon, or affixes to, or uses in advertising in connection with any unpatented article the word "patent" or any word or number importing the same is patented, for the purpose of deceiving the public; or Whoever marks upon, or affixes to, or uses in advertising in connection with any article the words "patent applied for," "patent pending," or any word importing that an application for patent has been made, when no application for patent has been made, or if made, is not pending, for the purpose of deceiving the public – Shall be fined not more than $500 for every such offense. <u>Only the United States may sue for the penalty authorized by this subsection.</u> (b) A person who has suffered a competitive injury as a result of a violation of this section may file a civil action in a district court of the United States for recovery of damages adequate to compensate for the injury. (c) <u>The marking of a product, in a manner described in subsection (a), with matter relating to a patent that covered that product but has expired is not a violation of this section.</u> (Emphasis added.)

First, AIA § 16(b) repealed the language that provided authorization for false marking *qui tam* actions from pre-AIA 35 U.S.C. § 292(b). Indeed, the "[a]ny person may sue for the penalty" language of pre-AIA 35 U.S.C. § 292(b) authorized private citizens who had not been directly harmed by a false marking offense committed by another party to sue that party on behalf of the U.S. government. Post-AIA 35 U.S.C. § 292(b) replaces this language with the phrase "[a] person who has suffered a competitive injury as a result of a violation of this section may file a civil action . . . ." A conforming limitation stating that "[o]nly the United States may sue for the penalty authorized by [post-AIA 35 U.S.C. § 292(a)]" was also added to post-AIA 35 U.S.C. § 292(a). As a result of these amendments, private citizens who have not actually suffered harm from another party's alleged false marking offense are no longer authorized to file suit on behalf of the U.S. government. *See Hall v. Bed Bath & Beyond, Inc.*, 705 F.3d 1357, 1373 (Fed. Cir. 2013) ("A competitive injury under [post-AIA 35 U.S.C.] *§ 292* requires an actual, tangible economic injury."). Instead, only the U.S. government and private citizens who have suffered a competitive injury as a result of a party's alleged false marking may file a civil action under post-AIA 35 U.S.C. § 292.

The second amendment made by AIA § 16(b) can be seen in the addition of post-AIA 35 U.S.C. § 292(c). That section states that "[t]he marking of a product, in a manner described in [post-AIA 35 U.S.C. § 292](a), with matter relating to a patent that covered that product but has expired is not a violation of this section." In other words, unlike pre-AIA 35 U.S.C. § 292(a), which permitted false marking claims to be brought on the basis that an article had been marked with the number of an expired patent, post-AIA 35 U.S.C. § 292(c) explicitly prohibits such false markings from serving as the basis of a false marking claim brought under post-AIA 35 U.S.C. § 292(a).

iii. Legislative Goals Addressed by Post-AIA 35 U.S.C. § 292

Congress' major objective in amending pre-AIA 35 U.S.C. § 292 was to put an end to the false-marking *qui tam* actions that were clogging up the dockets of federal courts across the nation. In addition, Congress made a policy decision to eliminate false marking suits that were premised on a party's marking of an article with the number of an expired patent, finding that the potential harm caused by such false markings was far outweighed by the expense manufacturers would incur to avoid them. The final Committee Report's background section provides helpful insight into Congress' reasons for amending pre-AIA 35 U.S.C. § 292. That report states:

> [A] recent survey of such [false-marking qui tam] suits found that a large majority involved valid patents that covered the products in question but had simply expired. For many products, it is difficult and expensive to change a mold or other means by which a product is marked as patented, and marked products continue to circulate in commerce for some period after the patent expires. It is doubtful that the Congress that originally enacted [35 U.S.C. § 292] anticipated that it would force manufacturers to immediately remove marked products from commerce once the patent expired, given that the expense to manufacturers of doing so will generally greatly outweigh any conceivable harm of allowing such products to continue to circulate in commerce.

Joe Matal, *A Guide to the Legislative History of the America Invents Act: Part II of II*, 21 FED. CIR. B.J. 539, 587 (2012) (citing H.R. REP. NO. 112-98, at 53 (2011)); *see also id.* at 588 (citing 157 CONG. REC. S5321, 2011 (statement of Sen. Kyl)) ("[False-marking *qui tam*] suits represent a tax that patent lawyers are imposing on domestic manufacturing—a shift in wealth to lawyers that comes at the expensive of manufacturing jobs . . .[T]his bill prevents such abuses by repealing the statute's qui tam action while still allowing parties who have [suffered] actual injury from false marking to sue and allowing the United States to enforce a $500-per-product fine where appropriate.").

iv. Effective Date of the AIA § 16(b)'s Amendments to Pre-AIA 35 U.S.C. § 292

Section 16(b)(4) of the AIA states that the changes to the false marking statute "shall apply to all cases, without exception, that are pending on, or commenced on or after, the date of the enactment of this Act." Leahy-Smith America Invents Act of 2011, Pub. L. No. 112-29, 125 Stat. 284, 329. The AIA was enacted on September 16, 2011. *Id.* at 284. Thus, any false-marking *qui tam* suit that was pending on September 16, 2011, or that was filed on or after that date, should be dismissed pursuant to AIA § 16(b)(4) and post-AIA 35 U.S.C. § 292(b). Similarly, pursuant to post-AIA 35 U.S.C. § 292(c), any false marking claim that is premised upon a party's alleged marking of an article with the number of an expired patent should be dismissed so long as the claim was pending on September 16, 2011, or was filed on or after that date.

v. Practical Implications of the AIA § 16(b)'s Amendments to Pre-AIA 35 U.S.C. § 292

The AIA's repeal of false-marking *qui tam* actions brought a swift end to the so-called "'cottage industry' of false marking litigation" that the Federal

Circuit was warned about in the *Forest Group* case. *Forest Group, Inc.*, 590 F.3d at 1303. Indeed, AIA § 16(b)(4) makes clear that the amendments to the false marking statute are retroactive by stating that those amendments "shall apply to all cases, without exception, that are pending on, or commenced on or after, the date of the enactment of this Act." Leahy-Smith America Invents Act of 2011, Pub. L. No. 112-29, 125 Stat. 284, 329. This means that any false-marking *qui tam* claims that were pending in either a trial court or at the Federal Circuit as of September 16, 2011 would be dismissed and no other false marking *qui tam* claims could be filed on or after that date. *See* 157 Cong. Rec. S1372, 2011 (statement of Sen. Kyl) (explaining that the effective-date provision of the false-marking provision of the AIA made the repeal of *qui tam* actions "fully retroactive," and ensured that the provision would apply "to cases pending at any level of appeal or review."); *see also America Invents Act and the Patent Marking Statute*, Stroock Special Bulletin (2012) (pointing out that after the AIA was implemented, some courts have *sua sponte* dismissed actions in which a false marking plaintiff had not suffered a competitive injury).

Furthermore, given that private citizens who have actually suffered a competitive injury are now the only private citizens who are authorized to bring false marking actions[e] under post-AIA 35 U.S.C. § 292(b), and that those private citizens are only entitled to recover compensatory damages[f] for their injuries,[g] very few false marking claims likely will be brought in the future. *See, e.g.*, H.R. Rep. No. 112-98, at 53 (2011) ("Though one might assume that section 292 actions are targeted at parties that assert fictitious patents in order to deter competitors, such a scenario is almost wholly unknown to false-marking litigation."); 157 Cong. Rec. S5320,

[e]Note that the Federal Circuit recently held that certain potential competitors may be able to demonstrate that they have suffered a "competitive injury" and would therefore have standing to sue for a violation of post-AIA 35 U.S.C. § 292(a). *See Sukumar v. Nautilus, Inc.*, 785 F.3d 1396, 1402 (Fed. Cir. 2015) ("Therefore, an injury is only a 'competitive injury' if it results from competition, and a potential competitor is engaged in competition if it has attempted to enter the market, which includes intent to enter the market and action to enter the market. And, for the sake of completeness, an entity has standing under *§ 292(b)* if it can demonstrate competitive injury that was caused by the alleged false marking.").

[f]Compensatory damages merely provide a plaintiff with a monetary amount that is necessary to replace what the plaintiff lost as a result of the defendant's misconduct. For example, if a plaintiff in a false marking suit proved that he/she lost profits of $1000 as a result of a patentee falsely marking his/her products with inaccurate patent information, that plaintiff would be entitled to recover $1000 and nothing more.

[g]Unlike private citizens who bring false marking claims, the U.S. government is still entitled to recover $500 per false marking offense if it succeeds in proving that another party has violated post-AIA 35 U.S.C. § 292(a).

2011 (statement of Sen. Kyl) ("Currently, such [false-marking] suits are often brought by parties asserting no actual competitive injury from the marking—or who do not even patent or manufacture anything in a relevant industry."). Patentees should, nevertheless, continue to be diligent in their efforts to accurately mark their patented products in light of the fact that their damages for patent infringement may be limited without such markings and that they still may be sued by competitors and/or the U.S. government if they falsely mark their articles.

III) AIA § 17: EFFECTS OF THE LACK OF ADVICE OF COUNSEL ON CLAIMS OF WILLFUL AND INDUCED INFRINGEMENT

Patentees have often pointed to an accused infringer's failure to obtain a non-infringement, unenforceability or invalidity opinion as evidence that the infringer acted in bad faith, or with the requisite intent, to willfully infringe or induce infringement of their patents. Though the Federal Circuit has long wrestled with how such evidence can be used to prove willful or induced infringement, the Court has generally found that such evidence is probative in determining whether a patent has been willfully infringed or whether an accused infringer has induced infringement of a patent. As discussed in Section III.C., AIA § 17 has altered the status quo set forth in the Federal Circuit's jurisprudence by no longer allowing such evidence to be used in proving such claims.

A. Background on Using the Lack of Advice of Counsel to Prove Willful Infringement Prior to the AIA

Similar to many traffic offenses (e.g., speeding, failure to use a turn signal, and failure to yield), patent infringement is a strict liability offense.[h] *In re Seagate Tech., LLC*, 497 F.3d 1360, 1368 (Fed. Cir. 2007) (en banc). In fact,

---

[h]Strict liability offenses impose legal liability on a person who commits the offense, regardless of whether the person had any intent to commit the offense. In other words, so long as it can be proved that the strict liability offense was committed by a particular person, that person will be liable for the penalty associated with the offense even if he/she did not intend to commit the offense. With regard to patent infringement, a patentee need only prove that an accused infringer infringed his/her patent. The patentee does not have to prove, for instance, that the infringer knew about his/her patent, or that the infringer had any intent to infringe his/her patent, in order to recover damages (e.g., a reasonable royalty or lost profits) or seek an injunction.

one of the only times the nature of an infringer's actions in infringing a patent becomes relevant is when a patentee is trying to prove that the infringer willfully infringed its patent for purposes of obtaining enhanced damages.[i] *See* 35 U.S.C. § 284 ("[T]he court may increase the damages up to three times the amount found or assessed . . . ."); 35 U.S.C. § 285 ("The court in exceptional cases may award reasonable attorney fees to the prevailing party"); *Aro Mfg. Co., Inc. v. Convertible Top Replacement Co., Inc.*, 377 U.S. 476, 508 (1964) (explaining that a patentee "could in a case of willful or bad-faith infringement recover punitive or 'increased' damages under [35 U.S.C. § 284's] trebling provision"); *In re Seagate Tech., LLC*, 497 F.3d at 1368 ("[35 U.S.C. § 284], similar to its predecessors, is devoid of any standard for awarding [enhanced damages]. Absent a statutory guide, we have held that an award of enhanced damages requires a showing of willful infringement."); *Modine Mfg. Co. v. The Allen Group, Inc.*, 917 F.2d 538, 543 (Fed. Cir. 1990) (explaining that a trial court has the authority to award attorneys fees under 35 U.S.C. § 285 in "exceptional cases" and that an express finding of willful infringement is a sufficient basis for finding that a case is "exceptional").

Although the Federal Circuit since its inception in 1982 has wrestled with the issue of what standard should be used to determine whether an accused infringer has willfully infringed a patent, proof of willful infringement has always required evidence that the accused infringer was aware of the patent at-issue and that the accused infringer did not take reasonable steps to avoid infringing the patent. *See In re Seagate Tech., LLC*, 497 F.3d at 1371 (explaining that infringement of a patent is willful when an infringer is aware of the patent, but nonetheless "act[s] despite an objectively high likelihood that its actions constitute[] infringement of a valid patent."); *Underwater Devices Inc. v. Morrison-Knudsen Co., Inc.*, 717 F.2d 1380,

---

[i]Other circumstances in which the nature of an infringer's actions becomes relevant include those involving allegations of induced infringement and contributory infringement. *See DSU Med. Corp. v. JMS Co., Ltd.*, 471 F.3d 1293, 1306 (Fed. Cir. 2006) (en banc in relevant part) (explaining that inducement "requires more than just intent to cause the acts that produce direct infringement" because it also requires "that the alleged infringer knowingly induced infringement and possessed specific intent to encourage another's infringement"); *Cross Med. Prods., Inc. v. Medtronic Sofamor Danek, Inc.*, 424 F.3d 1293, 1312 (Fed. Cir. 2005) (quoting *Golden Blount, Inc. v. Robert H. Peterson Co.*, 365 F.3d 1054, 1061 (Fed. Cir. 2004)) ("In order to succeed on a claim of contributory infringement, in addition to proving an act of direct infringement, plaintiff must show that defendant 'knew that the combination for which its components were especially made was both patented and infringing' and that defendant's components have 'no substantial non-infringing uses.'").

1389-90 (Fed. Cir. 1983) (finding willful infringement where an accused infringer had actual notice of the patentee's patents and failed to exercise its affirmative duty to seek and obtain competent legal advice with regard to those patents *before* initiating its infringing activities). In addition to dealing with defining the type of behavior that warrants a finding of willful infringement, the Federal Circuit has also grappled with the issue of what role, if any, an alleged infringer's failure to obtain advice of counsel should play when assessing the reasonableness of its actions with regard to an asserted patent. *See, e.g., Knorr-Bremse Systeme Fuer Nutzfahrzeuge GmbH v. Dana Corp.,* 383 F.3d 1337, 1345-46 (Fed. Cir. 2004) (en banc) (holding that an accused infringer's failure to obtain legal advice does not give rise to an adverse inference with respect to willful infringement); *Kloster Speedsteel AB v. Crucible Inc.,* 793 F.2d 1565, 1580 (Fed. Cir. 1986) (holding that an accused infringer's failure to produce advice from counsel "would warrant the conclusion that it either obtained no advice of counsel or did so and was advised that its [activities] would be an infringement of valid U.S. Patents.").

When the Court first considered these issues in *Underwater Devices Inc.,* it held that the issue of whether an alleged infringer had willfully infringed a patent should be decided using a negligence-type standard. 717 F.2d at 1390; *see also In re Seagate Tech., LLC,* 497 F.3d at 1371 ("In contrast, the duty of care announced in Underwater Devices sets a lower threshold for willful infringement that is more akin to negligence."). The Court specifically explained that where "a potential infringer has actual notice of another's patent rights, he has an affirmative duty to exercise due care to determine whether or not he is infringing." *Underwater Devices Inc.,* 717 F.2d at 1389. Moreover, according to the Court, that affirmative duty includes, among other things, "the duty to seek and obtain competent legal advice from counsel [e.g., to seek a legal opinion from counsel] *before* the initiation of any possible infringing activity." *Id.* at 1390 (emphasis in original). Thus, because the accused infringer in *Underwater Devices Inc.* had actual notice of the patents-in-suit but delayed obtaining an opinion of counsel until long *after* it commenced its infringing activities, and even *after* the patent suit was filed, the Federal Circuit found that it proceeded with its infringing activities in bad faith and willfully infringed the patent at-issue. *Id.* at 1389-90.

Underwater Devices was decided at a time "when widespread disregard of patent rights was undermining the national innovation incentive." *Knorr-Bremse,* 383 F.3d at 1343 (citing *Advisory Committee on Industrial Innovation Final Report,* Dep't of Commerce (Sept. 1979)). Indeed, the infringer's attorney in *Underwater Devices* had blatantly disregarded the patentee's rights by advising his client to "continue to refuse to even discuss the payment of a royalty" even though no analysis of the patent at-issue had been completed, and by advising his client that the patentee likely would not risk filing suit against him because "courts, in recent years, have -- in patent infringement cases -- found the patents claimed to be infringed upon invalid in approximately 80% of the cases." 717 F.2d at 1385.

As a result of the affirmative duty of care that was imposed on alleged infringers in *Underwater Devices*, many parties defended against willful infringement claims by asserting "advice of counsel" defenses. *In re Seagate Tech, LLC*, 497 F.3d at 1369. Under the "advice of counsel" defense, an accused willful infringer would try to establish that because of reasonable reliance on advice from counsel, its continued accused infringing activities were done in good faith. *Id.* The advice from counsel typically came in the form of an opinion that concluded that the patent at-issue was invalid, unenforceable and/or not infringed by the accused willful infringer's proposed activities. *Id.*

The biggest problem with the "advice of counsel" defense was that it often forced accused infringers to choose between: (1) asserting attorney-client privilege for any communications they had with the counsel who prepared the opinions on which they wished to rely, and (2) avoiding a finding of willfulness if infringement was found. *Quantum Corp. v. Tandon Corp.*, 940 F.2d 642, 644 (Fed. Cir. 1991).[j] Indeed, after *Underwater Devices*, courts began making negative inferences against accused willful infringers

---

[j]The attorney-client privilege encourages clients to disclose all pertinent information relating to a legal matter to their attorneys by protecting such disclosures from discovery during litigation. *See Upjohn Co. v. U.S.*, 449 U.S. 383, 389 (1981) ("The attorney-client privilege is the oldest of the privileges for confidential communications known to the common law. Its purpose is to encourage full and frank communication between attorneys and their clients and thereby promote broader public interests in the observance of law and administration of justice.") (citation omitted). Attorney-client privilege belongs to the client, who alone may waive it. *Am. Standard Inc. v. Pfizer Inc.*, 828 F.2d 734, 745 (Fed. Cir. 1987). The Federal Circuit has held that "[o]nce a party announces that it will rely on advice of counsel . . . in response to an assertion of willful infringement, the attorney-client privilege is waived." *In re EchoStar Commc'ns Corp.*, 448 F.3d 1294, 1299 (Fed. Cir. 2006). Moreover, once the attorney-client privilege is waived as to one communication, the waiver applies to all other communications relating to the same subject matter. *Fort James Corp. v. Solo Cup Corp.*, 412 F.3d 1340, 1349 (Fed. Cir. 2005). "This broad scope [of waiver] is grounded in principles of fairness and serves to prevent a party from simultaneously using the privilege as both a sword and a shield; that is, it prevents the inequitable result of a party disclosing favorable communications while asserting the privilege as to less favorable ones." *In re Seagate Tech., LLC*, 497 F.3d at 1372 (citing *In re Echostar Commc'ns Corp.*, 448 F.3d at 1301; *Fort James Corp.*, 412 F.3d at 1349).

who did not produce opinions of counsel. *See, e.g., Fromson v. Western Litho Plate & Supply Co.*, 853 F.2d 1568, 1572-73 (Fed. Cir. 1988) ("Where the infringer fails to introduce an exculpatory opinion of counsel at trial, a court must be free to infer that either no opinion was obtained or, if an opinion were obtained, it was contrary to the infringer's desire to initiate or continue its use of the patentee's invention."); *Kloster Speedsteel AB*, 793 F.2d at 1580 (observing that the infringer "has not even asserted that it sought advice of counsel when notified of the [patent] claims and [the patentee's] warning, or at any time before it began this litigation," and holding that the infringer's "silence on the subject, in alleged reliance on the attorney–client privilege, would warrant the conclusion that it either obtained no advice of counsel or did so and was advised that its importation and sale of the accused products would be an infringement of valid U.S. patents."); *Shatterproof Glass Corp. v. Libbey-Owens Ford Co.*, 758 F.2d 613, 628 (Fed. Cir. 1985) ("A record devoid of opinions of counsel and silent on [the accused infringer's] reaction to the existence of the [patentee's] patents may indeed lead to negative inferences, and the case for willfulness was dependent on determinations of credibility and motivation which were placed in issue at trial, and which are the province of the trier of fact.").

Recognizing that the flagrant disregard for patent rights referenced in *Underwater Devices* was no longer the norm, the Federal Circuit in 2004 began to change course with regard to its willful infringement doctrine. *See Knorr-Bremse*, 383 F.3d at 1344 (noting that the "conceptual underpinnings" of the precedent allowing negative inferences to be drawn where no opinion of counsel is produced "have significantly diminished in force"). Sitting en banc,[k] the Court held that "no adverse inference shall arise from invocation of the attorney-client and/or work product privilege." *Id.* In other words, after *Knorr-Bremse*, it was no longer proper for a judge or jury to infer that an alleged infringer had obtained an unfavorable opinion that its activities would have infringed the patent at-issue based solely on the fact that the alleged infringer invoked the attorney-client privilege and/or work product immunity to prevent having to disclose such opinion. In addition, the Court held that it is not appropriate to draw an adverse inference with respect to willful infringement when a defendant has not obtained legal advice. *Id.* at 1345. On this point, the Court explained that "[t]he issue

[k]When a case is heard "en banc," it simply means that all of the judges, as opposed to a select panel of three judges, heard the case. En banc opinions carry more weight than do cases that are heard by a select panel of judges.

here is not of privilege, but whether there is a legal duty upon a potential infringer to consult with counsel, such that failure to do so will provide an inference or evidentiary presumption that such opinion would have been negative." *Id.*

In 2007, the Federal Circuit took its holding in *Knorr-Bremse* a step further when it explicitly overruled the negligence-type standard for finding willful infringement that was articulated in *Underwater Devices*, finding that the standard was inconsistent with the standards that were used by other courts to evaluate willfulness in other civil litigation contexts. *In re Seagate Tech., LLC*, 497 F.3d at 1371. More specifically, the Court explained:

> *[W]e overrule the standard set out in* Underwater Devices *and hold that proof of willful infringement permitting enhanced damages requires at least a showing of objective recklessness. Because we abandon the affirmative duty of due care, we also reemphasize that there is no affirmative obligation to obtain opinion of counsel ... Accordingly, to establish willful infringement, a patentee must show by clear and convincing evidence that the infringer acted despite an objectively high likelihood that its actions constituted infringement of a valid patent.*
>
> *Id.*

After *Seagate*, and under today's current law, proof of willful infringement requires clear and convincing evidence that: (1) "the infringer acted despite an objectively high likelihood that its actions constituted infringement of a valid patent" and (2) the objectively defined risk was "known or [was] so obvious that it should have been known to the accused infringer." *Id.* An accused infringer's state of mind is not relevant to the first prong of the two-part test set forth in *Seagate. Id.* If the threshold objective standard of the first prong is satisfied, however, "the patentee must also demonstrate that this objectively-defined risk (determined by the record developed in the infringement proceeding) was either known or so obvious that it should have been known to the accused infringer." *Id.*

Although the Federal Circuit made clear that accused infringers had no affirmative duty to obtain opinions of counsel before embarking on potentially infringing activities, questions as to how evidence of a defendant's failure to obtain an opinion of counsel could be used to prove willful infringement still remained after *Knorr-Bremse* and *In re Seagate. See, e.g., In re Seagate Tech., LLC*, 497 F.3d at 1371 ("We leave it to future cases to further develop the application of [the objective recklessness] standard."); *Knorr-Bremse*, 383 F.3d at 1346-47 (Dyk, J., concurring-in-part and dissenting-in-part) (pointing out that the majority did not resolve the question of "whether the trier of fact, particularly the jury, can or should be told whether or not counsel was consulted (albeit without any inference as to the nature of the advice received) as part of the totality of the circumstances relevant to the question of willful infringement.");[1] Marcus S. Friedman & Barry J. Marenberg, *A Sharp Turn in the IP Highway: The Federal Circuit Redefines Willful Patent Infringement*, 178 N.J. L.J. 29 (2004) (questioning whether alleged infringers can avoid a finding of willful infringement without an opinion of counsel). As a result, district courts' views on whether and how evidence of a failure to obtain an opinion of counsel should be used in willful infringement analyses have varied widely since those cases were decided. *See, e.g., Abbott Diabetes Care Inc., et al. v. Roche Diagnostics Corp. et al.*, No. CO5-03117 MJJ, 2007 U.S. Dist. LEXIS 31193, at *37 (N.D. Cal. Apr. 24, 2007) (declining to draw any inference regarding willful infringement from the fact that Roche and Bayer did not obtain and/or produce exculpatory opinions of counsel); *Third Wave Technologies, Inc. v. Strategene Corp.*, 405 F. Supp. 2d 991, 1016 (W.D. Wis. 2005) (noting that *Knorr-Bremse* "did not say that it was improper for a jury to infer from an infringer's failure to consult counsel that the infringer had no prior knowledge of its opponent's patents or that it had not acted properly in other respects").

B. *Seagate's* Effect on Proving Induced Infringement Prior to the AIA

Shortly before the Federal Circuit's release of the *Seagate* opinion, a jury in the Central District of California found that, among other offenses, Qualcomm had induced infringement of and willfully infringed several claims of three

---

[1]In *Underwater Devices*, the court explained that a finding of willful infringement should not be based on a single failure on the part of the defendant (e.g., a failure to obtain advice of counsel), but rather on the "totality of the circumstances presented." 717 F.3d at 1390; *see also Rolls-Royce Ltd. v. GTE Valeron Corp.*, 800 F.2d 1101, 1109 (Fed. Cir. 1986) (explaining that the affirmative duty imposed on infringers in *Underwater Devices* will normally entail the obtaining of competent legal advice of counsel before infringing or continuing to infringe, but that does not mean that the absence of an opinion of counsel alone *requires* a finding of willful infringement in every case). Later on, the Federal Circuit developed several factors that could be considered in the totality of circumstances analysis for willful infringement. *See Read Corp. v. Portec, Inc.*, 970 F.2d 816, 826-27 (Fed. Cir. 1992) (listing factors, including whether the infringer deliberately copied the ideas of another, the infringer's behavior in the litigation, the infringer's size and financial condition, duration of the defendant's misconduct, and whether the defendant attempted to conceal its misconduct as factors that can be considered when evaluating whether infringement was willful).

patents[m] belonging to Broadcom. *Broadcom Corp. v. Qualcomm Inc.*, 543 F.3d 683, 687 (Fed. Cir. 2008). Qualcomm filed post-trial motions that, among other things, sought to set aside the jury's finding on willfulness and requested a new trial on all infringement claims and willfulness issues. *Id.*; *Broadcom Corp. v. Qualcomm Inc.*, No. SACV 05-467-JVS(RNBx), 2007 U.S. Dist. LEXIS 86627, at *5 (C.D. Cal. Nov. 21, 2007). The district court initially denied all of Qualcomm's motions. *Broadcom Corp.*, 543 F.3d at 687. Ten days after the district court denied Qualcomm's post-trial motions, however, the Federal Circuit released its en banc opinion in *In re Seagate*. *Id.* Because of the potential impact of *In re Seagate* on the willfulness determination previously made by the jury,[n] the district court sua sponte[o] invited a motion for reconsideration of its denial of Qualcomm's request for a new trial on willfulness and of its award of enhanced damages. *Id.* Upon consideration of the parties' arguments, the district court vacated the willfulness verdict and granted a new trial on the willfulness issues. *Id.*; *Broadcom Corp.*, 2007 U.S. Dist. LEXIS 86627, at *7. The district court did not grant a new trial on the induced infringement claim, however, despite that the jury had been instructed to consider whether or not Qualcomm obtained the advice of a competent lawyer in its analysis of that claim. *Broadcom Corp.*, 2007 U.S. Dist. LEXIS 86627, at *8-9, 12.

In support of its appeal of the district court's denial of its post-trial motion for a new trial on induced infringement, Qualcomm tried unsuccessfully to extend the reach of *In re Seagate* to the doctrine of induced infringement.[p] Qualcomm specifically argued that the jury instructions,

---

[m]The patents pertained to chipsets used in mobile radio devices such as cell phone handsets. *Broadcom Corp. v. Qualcomm Inc.*, 543 F.3d 683, 686 (Fed. Cir. 2008).

[n]Because *In re Seagate* had not yet been decided at the time the jury considered willfulness, the jury was not instructed under the "objective recklessness" standard established by that case. *Broadcom Corp.*, 2007 U.S. Dist. LEXIS 86627, *6-7. The instructions given to the jury also did not make clear that Qualcomm had no duty to obtain an opinion of counsel and thus were not in line with *In re Seagate's* abolition of the affirmative duty to acquire such opinions that was articulated in *Underwater Devices*. *Id.* at *9.

[o]Sua sponte is Latin for "on its own accord."

[p]In order to prove that an alleged infringer has induced infringement, a patentee must establish "first that there has been direct infringement, and second that the alleged infringer knowingly induced infringement and possessed specific intent to encourage another's infringement." *ACCO Brands, Inc. v. ABA Locks Mfrs. Co.*, 501 F.3d 1307, 1312 (Fed. Cir. 2007) (quoting *Minn. Mining & Mfg. Co. v. Chemque, Inc.*, 303 F.3d 1294, 1304-05 (Fed. Cir. 2002)). To meet the specific intent element of induced infringement, a patentee must demonstrate that "[t]he defendant must have intended to cause the acts that constitute the direct infringement and must have known or should have known tha[t] its action would cause the direct infringement." *DSU Med. Corp.*, 471 F.3d at 1305.

which advised the jury that it could consider whether or not Qualcomm obtained the advice of a competent lawyer when considering whether Qualcomm knew or should have known that the induced actions would constitute infringement, were impermissible in light of the Federal Circuit's *In re Seagate* opinion. *Broadcom Corp.*, 543 F.3d at 699. Qualcomm further argued that because specific intent is a stricter standard than the "objective recklessness" standard adopted in *Seagate*, evidence not probative of willful infringement could not be probative of specific intent for purposes of proving induced infringement. *Id.* Thus, according to Qualcomm's reasoning, evidence of a failure to obtain an opinion of counsel was not relevant to the specific intent issue for purposes of finding induced infringement. *Id.*

In denying Qualcomm's request for a new trial on induced infringement, the Federal Circuit explained that *In re Seagate* did not alter the state of mind requirement for induced infringement. *Id.* Confirming that the en banc holding in *DSU Med. Corp.* remained the relevant authority on induced infringement, the Federal Circuit explained:

> [W]e noted in DSU that [specific] intent may be established where an alleged infringer who "knew or should have known *his actions would induce actual infringements*," is shown to have induced infringing acts through his actions . . . That is, the "affirmative intent to cause direct infringement" . . . required by DSU may be shown -- just as the jury was instructed in DSU itself -- by establishing first that the defendant "intended to cause the acts that constitute the direct infringement," and second that the defendant "kn[ew] or should have known [that] its action would cause the direct infringement . . . Because opinion-of-counsel evidence, along with other factors, may reflect whether the accused infringer "knew or should have known" that its actions would cause another to directly infringe, we hold that such evidence remains relevant to the second prong of the intent analysis. Moreover, we . . . hold that the failure to procure such an opinion may be probative of intent in this context. It would be manifestly unfair to allow opinion-of-counsel evidence to serve an exculpatory function, as was the case in DSU itself . . . and yet not permit patentees to identify failures to procure such advice as circumstantial evidence of intent to infringe.
> **Id. (citing DSU Med. Corp., 471 F.3d at 1305-07) (emphasis in original).**

In sum, *Broadcom Corp. v. Qualcomm Inc.* confirmed that patentees were entitled to rely on evidence of an accused infringer's failure to obtain an opinion of counsel to prove both prongs of the specific intent element of induced infringement.

C. AIA § 17(a): Enactment of 35 U.S.C. § 298

AIA § 17(a) both clarified the law on how evidence of a failure to obtain an opinion of counsel could be used to prove willful infringement and changed the law on how such evidence could be used to support a claim of induced infringement. AIA § 17(a) specifically amended the Patent Act by adding a new statute, 35 U.S.C. § 298. Leahy-Smith America Invents Act of 2011, Pub. L. No. 112-29, 125 Stat. 284, 329. 35 U.S.C. § 298 reads:

> The failure of an infringer to obtain the advice of counsel with respect to any allegedly infringed patent, or the failure of the infringer to present such advice to the court or jury, may not be used to prove that the accused infringer willfully infringed the patent or that the infringer intended to induce infringement of the patent.
>
> **Id.**

Accordingly, although it was unclear whether and how evidence of a failure to obtain advice of counsel (e.g., an opinion of counsel) could be used to prove willful infringement after *Knorr-Bremse* and *In re Seagate*, new 35 U.S.C. § 298 makes clear that such evidence can no longer be utilized for this purpose. New 35 U.S.C. § 298 also abrogates the holding in *Broadcom Corp. v. Qualcomm Inc.* by stating that such evidence "may not be used to prove that the accused infringer . . . intended to induce infringement of the patent."

D. Legislative Goals Addressed by AIA § 17(a)

AIA § 17(a) enacted a new statute, 35 U.S.C. § 298, "to protect attorney-client privilege and to reduce pressure on accused infringers to obtain opinions of counsel for litigation purposes." H.R. Rep. No. 112-98, at 53 (2011). The new statute reflects Congress' policy choice that the probative value of evidence that a party failed to obtain advice of counsel is outweighed by the harm that coercing a waiver of attorney-client privilege inflicts on the attorney-client relationship. *Id.* Senator Kyl commented on the policies embodied in new 35 U.S.C. § 298 during the March 2011 debates on the AIA, stating:

> Permitting adverse inferences from a failure to procure an opinion or waive privilege undermines frank communication between clients and counsel. It also feeds the cottage industry of providing such opinions—an industry that is founded on an unhealthy relationship between clients and counsel and which amounts to a deadweight loss to the patent system. Some lawyers develop a lucrative business of producing these opinions, and inevitably become aware that continued requests for their services are contingent on their opinions' always coming out the same way—that the patent is invalid or not infringed. Section 298 reflects legislative skepticism of the probative value of such opinions.
>
> **Joe Matal, *A Guide to the Legislative History of the America Invents Act: Part II of II*, 21 Fed. Cir. B.J. 539, 589-90 (2012) (citing 157 Cong. Rec. S1374, 2011).**

E. Effective Date of AIA § 17(a)

When AIA § 17(a) was originally passed, it was not given its own effective date. Leahy-Smith America Invents Act of 2011, Pub. L. No. 112-29, 125 Stat. 284, 329. As a result, had no further legislation been passed addressing the effective date of this section, AIA § 17 would have been governed by the AIA's default effective date set forth in AIA § 35. Joe Matal, *A Guide to the Legislative History of the America Invents Act: Part II of II*, 21 FED. CIR. B.J. 539, 590 (2012); *see also* Leahy-Smith America Invents Act of 2011, Pub. L. No. 112-29, § 35, 125 Stat. 284, 341 ("Except as otherwise provided in this Act, the provisions of this Act shall take effect upon the expiration of the 1-year period beginning on the date of the enactment of this Act and shall apply to any patent issued on or after that effective date."). Congress provided AIA § 17 with its own effective date, however, when it passed the Leahy-Smith America Invents Technical Corrections on January 14, 2013. Specifically, section 1(a) of the Leahy-Smith America Invents Technical Corrections states that "section 298 of title 35, United States Code, shall apply to any civil action commenced on or after the date of the enactment of this Act." Pub. L. No. 112-274, 126 Stat. 2456. Accordingly, pursuant to 35 U.S.C. § 298, patentees should be prohibited from using evidence of a failure to obtain advice of counsel to prove either willful or induced infringement in any patent suit that is filed on or after January 14, 2013.

F. Practical Implications of AIA § 17(a)

Although Congress has made clear that evidence of a failure to obtain advice of counsel cannot be used to prove willful or induced infringement, it is important to understand that AIA § 17(a) does not prevent accused infringers from voluntarily relying on opinions of counsel to defend against such infringement allegations.[q] *See, e.g., Finisar Corp. v. DirecTV Group, Inc.,*

---

[q]While it is clear that accused infringers may rely on advice of counsel defenses when defending against willful or induced infringement claims, it is not clear how 35 U.S.C. § 298 will impact the type of evidence patent holders may use to rebut such defenses. *See Ultratec, Inc. v. Sorensen Commc'ns, Inc.*, No. 13-cv-346-bbc, 2014 U.S. Dist. LEXIS 141428, at *7 (W.D. Wis. Oct. 3, 2014) ("Plaintiffs do not object to defendants' motion in principle but argue that the statutory bar [imposed by 35 U.S.C. § 298] should not be interpreted so narrowly as to preclude them from rebutting any evidence defendants seek to introduce that implies that defendants consulted with counsel or had some legal basis for believing the patents to be invalid . . . Generally speaking, I agree with plaintiffs that the protection granted by *35 U.S.C. § 298* dissolves in the event defendants 'open the door' by attempting to refute a claim of willful infringement by implying that they relied on the advice of counsel.").

523 F.3d 1323, 1339 (Fed. Cir. 2008) (holding that "a competent opinion of counsel concluding either [non-infringement or invalidity] would provide a sufficient basis for [the defendant] to proceed without engaging in objectively reckless behavior with respect to the [asserted] patent"); *Comark Commc'ns, Inc. v. Harris Corp.*, 156 F.3d 1182, 1191 (Fed. Cir. 1998) (explaining that obtaining an objective opinion letter from counsel provides the basis for a defense against willful infringement); *Fujitsu Ltd. v. Belkin Int'l, Inc.*, No. 10-cv-03972-LHK, 2012 U.S. Dist. LEXIS 142102, at *121-22 (N.D. Cal. Sept. 28, 2012) (explaining that reliance on competent opinions of counsel that an asserted patent is invalid is relevant to a defense against induced infringement); *see also In re Seagate Tech., LLC*, 497 F.3d at 1369 (explaining that an accused infringer's reasonable reliance on advice of counsel that the asserted patent is invalid, unenforceable or not infringed is "crucial to the analysis" of willfulness, but is not necessarily dispositive); *Acumed LLC v. Stryker Corp.*, 483 F.3d 800, 810 (Fed. Cir. 2007) ("Favorable opinions of counsel normally present a well-grounded defense to willfulness, but the protection they afford is not absolute."). Thus, when a company begins making plans to practice a new technology, the best practice is still for that company to obtain an opinion from a qualified patent attorney to determine whether the new technology would infringe any claims of a valid, enforceable patent before commencing any work on the new technology.

To provide any insurance against a claim of willful or induced infringement, however, the opinion from the patent attorney must be <u>competent</u>[r] or it is of little value in showing that the company possessed a good faith belief that its technology did not infringe a valid patent. *Comark Commc'ns, Inc.*, 156 F.3d at 1191; *see also Jurgens v. CBK, Ltd.*, 80 F.3d 1566, 1572-73 (Fed. Cir. 1996) ("If infringers could rely on any opinion to defeat willful infringement, no matter how incompetent, insulation from increased damages would be complete."); *Read Corp.*, 970 F.2d at 828-29, *abrogated on other grounds by* Markman v. Westview Instruments, Inc., 52 F.3d 967 (Fed. Cir. 1995) ("Those cases where willful infringement is found despite the presence of an opinion of counsel generally involve situations where opinion of counsel was either ignored or found to be incompetent."). There must also be evidence that the patent attorney's

---

[r]Because the patentee must prove willfulness by clear and convincing evidence, the patentee bears the burden of showing that the opinion of counsel on which the defendant relies is incompetent. *In re Katz Interactive Call Processing Patent Litig.*, Nos. 07-ML-1816-B-RGK and 07-CV-2322-RGK, 2009 U.S. Dist. LEXIS 72134, at *93 (C.D. Cal. May 1, 2009).

opinion was communicated to the company and that the company relied on the opinion in carrying out its allegedly infringing activities in order for an opinion to exculpate the company from claims of willful or induced infringement. *See Fujitsu Ltd.*, 2012 U.S. Dist. LEXIS 142102, at *113 (finding that there was a genuine issue of fact as to whether an accused infringer relied on a patent attorney's opinions at all where the accused infringer provided no evidence that anyone at the company had reviewed the attorney's various opinions or independently determined that it was reasonable to rely on them and the accused infringer's 30(b)(6) designee on the topic of opinions of counsel could not confirm when the accused infringer received the opinion letter and had never seen the opinion letter himself).

The competency requirement applies to both the qualifications of the attorney giving the opinion and to the content of the opinion itself. *Jurgens*, 80 F.3d at 1572. The federal courts have delineated several factors for determining whether advice is competent, and whether it was reasonable for an accused infringer to rely on the advice. These factors include:[5]

(1) The Background Research Performed by the Attorney Providing the Opinion

The background research that should be performed by a patent attorney providing an opinion will vary from case to case. In general, however, patent attorneys should review the file history (i.e., the prosecution history), the specification, the claims, the abstract and any related patent applications (i.e., priority applications, parent applications and continuing applications) for the patent at-issue. Additionally, if the patent has already been litigated or has been involved in a post-grant proceeding before the PTAB by the time an attorney is providing an opinion, the attorney should review any orders or rulings associated with that patent (e.g., claim construction rulings). *See Krippelz v. Ford Motor Co.*, 636 F. Supp. 2d 669, 676 (N.D. Ill. 2009) ("[R]easonable defenses are limited to those consistent with [the court's *Markman*] ruling"). These materials should be reviewed regardless of whether the patent attorney is providing a non-infringement opinion, an invalidity opinion or an unenforceability

---

[5] The six factors listed here were delineated in *Chiron Corp. v. Genentech, Inc.*, 268 F. Supp. 2d 1117, 1121 (E.D. Cal. 2002).

opinion with regard to the patent.[t] In addition, the opinion should provide summaries of the materials that were reviewed while the opinion was being prepared so that it is clear that the patent attorney performed the requisite background research and had a reasonable understanding of what he/she was reviewing.

When providing an invalidity opinion, the patent attorney should also review all of the prior art that was cited during the prosecution of the patent at-issue.[u] *See, e.g., Jurgens*, 80 F.3d at 1572 (explaining that the steps "'normally considered to be necessary and proper in preparing an opinion' include a thorough review of the cited prior art and prosecution history.") (citation omitted). The opinion should set forth summaries of what is taught by at least those prior art references that were actually cited by the Examiner in his/her Office Actions during the prosecution of the patent-at-issue. Furthermore, if the attorney relies on any prior art references that were not cited during prosecution, the opinion should set forth a summary of those references that explains how the references are relevant to the patent claims at-issue.

In addition, when providing a non-infringement opinion, a patent attorney should perform sufficient research on his/her client's proposed new technology to determine whether or not that new technology meets all of the claim limitations of any of the patent claims for which he/she is providing the opinion. The patent attorney should also summarize any materials that he/she reviewed in connection with his/her infringement analysis (e.g., test results, product samples, or documents that provide key information about the characteristics of the new technology). The type of research a patent attorney will need to perform in order to evaluate infringement will vary based on the patent claims that are being analyzed in the non-infringement opinion. In some instances, the patent attorney may need to ask his/her client to provide certain test results for his/her technology to

---

[t]Notably, an opinion need not set forth both non-infringement and invalidity analyses in order to negate a finding of willful infringement. *See Graco, Inc. v. Binks Mfg. Co.*, 60 F.3d 785, 793 (Fed. Cir. 1995) ("There is no requirement that an opinion *must* address validity to negate a finding of willful infringement.") (emphasis in original). Instead, a competent opinion of counsel that concludes that either (1) an accused infringer did not infringe the patent at-issue, or (2) that the patent-at-issue was invalid, would provide a sufficient basis for the alleged infringer to proceed in practicing its new technology without engaging in objectively reckless behavior with regard to that patent. *Finisar Corp.*, 523 F.3d at 1339.

[u]A complete list of the prior art that was cited by both the Patent Examiner and the applicant during prosecution appears on the face of each patent.

determine whether the technology meets certain characteristics recited in the claims. For instance, if a client is seeking to market a new pharmaceutical product, that client may need to provide certain pharmacokinetic data to the patent attorney in order for that attorney to determine whether it meets the claim limitations of the patent claims at-issue. If the attorney is trying to determine whether the client would induce infringement of a patent, the client may need to provide copies of any written materials or instructions (e.g., package inserts) that will be provided to consumers in connection with its proposed new technology.

(2) Whether the Opinion Relied on Was Written or Oral

When deciding whether to practice new technology, companies should seek written opinions that analyze whether the new technology will infringe any valid, enforceable patent claims. Indeed, a written opinion will provide a company with better evidence that it had a reasonable defense to a charge of infringement than would an oral opinion. *See, e.g., Minn. Mining & Mfg. Co.*, 976 F.2d at 1580 (expressing skepticism about the competence of oral opinions by in-house counsel because of difficulties of proof and credibility issues).

(3) The Objectivity of the Opinion Being Relied On

Subjective, self-serving opinions are of little or no value when trying to defend against allegations of willful or induced infringement because such opinions are unreliable and lack credibility. For this reason, companies should generally seek opinions from outside counsel because outside counsel are less likely to be influenced by the business objectives and corporate pressures that may affect in-house counsel. *See SRI Int'l, Inc. v. Advanced Tech. Laboratories, Inc.*, 127 F.3d 1462, 1467 (Fed. Cir. 1997) (noting that while there is no per se rule against relying on the advice of in-house counsel, counsel's objectivity is an important factor in determining whether it was reasonable for an infringer to rely on an opinion of counsel); *Johns Hopkins Univ. v. Cellpro, Inc.*, 978 F. Supp. 184, 194 (D. Del. 1997) (considering the fact that an infringer's in-house patent lawyer reviewed and revised a draft opinion from outside counsel in determining that infringement was willful).

Additionally, even if outside counsel provides an opinion, the company should evaluate whether the reasoning and analysis in the opinion is consistent with the facts and applicable law, and with the positions the company has taken in any relevant prosecution, litigation and/or post-grant proceedings, before relying on the opinion. *See Liquid Dynamics Corp. v. Vaughan Co.,*

Inc., 449 F.3d 1209, 1226 (Fed. Cir. 2006) (affirming a finding of willfulness where the patentee "presented flaws in [an] opinion's factual basis"); *Minn. Mining & Mfg. Co.*, 976 F.2d at 1580-82 (affirming a finding of willfulness based in part on the fact that the in-house attorney's opinion regarding prior art was inconsistent with a position the same attorney had taken in prosecuting the infringer's patent); *Read Corp.*, 970 F.2d at 829 n.9 ("A good test that the advice given is genuine and not merely self-serving is whether the asserted defenses are backed up with viable proof during trial which raises substantial questions."); *Power Integrations, Inc. v. Fairchild Semiconductor Int'l, Inc.*, 725 F. Supp. 2d 474, 478 (D. Del. 2010) (finding that an opinion offered by the accused infringer was unreliable because it was plainly contrary to the facts regarding the structure of the accused infringer's devices). Further, opinions generally should not unequivocally state that a patent is invalid or unenforceable, or that the patent is not infringed by the client's proposed technology. *See Read Corp.*, 970 F.2d at 829, n.9 ("An opinion of counsel, of course, need not unequivocally state that the client will not be held liable for infringement. An honest opinion is more likely to speak of probabilities than certainties.").

(4) Whether the Attorney Rendering the Opinion Was a Patent Attorney

As is probably evident by this point in this text, patent law is highly complex and specialized. Accordingly, an opinion prepared by a registered patent attorney will likely carry more weight, and have a better probability of being adjudged competent, than one prepared by a general corporate attorney or someone who does not specialize in patent law.

(5) Whether the Opinion Was Detailed or Merely Conclusory

To effectively negate a finding of willful infringement, or specific intent for the purpose of proving induced infringement, an opinion "must be thorough enough, as combined with other factors, to instill a belief in the infringer that a court might reasonably hold the patent is invalid, not infringed, or unenforceable." *Ortho Pharm. Corp. v. Smith*, 959 F.2d 936, 944 (Fed. Cir. 1992); *see also SRI Int'l, Inc.*, 127 F.3d at 1466 (affirming a rejection of an advice of counsel defense because the advice was "conclusory and woefully incomplete . . . lacking both legal and factual analysis . . . [and] insufficient to meet the standard of due care appropriate to serve as an exculpatory opinion of counsel.") (internal quotation marks omitted); *Underwater Devices Inc.*, 717 F.2d at 1390, *overruled on other grounds by* In re Seagate Tech., LLC,

497 F.3d 1360 (Fed. Cir. 2007) (affirming a rejection of an advice of counsel defense because the opinion memo contained only "bald, conclusory and unsupported remarks regarding validity and infringement of [the patents at-issue]"); *Bear U.S.A., Inc. v. A.J. Sheepskin & Leather Outerwear, Inc.*, 909 F. Supp. 896, 907 (S.D.N.Y. 1995) (rejecting an advice of counsel defense because the opinion letter could "be described, at best, as superficial, containing no analysis explaining the conclusion reached . . . .").

A thorough opinion is one that sets forth proposed constructions for each claim element of each of the patent claims being analyzed, explains why such claim constructions are reasonable in light of the intrinsic and extrinsic evidence and any applicable case law, and provides sound reasoning, based on applicable case law and the facts of the case, for why each of the patent claims at-issue are not infringed, invalid, and/or unenforceable.[v] *See Read Corp.*, 970 F.2d at 829 (explaining that a written opinion may be incompetent on its face by reason of its containing merely conclusory statements without discussion of facts or obviously presenting only a superficial or off-the-cuff analysis); *AAT Bioquest, Inc. v. Tex. Fluorescence Laboratories, Inc.*, No. 14-cv-03909-DMR, 2015 U.S. Dist. LEXIS 160369, at *34-35 (N.D. Cal. Nov. 30, 2015) (finding that an infringer's reliance on its counsel's opinions was objectively unreasonable because the opinions bore significant indicia of unreliability due to the fact that they did not cite any case law and that the attorney who provided the opinions failed to thoroughly investigate the defense to infringement that was set forth in the opinions). In addition, if an opinion sets forth non-infringement positions, that opinion should detail why the proposed technology does not infringe the patent claims at-issue either literally or under the doctrine of equivalents. Furthermore, if an opinion sets forth invalidity arguments based on prior art, and the prior art was already considered by the U.S. PTO when the patent underwent examination, the opinion should discuss why the U.S. PTO's

---

[v]A thorough opinion should set forth an invalidity, unenforceability, and/or non-infringement analysis for each and every claim of the patent that is being evaluated. The only exception to this is that a non-infringement opinion can group together an independent claim with those claims that depend from it if the attorney determines that the client's new technology does not infringe the independent claim. *See Wahpeton Canvas Co., Inc. v. Frontier, Inc.*, 870 F.2d 1546, 1552 (Fed. Cir. 1989) ("It is axiomatic that dependent claims cannot be found infringed unless the claims from which they depend have been found to have been infringed . . . ."). The opinion should also articulate any additional reasons, to the extent they exist, that the new technology does not infringe the dependent claims even if the opinion articulates reasoning as to why the independent claim from which the dependent claims depend is not infringed.

analysis was incorrect and/or should proffer different invalidity arguments than those set forth by the U.S. PTO.

(6) Whether Material Information Was Withheld From the Attorney Who Gave the Opinion

Withholding information from an attorney who has been retained to provide an opinion on the validity, infringement and/or enforceability of a patent can be relevant to a willful or induced infringement inquiry in two ways. First, withheld or false information may affect the reliability of an attorney's advice to the extent that the advice is premised on false or misleading information, thereby negating any argument that reliance on the advice was reasonable. *Chiron Corp.*, 268 F. Supp. 2d at 1123. In other words, it would be absurd for an alleged infringer to argue that it was objectively reasonable in relying on an attorney's opinion when that alleged infringer knew that the opinion was based on false and/or incomplete information. Second, withholding material information is evidence of an infringer's bad faith, from which it can be inferred that the infringer did not intend to rely, and did not in fact rely, on the opinion that was rendered. *Id.*; *see also Comark Commc'ns, Inc.*, 156 F.3d at 1191 ("Whenever material information is intentionally withheld, or the best information is intentionally not made available to counsel during the preparation of the opinion, the opinion can no longer serve its prophylactic purpose of negating a finding of willful infringement.").

Although it may seem unlikely that a client would withhold information from its attorney, there are several scenarios in which this can occur.[w] One common situation in which many clients are tempted to withhold information from their attorneys occurs when a client receives conflicting test results. Indeed, clients often have to conduct certain testing (e.g., obtaining IR, NMR, or X-ray diffraction spectra) to determine whether a proposed new product would infringe a patent claim. Such tests do not always yield unequivocal results. In these instances, clients are often tempted

---

[w]Note that clients are most likely to encounter issues regarding withholding material information in the context of non-infringement opinions because patent attorneys typically rely on publicly available information, as opposed to information from their clients, when providing invalidity and unenforceability opinions. Patent attorneys must rely heavily on their clients to provide information about the new technology they intend to practice when preparing non-infringement opinions, however, because the attorney must compare the client's proposed new technology to what is claimed in any patents that might read on the technology.

to disclose to their attorney those test results that demonstrate that their new technology does not meet the limitations of the patent claims at-issue and withhold those results that show infringement of the claims. This is a bad strategy, however, because if the client is later sued for infringing the patent at-issue, the patentee can obtain samples of the client's new product and test that product itself. The patentee can also obtain all of the client's test results through discovery. If any of the client's test results demonstrate infringement, the client could then be liable for willful infringement. *See Abbott Laboratories v. Sandoz, Inc.*, 566 F.3d 1282, 1299 (Fed. Cir. 2009) (explaining that *de minimis* infringement can still be infringement under 35 U.S.C. § 271(a)).

Another common scenario in which clients are apt to inadvertently withhold information from their attorneys is when they are involved in multiple, parallel proceedings. For example, a client might challenge the validity of a patent in an inter partes review before the PTAB and in a district court litigation. When this happens, there is a great risk that the client will withhold information from one or more of its attorneys because the client often engages different attorneys to work on the different proceedings, and the proceedings do not usually advance at the same rate. This can be disastrous where, for example, a non-infringement or invalidity opinion is based on a proposed claim construction that is later rejected by either the PTAB or the district court proceeding and the opinion is never revised to reflect the correct claim construction.

Besides the six factors discussed previously, courts often look at the timing of when an opinion was rendered when assessing whether an infringer engaged in objectively reckless behavior or had specific intent in inducing infringement. With regard to timing, the best practice is to seek an opinion of counsel <u>before</u> commencing any activities that may infringe another's patent. Indeed, if an infringer commences its infringing acts before seeking an opinion, the delay in seeking the opinion will undercut any arguments that he/she relied on a reasonable defense to infringement in acting. *See Aspex Eyewear Inc. v. Clariti Eyewear, Inc.*, 605 F.3d 1305, 1313 (Fed. Cir. 2010) (noting that timely consultation with counsel may be evidence that an infringer did not engage in objectively reckless behavior); *Adidas Am., Inc. v. Payless Shoesource, Inc.*, 546 F. Supp. 2d 1029, 1049 (D. Oregon 2008) (finding that an alleged infringer's failure to obtain advice of counsel before engaging in its allegedly infringing conduct undermined its contention that it acted in good faith reliance on the advice of its counsel and raised the inference that the opinions were designed simply to bolster its advice of counsel defense); *Chiron Corp.*, 268 F. Supp. 2d at 1123 n.4 (quoting O'Malley, et al., *Federal Jury Practice & Instructions* § 19.08, at 885 (5th ed. 2000)) (noting that like the advice of counsel defense in the criminal law context, the advice of counsel defense is available in the infringement arena "if, *'before* [acting or failing to act], [the d]efendant, while acting in *good faith and for the purpose of securing advice on the lawfulness of [his] possible future conduct,* sought and obtained the advice of an attorney whom [he] considered to be competent, and made a full and accurate report or disclosure to this attorney of *all* important and material facts of which [he] had knowledge or had the means of knowing, and then acted strictly in accordance with the advice [his] attorney gave following this full report or disclosure, then [the d]efendant would not be willfully or deliberately doing wrong in [performing or omitting] some act the law [forbids or requires.]'") (emphasis and alterations in original).

Those seeking opinions of counsel should not only be cognizant of the aforementioned factors but should also check any opinion obtained from counsel against these factors to ensure that the opinion will stand up to a competency challenge in court if it is relied upon to defeat a claim of willful and/or induced infringement.

IV) AIA § 19(d): JOINDER OF PARTIES IN PATENT INFRINGEMENT SUITS

Plaintiffs who wish to sue multiple defendants in a single patent infringement suit must comply with Fed. R. Civ. P. 20(a)(2), which sets forth a two-part test for when joinder of defendants is proper in civil actions. Fed. R. Civ. P. 20(a)(2) specifically states that multiple persons may be joined as defendants in one action if:

(A) any right to relief is asserted against them jointly, severally, or in the alternative with respect to or arising out of the same transaction, occurrence, or series of transactions or occurrences; and

(B) any question of law or fact common to all defendants will arise in the action.

The reason that the federal civil procedure provides joinder rules, such as Fed. R. Civ. P. 20(a)(2), is to draw all factually-related claims and parties into a single lawsuit to promote convenience and efficiency to both the court and the parties. 6A Charles Alan Wright et al., FEDERAL PRACTICE AND PROCEDURE § 1652 (3d ed. 1998); *see also United Mine Workers of Am. v. Gibbs*, 383 U.S. 715, 724 (1966) ("Under the Rules, the impulse is toward entertaining the broadest possible scope of action consistent with fairness to the parties; joinder of claims, parties and remedies is strongly encouraged.").

As explained in Section IV.A., federal courts were split on how to apply Fed. R. Civ. P. 20(a)(2) in patent infringement suits brought by patent trolls prior to the AIA. AIA § 19(d) was therefore enacted to clarify the law and bring uniformity to joinder practice in patent suits. AIA 19(d) is discussed in detail in Section IV.B.–E.

A. Joinder of Defendants in Patent Infringement Suits Prior to the AIA

One of the major problems confronting Congress during the time the AIA was being debated was the ever-rising number of patent infringement suits that were being filed by patent trolls. *See* Ahmed J. Davis & Karolina Jesien, *The Balance of Power in Patent Law: Moving Towards Effectiveness in Addressing Patent Troll Concerns*, 22 FORDHAM INTELL. PROP. MEDIA & ENT. L.J. 836 n.2 (2012) (explaining that non-practicing entities brought approximately one hundred lawsuits targeting five hundred operating companies in 2001, while in 2010, the numbers increased to more than five hundred lawsuits targeting over 2300 operating companies). One of the tactics patent trolls liked to use was to file one patent infringement suit against multiple, unrelated defendants. Some patent trolls have even sued over 50 unrelated defendants in one action. *See* Erick Robinson, *New Patent Reform Law Could Reduce Lawsuits by Non-Practicing Entities* (2011), https://opensource.com/law/11/9/new-patent-reform-law-could-reduce-lawsuits-non-practicing-entity (explaining that different courts have had differing tolerances for the joinder tactics employed by patent trolls, but many have allowed cases with 50 or more defendants to be joined). Suing multiple defendants in one action allows patent trolls to simultaneously minimize their litigation costs (e.g., court filing fees and discovery costs) and maximize their chances of obtaining a settlement either by forcing competitors and other unfriendly parties to cooperate with one another to defend the case or by forcing companies to spend even more to defend against the suit on their own.[x] *Id.* One of the only procedural hurdles patent trolls faced in exploiting this tactic was demonstrating that joinder of the defendants named in their complaints was

---

[x]Cooperating with competitors is never ideal for any company because it often forces the company to exchange confidential documents with their competitors and to also try to develop unified strategies with companies that have competing interests.

proper under Fed. R. Civ. P. 20(a)(2).[y] While some jurisdictions permitted patent trolls to join unrelated defendants who were alleged to have infringed a patent based on separate acts in one suit, other jurisdictions found that the two-part test set forth in Fed. R. Civ. P. 20(a)(2) could not be met by defendants who had such an attenuated relationship.

Of all of the jurisdictions in the U.S., the federal district courts in Texas and Louisiana were most receptive to the tactic of suing multiple unrelated defendants in one patent infringement suit. Indeed, these district courts took a very relaxed approach to the two-part test for joinder set forth in Fed. R. Civ. P. 20(a)(2). Under their pre-AIA interpretation of Fed. R. Civ. P. 20(a)(2), joinder of two or more defendants in a single patent suit was proper, regardless of whether the defendants were related, so long as the defendants were being sued for infringing the same patent or patents and there was no evidence that one defendant's allegedly infringing method(s) and/or product(s) was dramatically different from another defendant's allegedly infringing method(s) and/or product(s). *See, e.g., Alford Safety Services, Inc. v. Hot-Hed, Inc.*, No. 10-1319, 2010 U.S. Dist. LEXIS 98152, at *26-27 (E.D. La. Aug. 23, 2010) (finding that joinder of two unrelated defendants was proper because even though the designs of their accused infringing products might have differed, the underlying issue of infringement of the same patents was the same); *Mannatech, Inc. v. Country Life, LLC*, No. 3:10-CV-533-O, 2010 U.S. Dist. LEXIS 75353, at *4-5 (N.D. Tex. July 26, 2010) (holding that joinder of unrelated defendants was proper because the fact that all of the defendants' dietary supplement products were alleged to infringe the same patent was sufficient to meet the "same transaction or occurrence" prong of the joinder analysis and the validity and scope of the asserted patent was a question of law that was common to all defendants in the case); *Adrain v. Genetec Inc.*, No. 2:08-CV-423, 2009 U.S. Dist. LEXIS

---

[y]One of the other procedural hurdles patent trolls had to deal with was showing that venue was proper for all of the defendants. For example, if a patent troll brought suit against a dozen defendants in the Eastern District of Texas, and one or more of those defendants were not based in Texas, the defendants could file a motion to transfer venue to try to have the case heard in a different district court. *See* 28 U.S.C. § 1404(a) ("For the convenience of parties and witnesses, in the interest of justice, a district court may transfer any civil action to any other district or division where it might have been brought."). To determine whether a case should be transferred to another district court for the sake of convenience, courts evaluate several factors, including: (1) the plaintiff's chosen forum, (2) the defendant's preference, (3) where the claim arose, (4) the convenience of the parties, (5) the convenience of the witnesses, and (6) the location of books and records. *Teleconference Sys. v. Proctor & Gamble Pharms., Inc.*, 676 F. Supp. 2d 321, 329 (D. Del. 2009).

86855, at *6-7 (E.D. Tex. Sept. 22, 2009) (finding that joinder of unrelated defendants was proper because the defendants sold similar license plate recognition systems that allegedly infringed the same claims of the asserted patent, thereby satisfying the "same transaction or occurrence test" and the "common question of law or fact" prong was satisfied due to the fact that the validity and scope of the asserted patent was a question common to all of the defendants in this case); *MyMail, Ltd. v. America Online, Inc.*, 223 F.R.D. 455, 456-57 (E.D. Tex. 2004) (denying defendants' motion to correct misjoinder under Fed. R. Civ. P. 20(a), where all of the defendants were alleged to have infringe the same patent and the record before the Court did not show any dramatic differences between the allegedly infringing products and methods such that determining infringement in one case would be less proper or efficient than determining infringement in multiple cases).

With regard to the first prong of the two-part test under Fed. R. Civ. P. 20(a)(2), the district courts in Texas followed the Eighth Circuit's "logical relationship test" to determine whether claims "arose from the same transaction or occurrence." *Adrain*, 2009 U.S. Dist. LEXIS 86855, at *6-7 (citing *Mosley v. General Motors Corp.*, 497 F.2d 1330, 1332-33 (8th Cir. 1974)); *see also Mannetech, Inc.*, 2010 U.S. Dist. LEXIS 75353, at *4 (citing *Mosley*, 497 F.2d at 1332-33) (explaining that under the logical relationship test, "all logically related events entitling a person to institute a legal action against another generally are regarded as comprising a transaction or occurrence"). According to this test, a "logical relationship" between alleged acts of infringement existed if there was "some nucleus of operative facts or law." *Adrain*, 2009 U.S. Dist. LEXIS 86855, at *6. Moreover, so long as there was no evidence that each of the defendants' allegedly infringing methods or products were dramatically different from one another, the courts held that there was a nucleus of fact or law such that the same transaction or occurrence test of Fed. R. Civ. P. 20(a)(2) was satisfied. *See Mannatech, Inc.*, 2010 U.S. Dist. LEXIS 75353, at *4-5 (finding that the defendants' alleged infringing acts arose out of the same transaction or occurrence where all of their allegedly infringing products were dietary supplement products that allegedly embodied the invention disclosed and claimed in the patent at-issue); *Adrain*, 2009 U.S. Dist. LEXIS 86855, at *7 (explaining that the similarity of defendants' accused license plate recognition systems was sufficient to satisfy the nucleus of fact or law test for purposes of joinder); *MyMail Ltd.*, 223 F.R.D. at 457 (finding that there was a nucleus of facts or law that satisfied the same transaction or occurrence test where the record did not show that the defendants' methods or products were dramatically different from one another).

Regarding the second prong of the two-part test under Fed. R. Civ. P. 20(a)(2), the Texas courts held that there were common issues of law and fact that were sufficient to satisfy the requirements for joinder where the same patent or patents were being asserted against all of the defendants. *See Mannatech*, 2010 U.S. Dist. LEXIS 75353, at *5 (finding that the second requirement of Fed. R. Civ. P. 20(a)(2) had been satisfied because the validity and scope of the '807 patent was a question of law common to all defendants in the case); *Adrain*, 2009 U.S. Dist. LEXIS 86855, at *7-8 (holding that the second prong of the two part-test under Fed. R. Civ. P. 20(a) (2) had been satisfied because the validity and scope of the '669 patent was a question common to all of the defendants in the case).

Most of the jurisdictions outside of Texas and Louisiana took a much stricter approach to the "same transaction or occurrence" prong of the two-part test of Fed. R. Civ. P. 20(a)(2) and did not allow multiple defendants to be joined just because the same patent was asserted against them or because they made or used similar products or processes. *See, e.g., Rudd v. Lux Prods. Corp.*, No. 09-CV-6957, 2011 U.S. Dist. LEXIS 4804, at *7 (N.D. Ill. Jan. 12, 2011) ("After researching the issue, the Court determines that [the Eastern District of Texas's] approach [to Rule 20] is in the minority."); *WiAV Networks, LLC v. 3COM Corp. et al.*, No. C 10-03448 WHA, 2010 U.S. Dist. LEXIS 110957, at *19 (C.D. Cal. Oct. 1, 2010) (quoting *Spread Spectrum Screening, LLC v. Eastman Kodak Co.*, 2010 U.S. Dist. LEXIS 90549, 2010 WL 3516106, at *2 (N.D. Ill. Sept. 1, 2010)) ("[N]umerous courts have found that 'joinder is often improper where [multiple] competing businesses have allegedly infringed the same patent by selling different products.'" (second alteration in original)); *see also Golden Scorpio Corp. v. Steel Horse Bar & Grill*, 596 F. Supp. 2d 1282, 1285 (D. Ariz. 2009) (holding that defendants who independently infringed the same trademark are not part of the same transaction or occurrence); *Arista Records LLC v. Does 1-4*, 589 F. Supp. 2d 151, 154-55 (D. Conn. 2008) (holding that defendants who independently infringed the same copyright are not part of the same transaction or occurrence). As a result, the district courts in Texas became a very popular venue for patent trolls that were seeking to sue more than one defendant for patent infringement.

B. AIA § 19(d)'s Changes to Joinder of Defendants in Patent Infringement Suits

AIA § 19(d) adds a new statute, 35 U.S.C. § 299, to the patent laws. Leahy-Smith America Invents Act of 2011, Pub. L. No. 112-29, 125 Stat. 284, 332-33; *see also* Leahy-Smith America Invents Technical Corrections Act of 2013, Pub. L. No. 112-274, § 1(c), 126 Stat. 2456 (replacing "or counterclaim defendants only if" with "only if" in the version of post-AIA 35 U.S.C. § 299(a) that was passed in the Leahy-Smith America Invents Act of 2011). New 35 U.S.C. § 299 reads as follows:

> (a) JOINDER OF ACCUSED INFRINGERS.-- With respect to any civil action arising under any Act of Congress relating to patents, other than an action or trial in which an act of infringement under section 271(e)(2) has been pled, parties that are accused infringers may be joined in one action as defendants or counterclaim defendants, or have their actions consolidated for trial, only if--
> (1) any right to relief is asserted against the parties jointly, severally, or in the alternative with respect to or arising out of the same transaction, occurrence, or series of transactions or occurrences relating to the making, using, importing into the United States, offering for sale, or selling of the same accused product or process; and
> (2) questions of fact common to all defendants or counterclaim defendants will arise in the action.
> (b) ALLEGATIONS INSUFFICIENT FOR JOINDER.-- For purposes of this subsection, accused infringers may not be joined in one action as defendants or counterclaim defendants, or have their actions consolidated for trial, based solely on allegations that they each have infringed the patent or patents in suit.
> (c) WAIVER.-- A party that is an accused infringer may waive the limitations set forth in this section with respect to that party.

New 35 U.S.C. § 299(a) should look familiar because that section essentially repeats the two-part test for joinder that is set forth in Fed. R. Civ. P. 20(a)(2). The only differences between 35 U.S.C. § 299(a) and Fed. R. Civ. P. 20(a)(2) are that: (1) 35 U.S.C. § 299(a) adds the language "relating to the making, using, importing into the United States, offering for sale, or selling of the same accused product or process" and (2) 35 U.S.C. § 299(a) applies to both joinder of defendants and counterclaim defendants whereas Fed. R. Civ. P. 20(a)(2) applies only to the joinder of defendants.[z] The added language in

---

[z] New 35 U.S.C. § 299(a) brings more significant changes to consolidation of patent suits for trial under Fed. R. Civ. P. 42 than it does to joinder of defendants under Fed. R. Civ. P. 20(a)(2). Indeed, Fed. R. Civ. P. 42(a) provides that "[i]f actions before the court involve a common question of law or fact, the court may . . . consolidate the actions." 35 U.S.C. § 299(a) replaces this simple test, however, with the two-part analysis that is required for joinder under Fed. R. Civ. P. 20(a)(2). Thus, 35 U.S.C. § 299(a) makes it considerably more difficult for litigants to consolidate multiple patent suits for trial because they now have to show that the actions they wish to consolidate involve a common question of fact and arise out of the same transaction, occurrence, or series of transactions or occurrences.

35 U.S.C. § 299(a) simply makes clear that it is the alleged infringing acts, and not the patent that is being infringed, that must meet the "same transaction or occurrence" test for joinder.

The most significant changes made to joinder in patent suits are seen in section (b) of 35 U.S.C. § 299. That section explicitly abrogates the pre-AIA case law of the Texas and Louisiana district courts discussed in Section IV.A. by stating that "accused infringers may not be joined in one action as defendants or counterclaim defendants . . . based solely on allegations that they have each infringed the patent or patents in suit." Thus, the fact that all defendants are sued for infringement of the same patent or patents is no longer a sufficient reason to join those defendants in one suit. 35 U.S.C. § 299(b) also states that "accused infringers may not . . . have their actions consolidated[a] for trial[] based solely on allegations that they each have infringed the patent or patents in suit." Notably, new 35 U.S.C. § 299(b) does not limit a plaintiff's ability to request consolidation for other stages of litigation (e.g., discovery or claim construction).

Importantly, new 35 U.S.C. § 299 does not impact patent infringement suits brought under the Hatch-Waxman Act or suits that are brought before the International Trade Commission. Indeed, 35 U.S.C. § 299(a) explicitly states that the statute applies "to any civil action *arising under any Act of Congress relating to patents, other than an action or trial in which an act of infringement under section 271(e)(2)* has been pled." (emphasis added). Section 271(e)(2) is the section of the patent infringement statute that provides the basis for infringement claims under the Hatch-Waxman Act. Additionally, proceedings before the International Trade Commission are not "civil actions arising under any Act of Congress relating to patents." *See* Thomas Martin, *ITC is a More Desirable Venue in the Wake of AIA*, Law360, Feb. 11, 2013 ("Section 299 applies only to 'civil actions,' which are heard by Article III, or judicial, tribunals. By contrast, the ITC is an Article I, or legislative, tribunal that hears only those actions enumerated by Congress in the Tariff Act.") (citations omitted). Accordingly, these proceedings are not subject to 35 U.S.C. § 299.

---

[a]Fed. R. Civ. P. 42 permits parties to consolidate multiple suits for various reasons. Notably, suits can be consolidated for certain stages of litigation but not for others. For example, suits might be consolidated for a hearing (e.g., a Markman hearing) but not for trial. Fed. R. Civ. P. 42(a). Furthermore, the fact that a case has been severed due to improper joinder under Fed. R. Civ. P. 20(a)(2) does not prohibit a court from consolidating two or more of the suits resulting from the severance. Charles Alan Wright, Arthur R. Miller & Edward H. Cooper, Federal Practice and Procedure § 2382 (3d ed. 2011).

C. Legislative Goals Addressed by AIA § 19(d)

35 U.S.C. § 299 was enacted to overturn the interpretation of Fed. R. Civ. P. 20(a)(2) that was advanced by the district courts in Texas and Louisiana and "effectively conform[] these courts' jurisprudence to that followed by a majority of jurisdictions." *See* H.R. REP. NO. 112-98, at 55 n.61 (2011) ("Section 299 legislatively abrogates the construction of Rule 20(a) adopted in *MyMail, Ltd. v. America Online, Inc.,* 223 F.R.D. 455 (E.D. Tex. 2004); *Sprint Communications Co. v. Theglobe.com, Inc.,* 233 F.R.D. 615 (D. Kan. 2006); *Adrain v. Genetec Inc.,* 2009 WL3063414 (E.D. Tex. September 22, 2009); *Better Educ. Inc. v. Einstruction Corp.,* 2010 WL 918307 (E.D. Tex. March 10, 2010); *Mannatech, Inc. v. Country Life,* LLC, 2010 WL 2944574 (N.D. Tex. July 26, 2010); *Alford Safety Services, Inc., v. Hot-Hed, Inc.,* 2010 WL 3418233 (E.D. La. August 24, 2010); and *Eolas Technologies, Inc. v. Adobe Systems, Inc.,* 2010 WL 3835762 (E.D. Tex. Sep[]tember 28, 2010)—effectively conforming these courts' jurisprudence to that followed by a majority of jurisdictions.").

The House Committee Report on new 35 U.S.C. § 299 cites *Rudd v. Lux Products Corp.,* 2011 U.S. Dist. LEXIS 4804 (N.D. Ill. Jan. 12, 2011) as a case that represents the majority view on interpreting Fed. R. Civ. P. 20(a)(2). In that case, the district court explained that "a party fails to satisfy *Rule 20(a)'s* requirement of a common transaction or occurrence where unrelated defendants, based on different acts, are alleged to have infringed the same patent… Moreover, allegations that unrelated defendants design, manufacture, and sell similar products do not satisfy *Rule 20(a)'s* requirement." 2011 U.S. Dist. LEXIS 4804, at *8-9 (N.D. Ill. Jan. 12, 2011) (citations omitted) (emphasis in original).

By rejecting the lenient interpretation of Fed. R. Civ. P. 20(a)(2) that was advanced by the Texas and Louisiana district courts, Congress' clear intent was to end patent trolls' practice of suing multiple unrelated defendants in a single patent suit. Joe Matal, *A Guide to the Legislative History of the America Invents Act: Part II of II,* 21 FED. CIR. B.J. 539, 592 (2012); *see also* H.R. REP. NO. 112-98, at 54 (2011) ("The [AIA] also addresses problems occasioned by the joinder of defendants (sometimes numbering in the dozens) who have tenuous connections to the underlying disputes in patent infringement suits."). Moreover, in an effort to close any loopholes that might have been exploited by patent trolls, Congress expanded the restriction on joinder to also apply to consolidation of cases at trial. 35 U.S.C. § 299(b). Senator Kyl explained that:

> If a court that was barred from joining defendants in one action could instead simply consolidate their cases for trial under [Fed. R. Civ. P.] 42, section 299's purpose of allowing unrelated patent defendants to insist on being tried separately would be undermined.
> **157 CONG. REC. S5429, 2011.**

D. Effective Date of AIA § 19(d)

AIA § 19(e) provides that the amendments made by section 19 "shall apply to any civil action commenced on or after the date of the enactment of this Act." Leahy-Smith America Invents Act of 2011, Pub. L. No. 112-29, 125 Stat. 284, 333. The AIA was enacted on September 16, 2011. *Id.* at 284. Thus, 35 U.S.C. § 299's new standard for joinder will apply to any patent suit that is filed on or after September 16, 2011.

E. Practical Effects of AIA § 19(d)

While Congress' enactment of 35 U.S.C. § 299 has had an impact on patent trolls' litigation tactics, the statute has done little to curb the total amount of litigation brought by such entities. Indeed, since the AIA was passed, the number of multi-defendant suits filed by patent trolls has decreased but the number of single-defendant cases has increased. *See* Macedo et al., *AIA's Impact On Multidefendant Patent Litigation: Part 2* (2012), http://www.law360.com/ip/articles/387458/aia-s-impact-on-multidefendant-patent-litigation-part-2 (explaining that immediately after the AIA was passed, many non-practicing entities "began to conform their practices: [i]nstead of instituting one massive multidefendant infringement action, they would institute a multitude of separate but nearly identical patent infringement complaints against unrelated entities in the same court."); Maya M. Eckstein et al., *The (Unintended) Consequences of the AIA Joinder Provision*, AIPLA Spring Meeting (2012) (explaining that "Chipworks, which tracks U.S. patent litigations on a quarterly basis, recently noted that patent infringement filings reached on all time high of 1100 patent cases in the fourth quarter of 2011" and that the reason for this increase was that "[n]on-practicing entities (NPEs), who had routinely filed large multi-defendant cases, took to filing multiple parallel cases against single defendants in the months following the passage of the AIA.") (citation omitted).

Moreover, some patent trolls have turned to consolidation and/or Multidistrict Litigation ("MDL")[bb] tactics to reap some of the same benefits they once derived from bringing multi-defendant suits. *See* Dongbiao Shen,

---

[bb]Multidistrict litigation ("MDL") allows parties to request that cases pending in multiple districts be transferred to a single district court so that they can be consolidated for pretrial purposes. A Judicial Panel that consists of seven circuit and/or district judges appointed by the Chief Justice of the United States meets once every two months to determine whether transfer to MDL is proper. 28 U.S.C. § 1407(d).

Misjoinder or Mishap: The Consequences of the AIA Joinder Provision, 29 BERKE-
LEY TECH. L.J. 546 (2014) ("Additionally, both patent holders and accused
infringers are employing multidistrict litigation ('MDL') to consolidate
multiple single-defendant suits pending in different districts."); *Multi-Defendant Joinder Under the America Invents Act: Much Ado About Nothing?* (2012),
http://www.quinnemanuel.com/the-firm/news-events/article-december-
2012-multi-defendant-joinder-under-the-america-invents-act-much-ado-
about-nothing/ (citing *SoftView LLC v. Apple Inc.*, 2012 WL 3061027, at
*11 (D. Del. Jul. 26, 2012); *C.R. Bard, Inc. v. Med. Components, Inc.*, 2012 WL
3060105, at *1-2 (D. Utah Jul. 25, 2012); *Rotatable Tech. LLC v. Nokia Inc.*,
No. 2:12-cv-265 (E.D. Tex. filed May 1, 2012), ECF No. 60; and *Roy-G-Biv
Corp. v. Abb Ltd.*, No. 6:11-cv-00622 (E.D. Tex. filed Nov. 15, 2011), ECF
No. 51 as examples of cases in which suits against multiple unrelated defendants have been consolidated); Maya M. Eckstein et al., *The (Unintended)
Consequences of the AIA Joinder Provision*, AIPLA Spring Meeting (2012)
("Consolidation could relieve plaintiffs of many of the burdens exacted
by the AIA, essentially allowing plaintiffs to prosecute multiple matters
together, to allow for a single scheduling and protective order, to present
witnesses for deposition only once, rather than in each separately-filed case,
and to have consolidated hearings on matters including claim construction
and summary judgment.").

35 U.S.C. § 299 does not restrict consolidation of patent suits under
Fed. R. Civ. P. 42 to the same degree that it restricts joinder of multiple
defendants in such suits. In fact, 35 U.S.C. § 299 restricts joinder during
all stages of litigation but only restricts consolidation during trial. *See* 35
U.S.C. § 299(b) ("For purposes of this subsection, accused infringers may
not be joined in one action as defendants or counterclaim defendants, or
have their actions consolidated for trial, based solely on allegations that
they each have infringed the patent or patents in suit.") (emphasis added).
Additionally, courts are generally eager to consolidate cases because doing
so relieves them of many burdens that are created by adjudicating several parallel matters involving the same patents and similar technology. *See*
Maya M. Eckstein et al., *The (Unintended) Consequences of the AIA Joinder
Provision*, AIPLA Spring Meeting (2012) ("Faced with 5, 10, 20 or more
cases filed by the same plaintiff alleging infringement of the same patents,
judges are likely to seek efficiencies for themselves, their clerks, and other
court personnel."). Thus, litigants have been permitted to consolidate patent suits for pre-trial purposes (e.g., claim construction and discovery) even
after 35 U.S.C. § 299 went into effect.

Furthermore, 35 U.S.C. § 299 has no impact on Multidistrict Litigation. Thus, patent trolls are free to utilize this tactic as a means to consolidate several suits for pretrial purposes. Indeed, 28 U.S.C. § 1407 permits "consolidated pretrial proceedings" when "civil actions involving one or more common questions of fact are pending in different districts." The statute aims to increase the "convenience of parties and witnesses" and "promote the just and efficient conduct of . . . actions" by eliminating inconsistent rulings and reducing the costs associated with litigating multiple similar suits in different districts.

Because patent trolls have found various ways to work around the barriers to joinder that were promulgated in new 35 U.S.C. § 299, it is unlikely that the new legislation will accomplish its intended goals of significantly reducing the number of suits brought by such entities.

CHAPTER SEVEN

AIA's Modifications to Patent Infringement Defenses

Contents

I) INTRODUCTION

The prior user rights and best mode defenses were seldom used prior to the AIA. An expanded prior user rights defense, however, was viewed as an essential component of a first–inventor–to–file system. As such, the AIA amended the statute governing the defense (i.e., 35 U.S.C. § 273) so that more patent infringement defendants could use the defense against a broader range of patent claims. Notwithstanding these amendments, some of the changes to 35 U.S.C. § 273 actually make it more difficult for defendants to establish that they have a prior user right to a claimed invention. The added hurdles to obtaining prior user rights in post–AIA 35 U.S.C. § 273 will likely prevent many alleged infringers from relying on the defense in the future.

In contrast to Congress' desire to expand the prior user rights defense, many legislators wanted to completely repeal the best mode requirement from 35 U.S.C. § 112, ¶ 1, so that it could neither serve as a basis for invalidating a

America Invents Act Primer. http://dx.doi.org/10.1016/B978-0-12-812096-5.00007-7

patent nor be a requirement for patentability. While Congress did not go so far as to repeal best mode from the patent statute, the AIA provides several statutory amendments that completely eliminate the possibility of invalidating a patent or rendering it unenforceable, either in a district court litigation or in a proceeding before the Patent Trial and Appeal Board, based on a failure to disclose the best mode. Given that best mode is still a requirement for patentability under post-AIA 35 U.S.C. § 112(a), and that failure to disclose the best mode could, in many instances, affect the scope of a patent and impact its validity on grounds other than best mode (e.g., enablement, novelty, and/or obviousness), prudent patent applicants should continue to include the best mode of carrying out their invention in their patent applications.

II) AIA § 5: DEFENSE TO INFRINGEMENT BASED ON PRIOR COMMERCIAL USE, i.e., THE "PRIOR USER RIGHTS DEFENSE"

A. Background on the Prior User Rights Defense

The prior user rights defense is a limited defense[a] to patent infringement that is based on principles of equity. *See, e.g.*, Keith M. Kupferschmid, *Prior User Rights: The Inventor's Lottery Ticket*, 21 AIPLA Q.J. 213, 216 (1993). Prior user rights only arise in certain limited circumstances. *Id.* In particular, the rights arise when:

1) one party (the patentee) obtains a patent on his/her invention;
2) another party (the prior user[b]) had commercially used the invention that the patentee patented at some point in time before the filing date or priority date of the patent mentioned in #1;
3) the use by the prior user in #2 does not constitute patent invalidating prior art[c];

[a]The prior user rights defense is limited because: (1) it is not available to everyone who is accused of infringing a patent, (2) it is a personal right that cannot be transferred or sold separately from the underlying business to which the rights pertain, and (3) restrictions exist for how and where the prior user may develop and/or use the patented invention for which the prior user acquired prior user rights. Kyla Harriel, *Prior User Rights in a First-to-Invent Patent System: Why Not?*, IDEA, 543, 546, 552 (1996); Gary L. Griswold & F. Andrew Ubel, *Prior User Rights—A Necessary Part of a First-to-File System*, 26 J. MARSHALL L. REV. 567, 571 (1993).

[b]The term "prior user" is synonymous with the term "prior commercial user." Similarly, the term "prior user right" is synonymous with the term "prior commercial user right."

[c]Because prior uses that cause an invention to be available to the public typically constitute prior art under pre- or post-AIA 35 U.S.C. §§ 102 and 103, prior user rights are typically associated with a party who uses the invention in secret. Keith M. Kupferschmid, *Prior User Rights: The Inventor's Lottery Ticket*, 21 AIPLA Q.J. 213, 216 n.4 (1993).

4) the prior user continues to use the invention after the patent mentioned in #1 issues; and

5) the patentee sues the prior user for patent infringement based on the prior user's continued use of the patented invention in #4.

Id. If a prior user successfully establishes to a court that he/she had a vested prior user right in the patented invention, the prior user may continue to exploit the invention in a manner that normally would constitute infringement without being liable to the patentee for such use. *Id.* In other words, prior user rights maintain the status quo that existed between the patentee and the prior user before the filing date or priority date of the patent so that the prior user can continue using and/or developing the technology that he/she was already commercially using before the filing date or priority date of the patent at-issue. Kyla Harriel, *Prior User Rights in a First-to-Invent Patent System: Why Not?*, IDEA, 543, 546 (1996).

Prior user rights issues typically arise with regard to process or method inventions.[d] As compared to patents covering products or compositions, process patents are extremely difficult to enforce[e] because businesses do not typically disclose the processes or methods they use to manufacture their products and it is often very difficult, if not impossible, to determine

---

[d]*See, e.g.*, 157 CONG. REC. S5430, 2011 (statement of Sen. Kyl) ("Generally, products that are sold to consumers will not need a [prior user rights] defense over the long term. As soon as the product is sold to the public, any invention that is embodied or otherwise inherent in that product becomes prior art and cannot be patented by another party, or even by the maker of the product after the grace period has expired."); *The Patent System Harmonization Act of 1992: J. Hearing on S. 2605 and H.R. 4978 Before the Subcomm. on Patents, Copyrights, and Trademarks of the S. Comm. on the Judiciary and the Subcomm. on Intellectual Prop. and Judicial Admin. of the H. Comm. on the Judiciary*, 102d Cong. 189, 196 (1992) (statement of Robert A. Armitage, on behalf of NAM) ("Prior user rights have effect only for inventions that have been commercialized in secret and almost always arise in connection with trade-secret manufacturing processes.").

[e]One of the reasons process patents are difficult to enforce is that Federal Rule of Civil Procedure 11 requires a patentee to conduct an adequate pre-filing investigation before filing a patent infringement suit. Namely, a patentee must compare the accused device or process to the patent claims. *See, e.g., Judin v. U.S.*, 110 F.3d 780, 784-85 (Fed. Cir. 1997) (imposing sanctions under Fed. R. Civ. P. 11 against a patentee who made no reasonable effort to ascertain whether the accused device satisfied the two key claim limitations prior to filing the infringement suit and offered no adequate explanation for failing to obtain, or attempting to obtain, a sample of the accused device so that its actual design and functioning could be compared with the claims of the patent). Conducting such pre-filing investigations is often more difficult for processes, which are typically practiced in secret, than for products, which are often readily available to the public.

how a product has been manufactured merely by looking at it. *See, e.g.,*
The Patent Prior User Rights Act and the Patent Reexamination Reform Act:
Hearing on S. 2272 and S. 2341 Before the Subcomm. on Patents, Copyrights,
and Trademarks of the S. Comm. on the Judiciary, 103d Cong. 23, 24 (1994)
(statement of Roger S. Smith, President, IPO) ("Many important technological achievements—notably processes—can only be effectively exploited through secret use. Processes are naturally practiced away from the
public's view in most cases. Patents covering them consequently are very
difficult to enforce, so process patents often do not provide meaningful
protection."). As a result, owners of process patents often have difficulty
determining whether others (e.g., their competitors) are infringing their
patents. Moreover, because the written description and enablement requirements of 35 U.S.C. § 112 require patent applicants to disclose their
inventions in sufficient detail to allow others to make and use their inventions, inventors of processes or methods are often reluctant to seek patent
protection because doing so could enable their competitors to capitalize on their inventions with little chance of being sued. Representative
Lamar Smith summarized well the patenting challenges manufacturers
face, explaining:

> For many manufacturers, the patent system presents a catch-22. If they patent a
> process, they disclose it to the world and foreign manufacturers will learn of it and,
> in many cases, use it in secret without paying licensing fees. The patents issued on
> manufacturing processes are very difficult to police, and oftentimes patenting the
> idea simply means giving the invention away to foreign competitors. On the other
> hand, if the U.S. manufacturer doesn't patent the process, then under the current
> system a later party can get a patent and force the manufacturer to stop using a
> process that they independently invented and used.
>
> **157 CONG. REC. H4483, 2011.**

The risks and challenges involved with process and method patents
cause many businesses to choose trade secret protection over patent protection for their innovative processes and methods. By protecting processes
and methods as trade secrets, businesses are not forced to disclose them to
the public, which greatly reduces the likelihood that their competitors will
capitalize on their inventions without paying for licenses. While this strategy
is often the best for protecting process and method inventions, problems can
arise when another person independently invents a process or method that
is subject to trade secret protection and that person patents the process or
method. In such instances, the business that chose to protect its invention as
a trade secret risks facing liability for patent infringement if it continues to

utilize the patented process or method.[f] The prior user rights defense aims to ameliorate this problem by allowing the business that opted to protect its innovative method or process as a trade secret to continue using the method or process without patent infringement liability.

B. Recent History of Prior User Rights in the U.S.

Pre-AIA 35 U.S.C. § 273, which is the statute that provides the prior user rights defense, was initially enacted as part of the 1999 American Inventors Protection Act (AIPA) in response to the Federal Circuit's decision in *State Street Bank & Trust Co. v. Signature Fin. Group, Inc.*, 149 F.3d 1368 (Fed. Cir. 1998), *cert. denied*, 525 U.S. 1093 (1999). U.S. PATENT AND TRADEMARK OFFICE, REPORT TO CONGRESS ON THE PRIOR USER RIGHTS DEFENSE 6 (2012), http://www.uspto.gov/sites/default/files/aia_implementation/20120113-pur_report.pdf. The *State Street Bank* case affirmed that business methods are patentable subject matter under 35 U.S.C. § 101. *State Street Bank & Trust Co.*, 149 F.3d at 1375-76. In doing so, the decision created a great deal of uncertainty for U.S. businesses as to whether they might be liable for patent infringement for their continued use of internal business processes they once thought were unpatentable. U.S. PATENT AND TRADEMARK OFFICE, REPORT TO CONGRESS ON THE PRIOR USER RIGHTS DEFENSE 6 (2012), http://www.uspto.gov/sites/default/files/aia_implementation/20120113-pur_report.pdf. 35 U.S.C. § 273 sought to strike a balance between patent owners and businesses that were concerned about liability for infringing business method patents by providing a narrow prior user rights defense that was only applicable to patents covering methods of doing or conducting business. *Id.*; *see also*

---

[f]Historically, this potential for patent infringement liability in the U.S. created a strong incentive for manufacturers to conduct their manufacturing processes overseas. *See, e.g., America Invents Act: Hearing on H.R. 1249 Before the Subcomm. on Intellectual Prop., Competition, and the Internet of the H. Comm. on the Judiciary*, 112th Cong. 51 (2011) (statement of David J. Kappos, Director, USPTO) ("[P]rior user rights have the advantage of being very pro-American manufacturing. Currently, there is actually an incentive for American businesses to locate their factories overseas . . . because . . . other countries have prior user rights."). Once 35 U.S.C. § 271(g) was enacted in 1998, however, conducting manufacturing processes overseas did not completely absolve manufacturers of patent infringement liability. *See* Gregory C. Gramenopoulos, *The Extraterritorial Reach of U.S. Patents: Implications for the Global Marketplace*, BNA International Piracy and Brand Awareness (May 2006) ("In 1998, Congress further added section 271(g) to Title 35 of the Patent Act. This subsection creates liability for importing into the United States or selling, offering to sell, or using in the United States, a product 'made by' a process covered by a U.S. patent.").

pre-AIA 35 U.S.C. § 273(b)(1) ("It shall be a defense to an action for infringement under section 271 of this title with respect to any subject matter that would otherwise infringe one or more claims for a method in the patent being asserted against a person, if such person had, acting in good faith, actually reduced the subject matter to practice at least 1 year before the effective filing date of such patent, and commercially used the subject matter before the effective filing date of such patent."); pre-AIA 35 U.S.C. § 273(a)(3) (defining the term "method" in pre-AIA 35 U.S.C. § 273 to mean "a method of doing or conducting business.").

Even though pre-AIA 35 U.S.C. § 273 was the first statute to explicitly provide a prior user rights defense, the defense arguably existed under pre-AIA 35 U.S.C. § 102(g) before the prior user rights statute was enacted. Indeed, pre-AIA 35 U.S.C. § 102(g)(2) provides that a person shall be entitled to a patent unless "before such person's invention thereof, the invention was made in this country by another inventor who had not abandoned, suppressed, or concealed it." As explained by Judge Newman of the Court of Appeals for the Federal Circuit, trade secrets, if commercially used, could be used to invalidate a patent under pre-AIA 35 U.S.C. § 102(g), thereby securing the right of the trade secret holder to continue using that trade secret. Judge Newman specifically stated:

> I have not seen anyone who was a prior user who has been stopped upon raising the [post-AIA] 102(g) defense and from that viewpoint[,] it seems that the prior user right is alive and well. Because someone has kept it as a trade secret has not succeeded, as far as I can tell, in avoiding the defense, because if it has been in commercial use, even if the process has been kept secret, it is considered a bar [to patentability]. If we go to a first-to-file system[,] we must face the important points that have been raised about forcing people into the patent system, even for marginal inventions technologically, in order to protect their prior user right. But if we stay with the current first-to-invent system, we would be changing direction if we felt that there should not be prior user right.

32 IDEA 7, 60 (1991-92) (reprinting transcript of conference held Apr. 27, 1991, hosted by the Franklin Pierce Law Center, in cooperation with the Kenneth J. Germeshausen Center for the Law of Innovation and Entrepreneurship, and the PTC Research Foundation); *see also Dunlop Holdings Ltd. v. Ram Golf Corp.*, 524 F.2d 33, 36-37 (7th Cir. 1975) (finding that an alleged infringer's nondisclosure of one of the ingredients that made his inventive golf ball durable did not amount to suppression or concealment of his durable golf ball invention for purposes of 35 U.S.C. § 102(g) where "[t]he evidence clearly demonstrate[d] that [the alleged infringer] endeavored to

market his [invention] as promptly and effectively as possible" and he publicly used the inventive golf ball before the date of invention of the relevant patent claims).

Pre-AIA 35 U.S.C. § 102(g) was repealed by the AIA as of March 16, 2013. Leahy-Smith America Invents Act of 2011, Pub. L. No. 112-29, §§ 3(b) and 3(n)(1), 125 Stat. 284, 285-87, 293. As a result, that statute is no longer available to prior users who are sued for infringing a patent, wherein all of the claims of the patent are subject to the first-inventor-to-file patent system (i.e., patents wherein all of the claims have an effective filing date that is on or after March 16, 2013). *Id.* at § 3(n)(1) and (2), 125 Stat. 284, 293; *see also* Thomas L. Irving & Deborah M. Herzfeld, *Top Five Dangers for the AIA Unwary*, Landslide (2013) (explaining that pre-AIA 35 U.S.C. § 102(g) still applies to patents and patent applications that contain at least one claim having an effective filing date that is on or after March 16, 2013 and at least one claim having an effective filing date that is before March 16, 2013). Furthermore, while 35 U.S.C. § 273 is still available in the first-inventor-to-file patent system, that statute was significantly revised by the AIA. Accordingly, the remainder of this section will provide a comparison of pre- and post-AIA 35 U.S.C. § 273.

C. Comparison of Pre- and Post-AIA 35 U.S.C. § 273

The AIA's amendments to pre-AIA 35 U.S.C. § 273 both expand and narrow the prior user rights defense that existed before the AIA went into effect. For instance, the AIA expands the prior user rights defense by broadening the types of patent claims that are subject to the defense and by making the defense available to a broader group of defendants. On the other hand, the AIA narrows the prior user rights defense by requiring that a defendant commercially use the patented subject matter as of an earlier time in order to qualify for the prior user rights defense and by prohibiting use of the defense against patents that were owned or subject to an obligation of assignment to certain institutions of higher education or technology transfer organizations at the time the invention was made. There are also numerous similarities between pre- and post-AIA 35 U.S.C. § 273. Both the differences and similarities are discussed in more detail in Sections II.C.i.–II.C.iii. Furthermore, for ease of reference, pre- and post-AIA 35 U.S.C. § 273(a) and (b)[g] are shown in the following table:

---

[g]Note that for the sake of brevity, only a portion of pre-AIA 35 U.S.C. § 273(b) is shown.

Pre-AIA 35 U.S.C. § 273(a) and (b)(1) and (2)	Post-AIA 35 U.S.C. § 273(a) and (b)

(a) DEFINITIONS.- For purposes of this section-
  (1) the terms "commercially used" and "commercial use" mean use of a method in the United States, so long as such use is in connection with an internal commercial use or an actual arm's-length sale or other arm's-length commercial transfer of a useful end result, whether or not the subject matter at issue is accessible to or otherwise known to the public, except that the subject matter for which commercial marketing or use is subject to a premarketing regulatory review period during which the safety or efficacy of the subject matter is established, including any period specified in section 156(g), shall be deemed "commercially used" and in "commercial use" during such regulatory review period;
  (2) in the case of activities performed by a nonprofit research laboratory, or nonprofit entity such as a university, research center, or hospital, a use for which the public is the intended beneficiary shall be considered to be a use described in paragraph (1), except that the use-
    (A) may be asserted as a defense under this section only for continued use by and in the laboratory or nonprofit entity; and
    (B) may not be asserted as a defense with respect to any subsequent commercialization or use outside such laboratory or nonprofit entity;
  (3) the term "method" means a method of doing or conducting business; and
	(a) IN GENERAL.— A person shall be entitled to a defense under section 282(b) with respect to subject matter consisting of a process, or consisting of a machine, manufacture, or composition of matter used in a manufacturing or other commercial process, that would otherwise infringe a claimed invention being asserted against the person if—
	  (1) such person, acting in good faith, commercially used the subject matter in the United States, either in connection with an internal commercial use or an actual arm's length sale or other arm's length commercial transfer of a useful end result of such commercial use; and
	  (2) such commercial use occurred at least 1 year before the earlier of either—
	    (A) the effective filing date of the claimed invention; or
	    (B) the date on which the claimed invention was disclosed to the public in a manner that qualified for the exception from prior art under section 102(b).

(*Continued on next page*)

Pre-AIA 35 U.S.C. § 273(a) and (b)(1) and (2)	Post-AIA 35 U.S.C. § 273(a) and (b)
(4) the "effective filing date" of a patent is the earlier of the actual filing date of the application for the patent or the filing date of any earlier United States, foreign, or international application to which the subject matter at issue is entitled under section 119, 120, or 365 of this title. (b) DEFENSE TO INFRINGEMENT.- (1) IN GENERAL.- It shall be a defense to an action for infringement under section 271 of this title with respect to any subject matter that would otherwise infringe one or more claims for a method in the patent being asserted against a person, if such person had, acting in good faith, actually reduced the subject matter to practice at least 1 year before the effective filing date of such patent, and commercially used the subject matter before the effective filing date of such patent. (2) EXHAUSTION OF RIGHT.- The sale or other disposition of a useful end product produced by a patented method, by a person entitled to assert a defense under this section with respect to that useful end result shall exhaust the patent owner's rights under the patent to the extent such rights would have been exhausted had such sale or other disposition been made by the patent owner.	(b) BURDEN OF PROOF.— A person asserting a defense under this section shall have the burden of establishing the defense by clear and convincing evidence.

i. Expansion of the Prior User Rights Defense Under the AIA

AIA § 5(a)'s amendments to pre-AIA 35 U.S.C. § 273 expand the prior user rights defense in two major ways. First, the AIA modified the statute so that the prior user rights defense could be used against a broader range of patent claims. In particular, pre-AIA 35 U.S.C. § 273(b)(1) states: "[i]t shall be a defense to an action for infringement under section 271 of this title with respect to any subject matter that would otherwise infringe one or more claims for a method in the patent being asserted against a person, if such person had, acting in good faith, actually reduced the subject matter to practice at least 1 year before the effective filing date of such patent, and commercially used the subject matter before the effective filing date of such patent."[h] Pre-AIA 35 U.S.C. § 273(a)(3) defines "method" as "a method of doing or conducting business." Thus, only business method patent claims were subject to the prior user rights defense under pre-AIA 35 U.S.C. § 273. *See Sabasta v. Buckaroos, Inc.*, 507 F. Supp. 2d 986, 1005 (S.D. Iowa 2007) ("While there is no case law directly addressing the scope of § 273 as it was enacted, the legislative history clearly indicates an intent for the [prior user rights defense] to have a limited scope, that is, the defense is designed to protect small businesses from patent infringement suits for methods of conducting business that use a novel process employing unpatentable subject matter, but that have 'useful, concrete and tangible result[s].' The fact that Buckaroos is in business and uses a process to manufacture ribbed pipe saddles does not, in light of the legislative history and the *State Street Bank* case, bring it within the intended purview of § 273.") (citations omitted).

In contrast to pre-AIA 35 U.S.C. § 273, post-AIA 35 U.S.C. § 273(a) explains that "[a] person shall be entitled to a [prior user rights defense] with respect to subject matter consisting of a process, or consisting of a

---

[h]It might seem like the post-AIA statute expands the prior user rights defense by omitting the requirement that the prior user actually reduce the patented invention to practice prior to the effective filing date of the patent at-issue. *Compare* pre-AIA 35 U.S.C. § 273(b)(1) (requiring that a prior user acted in good faith to: (1) actually reduce the subject matter of the patent claim(s) to practice at least 1 year before the effective filing date of the patent-at-issue; and (2) commercially use the subject matter of the patent claim(s) in the United States on a date that precedes the effective filing date of the patent-at-issue) *with* post-AIA 35 U.S.C. § 273(a)(1) and (2) (explaining that a defendant asserting the prior user rights defense must plead that he/she acted in good faith to "commercially use" the subject matter of the patent claim(s) in the United States but making no explicit requirement that the prior user reduce the invention to practice prior to the patenting of that invention). The legislative history for post-AIA 35 U.S.C. § 273 explains, however, that the express requirement to "reduce the subject matter of the patent claim(s) to practice" was only eliminated because "the use of a process, or the use of product in a commercial process, will always constitute a reduction to practice." 157 Cong. Rec. S5430, 2011 (statement of Sen. Kyl). Thus, while the post-AIA statute does not explicitly require reduction to practice, reduction to practice is implicit in the "commercial use" requirement.

machine, manufacture, or composition of matter used in a manufacturing or other commercial processes . . . ." The post-AIA prior user rights defense can therefore be used against more than just business method patent claims because both process patent claims (including, but not limited to, business method claims), as well as claims covering products that are used in manufacturing or commercial processes (e.g., machines, starting materials used in manufacturing or compositions of matter used in manufacturing goods) are subject to the defense. *See* 157 Cong. Rec. S5430, 2011 (statement of Sen. Kyl) ("Subsection (a) [of post-AIA 35 U.S.C. § 273] expands the defense beyond just processes to also cover products that are used in a manufacturing or other commercial process . . . Some products . . . consist of tools or other devices that are used only by the inventor inside his closed factory. Others consist of substances that are exhausted in a manufacturing process and never become accessible to the public. Such products will not become prior art. Revised section 273 therefore allows the defense to be asserted with respect to such products.").

The AIA also expands the prior user rights defense by making the defense available to additional types of alleged patent infringers. Under the pre-AIA statute, only the person who actually performed the acts necessary to establish the prior user rights defense, i.e., the prior user himself/herself, could rely on the defense unless that person transferred his/her entire enterprise or line of business to another person. Pre-AIA 35 U.S.C. § 273(b)(6). In other words, only direct infringers were eligible to use the defense under the pre-AIA statute. The post-AIA statute, however, makes the defense available to: (1) the person who actually performed the acts necessary to establish the prior user rights defense, (2) the person who directed the performance of those acts, and (3) an entity that controls, is controlled by, or is under common control with such person. Post-AIA 35 U.S.C. § 273(e)(1)(A).

Regarding the "directed the performance of" language, Senator Kyl explained,

> *The House bill originally allowed the defendant to assert the defense if he performed the commercial use or "caused" its performance. The word "caused," however, could be read to include even those uses that a vendor made without instructions or even the contemporaneous knowledge of the person asserting the defense. The final bill uses the word "directed," which limits the provision only to those third-party commercial uses that the defendant actually instructed the vendor or contractor to use. In analogous contexts, the word "directed" has been*

understood to require evidence that the defendant affirmatively directed the vendor or contractor in the manner of the work or use of the product.

157 Cong. Rec. S5430, 2011 (quoting and paraphrasing *Ortega v. Puccia*, 866 N.Y.S.2d 323, 328 (N.Y. App. 2008)).

Furthermore, Senator Kyl shed light on the meaning of "control," explaining:

> *Subsection (e)(1)(A)'s reference to entities that "control, are controlled by, or under common control with" the defendant borrows a term that is used in several federal statutes. See 12 U.S.C. 1841(k), involving bank holding companies, 15 U.S.C. 78c(a)(4)(B)(vi), involving securities regulation, 15 U.S.C. 6809(6), involving financial privacy, and 49 U.S.C. 30106(d)(1), involving motor vehicle safety. Black's Law Dictionary 378 (9th ed. 2009) defines "control" as the "direct or indirect power to govern the management and policies of a person or entity, whether through ownership of securities, by contract, or otherwise; the power or authority to manage, direct, or oversee."*

157 Cong. Rec. S5430-31, 2011.

Thus, under the new statute, both direct *and* indirect infringers are eligible to use the prior user rights defense. U.S. Patent and Trademark Office, Report to Congress on the Prior User Rights Defense 18 (2012), http://www.uspto.gov/sites/default/files/aia_implementation/20120113-pur_report.pdf.

ii. Narrowing of the Prior User Rights Defense Under the AIA

As compared to pre-AIA 35 U.S.C. § 273, post-AIA 35 U.S.C. § 273 restricts the prior user rights defense in two major ways. The first and most significant way the defense was limited by the AIA can be seen in the timing by which an alleged infringer must have used the invention at-issue in order to obtain prior user rights. Indeed, to utilize the prior user rights defense under the pre-AIA statute, an alleged infringer must have: (1) actually reduced the allegedly infringing subject matter to practice at least 1 year before the effective filing date of the patent he/she is alleged to have infringed; and (2) commercially used the subject matter at some point before the effective filing date of that patent.[i] Pre-AIA 35 U.S.C. § 273(b)(1). Post-AIA 35 U.S.C. § 273(a)(2), on the other hand, requires that in order to invoke the prior user rights defense, an alleged infringer must have commercially used the allegedly infringing subject matter "at least 1 year before the earlier of either (A) the effective filing

---

[i]Pre-AIA 35 U.S.C. § 273(a)(4) defines the term "effective filing date" as the "the earlier of the actual filing date of the application for the patent or the filing date of any earlier United States, foreign, or international application to which the subject matter at issue is entitled under section 119, 120, or 365 of this title."

date of the claimed invention; or (B) the date on which the claimed invention was disclosed to the public in a manner that qualified for the exception from prior art under [35 U.S.C. §] 102(b)."[j] By requiring that the commercial use must occur before the <u>earlier of</u> options (A) or (B), the post-AIA statute makes it harder for alleged infringers to rely on a prior commercial use where there has been a public disclosure that qualifies as an exception from prior art under post-AIA 35 U.S.C. § 102(b) because, in those instances, the prior commercial use must occur at a time that is much earlier than it would have under the pre-AIA statute.[k] The following timeline helps demonstrate this point:

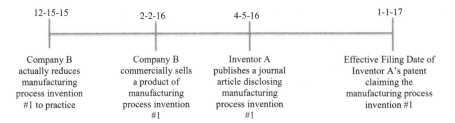

12-15-15	2-2-16	4-5-16	1-1-17
Company B actually reduces manufacturing process invention #1 to practice	Company B commercially sells a product of manufacturing process invention #1	Inventor A publishes a journal article disclosing manufacturing process invention #1	Effective Filing Date of Inventor A's patent claiming the manufacturing process invention #1

In the scenario shown in the previous timeline, if Inventor A sued Company B for infringement of his patent claiming the manufacturing

[j] The requirement that there be a time gap between the prior commercial use and the patent's effective filing date or a public disclosure was imposed to ensure that the person claiming a prior user right did not actually derive the invention from the patentee. Joe Matal, *A Guide to the Legislative History of the America Invents Act: Part II of II*, 21 FED. CIR. B.J. 539, 566 (2012).

[k] Some might argue that the AIA's requirement that the commercial use occur an entire year before the earlier of either the effective filing date of the claimed invention or the date on which a post-AIA 35 U.S.C. § 102(b)-excepted public disclosure of the claimed invention occurred essentially guts the prior user rights defense so that it is entirely worthless. *See, e.g.,* Joe Matal, *A Guide to the Legislative History of the America Invents Act: Part II of II*, 21 FED. CIR. B.J. 539, 566 (2012) ("Requiring that actual commercial use be established a full year before a patented invention's effective-filing date or [a post-AIA] § 102(b) public disclosure inevitably will result in situations in which the right to continued use of an invention is denied to a person who clearly was the first inventor."); *Prior Domestic Commercial Use Act of 1995: Hearing on H.R. 2235 Before the Subcomm. on Courts and Intellectual Prop. of the H. Comm. on the Judiciary*, 104th Cong. 13-14 (1995) (statement of Karl F. Jorda, Professor, Franklin Pierce Law Center) ("[The requirement that a commercial use occur an entire year before the effective date of the claimed invention] goes too far and guts this prior user defense. It is too radical and stands well-established patent and trade secret law principles on their heads . . . It also complicates this prior user right, which is already drastically limited and qualified, beyond reason. A first inventor is a first inventor and should be accorded the status of a first inventor especially in the one-year period prior to the entry of a rival inventor because it is in that period that the same invention is likely to be made by more than one inventor due to outside stimuli.").

process invention #1, Company B would be entitled to rely on the prior user rights defense under pre-AIA 35 U.S.C. § 273 but not under the post-AIA statute. Indeed, Company B's activities meet the requirements of pre-AIA 35 U.S.C. § 273(b)(1) because (1) Company B's actual reduction to practice occurred on December 15, 2015, which is a date that is more than 1 year prior to the January 1, 2017 effective filing date of Inventor A's patent, and (2) Company B commercially used the patented subject matter on February 2, 2016, which is a date that is prior to the January 1, 2017 effective filing date of Inventor A's patent. Company B could not rely on the prior user rights defense under post-AIA 35 U.S.C. § 273, however, because under that statute, Company B's commercial use would have had to occur at least 1 year before the earlier of either (A) the effective filing date of the claimed invention; or (B) the date on which the claimed invention was disclosed to the public in a manner that qualified for the exception from prior art under post-AIA 35 U.S.C. § 102(b). Inventor A's publication of the journal article disclosing the manufacturing process invention #1 on April 5, 2016 is a disclosure to the public that qualifies for an exception from prior art under post-AIA 35 U.S.C. § 102(b). *See* post-AIA 35 U.S.C. § 102(b)(1)(A) (explaining that a disclosure made 1 year or less before the effective filing date of a claimed invention shall not be prior art if the disclosure was made by the inventor). Accordingly, in order to qualify for the prior user rights defense, Company B's commercial use would have had to occur on or before April 5, 2015.

The second major way in which the prior user rights defense was limited by the AIA was through the addition of the university exception clause in post-AIA 35 U.S.C. § 273(e)(5). That clause reads as follows:

(5) UNIVERSITY EXCEPTION.--

(A) IN GENERAL.-- A person commercially using subject matter to which subsection (a) applies may not assert a defense under this section if the claimed invention with respect to which the defense is asserted was, at the time the invention was made, owned or subject to an obligation of assignment to either an institution of higher education (as defined in section 101(a) of the Higher Education Act of 1965 (20 U.S.C. 1001(a)), or a technology transfer organization whose primary purpose is to facilitate the commercialization of technologies developed by one or more such institutions of higher education.

(B) EXCEPTION.-- Subparagraph (A) shall not apply if any of the activities required to reduce to practice the subject matter of the claimed invention could not have been undertaken using funds provided by the Federal Government.

Pre-AIA 35 U.S.C. § 273 did not contain a university exception clause. The university exception clause prohibits alleged infringers from asserting the prior user rights defense against claimed inventions that were: (1) owned or subject to an obligation of assignment to an institution of higher education (as defined in section 1001(a) of the Higher Education Act of 1965)[l] at the time the invention was made, or (2) owned or subject to an obligation of assignment to a technology transfer organization whose primary purpose[m] is to facilitate the commercialization of technologies developed by one or more such institutions of higher education at the time the invention was made.

Regarding the exception set forth in post-AIA 35 U.S.C. § 273(e)(5)(B), Senator Kyl explained:

> Subparagraph (B), the exception to the university exception, is only intended to preclude application of subparagraph (A) when the federal government is affirmatively prohibited, whether by statute, regulation, or executive order, from funding research in the activities in question.
>
> **157 Cong. Rec. S5431, 2011.**

One criticism of the university exception clause is that it threatens to preclude the use of the prior user rights defense against many patents that are asserted by organizations other than the universities and technology transfer organizations it was enacted to protect.[n] Indeed, post-AIA 35 U.S.C. § 273(e)(5)(A) does not require that a patent be owned by an institution of higher education or a technology transfer organization at the time it is asserted against an alleged infringer. Instead, the patent must only be owned or subject to an obligation of assignment to one of those organizations "at the time the invention was made." Thus, if an institution of higher education or a technology transfer organization obtains a patent on one of its inventions, and then assigns the patent to another entity (e.g., a non-practicing entity) that later enforces the patent, the university exception clause will preclude an alleged infringer from relying on the prior user rights defense.

---

[l]20 U.S.C. § 1001(a) provides a list of criteria that must be met in order for a college or university to meet the definition of an "institution of higher education."

[m]The university exception clause is intended to include those non-profit technology transfer organizations that are entitled to receive assignments of inventions pursuant to 35 U.S.C. § 207(c)(7). 157 Cong. Rec. S5427-28, 2011 (statements of Sen. Kohl and Sen. Leahy).

[n]Prior to the AIA, universities had argued that they should be exempt from prior user rights because they could never establish themselves as prior users given that they do not manufacture or market their inventions. They essentially argued that they were hurt by a prior user rights system because while their patents were subject to a prior user rights defense, they were unable to claim the benefit of such defense themselves. Keith M. Kupferschmid, *Prior User Rights: The Inventor's Lottery Ticket*, 21 AIPLA Q.J. 213, 234 (1993).

iii. Similarities Between Pre- and Post-AIA 35 U.S.C. § 273

While the AIA brought about many significant changes to the prior user rights defense, many of the clauses of post-AIA 35 U.S.C. § 273 are virtually identical to the corresponding clauses in the pre-AIA statute. This section will give a brief description of each of these clauses.

1. Exhaustion of Patent Rights Clause

"The longstanding doctrine of patent exhaustion provides that the initial authorized sale of a patented item terminates all patent rights to that item." *Quanta Computer, Inc. v. LG Electronics, Inc.*, 553 U.S. 617, 625 (2008). The doctrine of patent exhaustion prevents a patentee from placing post-sale restrictions on the use of a patented article so long as the patentee authorized its sale. The exhaustion of patent rights clause is set forth in pre-AIA 35 U.S.C. § 273(b)(2) and post-AIA 35 U.S.C. § 273(d). The AIPA Committee Report for the pre-AIA statute best illustrates how patent exhaustion works within the context of prior user rights. That report explains, "[f]or example, if [under the doctrine of patent exhaustion,] a purchaser would have had the right to resell a product if bought from the patent owner, the purchaser has the same right if the product is purchased from a person entitled to a [prior user rights] defense." H.R. REP. No. 106-287, at 48 (1999) (citations omitted).

If patent exhaustion did not occur when a prior user sold a product made by a patented method or process to which they had acquired prior user rights, then their prior user rights would, in most cases, have little to no value.° Indeed, without patent exhaustion, someone who purchased a product made by someone with prior user rights to a patented manufacturing process for that product would likely infringe the patent on the manufacturing process once they sold, offered to sell, or used the product. *See* 35 U.S.C. § 271(g) ("Whoever without authority ... offers to sell, sells, or uses within the United States a product which is made by a process patented in the United States shall be liable as an infringer, if the ... offer to sell, sale, or use

---

°The Supreme Court recently confirmed that a sale of a product can exhaust patent claims to a method or process where the product "embodie[s] essential features of the patented invention" and the only reasonable and intended use for the product is to practice the patent. *Quanta Computer, Inc.*, 553 U.S. at 628, 631. One example of an instance in which the Supreme Court found that the sale of a product exhausted patent claims to a method was in *Ethyl Gasoline Corp. v. U.S.*, 309 U.S. 436, 446, 457 (1940). There the Court specifically found that the sale of a motor fuel produced under one patent also exhausted another patent that was directed to a method of using the fuel in combustion motors. *Id.*

of the product occurs during the term of such process patent."). Absent patent exhaustion then, a person with prior user rights would have no financial incentive to practice the patented process or method to which he obtained his rights because no one would want to buy the product of that process or method. *See Prior User Rights (Relative to Patents): Hearing Before the Subcomm. on Intellectual Prop. and Judicial Admin. of the H. Comm. on the Judiciary*, 103d Cong. 135, 140 (1994) (Letter of Harold C. Wegner, Prof. of Law and Director of the Intellectual Prop. Law Program, George Washington University) (explaining that without patent exhaustion, "a [person] who has a prior user right would never be able to use that prior user right because his customers would become patent infringers.").

A comparison of the pre- and post-AIA exhaustion of patent right clauses is set forth in the following table:

Pre-AIA 35 U.S.C. § 273(b)(2)	Post-AIA 35 U.S.C. § 273(d)
EXHAUSTION OF RIGHT.- The sale or other disposition of a useful end product produced by a patented method, by a person entitled to assert a defense under this section with respect to that useful end result shall exhaust the patent owner's rights under the patent to the extent such rights would have been exhausted had such sale or other disposition been made by the patent owner.	EXHAUSTION OF RIGHTS.-- Notwithstanding subsection (e)(1), the sale or other disposition of a useful end result by a person entitled to assert a defense under this section in connection with a patent with respect to that useful end result shall exhaust the patent owner's rights under the patent to the extent that such rights would have been exhausted had such sale or other disposition been made by the patent owner.

The only difference between the two clauses is that the latter clause adds the words "[n]otwithstanding subsection (e)(1)." This addition was included to make it clear that the patent exhaustion that attaches to a prior use supersedes the otherwise personal and nontransferable nature of the prior-user defense. Joe Matal, *A Guide to the Legislative History of the America Invents Act: Part II of II*, 21 FED. CIR. B.J. 539, 569 (2012). In other words, despite the fact that pre-AIA 35 U.S.C. § 273(b)(6) and post-AIA 35 U.S.C. § 273(e)(1)(A) and (B) limit who can use the prior user rights defense, and restrict the transferability of prior user rights, those clauses do not prevent those who purchase products from persons with prior user rights from defending against claims of patent infringement by arguing that the patent rights were exhausted by such purchases.

2. Transfer of Right Clause

Pre-AIA 35 U.S.C. § 273(b)(6) and post-AIA 35 U.S.C. § 273(e)(1)(B) place limitations on how prior user rights may be transferred from one person or business entity to another. A comparison of the two statutory sections appears as follows:

Pre-AIA 35 U.S.C. § 273(b)(6)	Post-AIA 35 U.S.C. § 273(e)(1)(B)
PERSONAL DEFENSE.- The defense under this section may be asserted only by the person who performed the acts necessary to establish the defense and, except for any transfer to the patent owner, the right to assert the defense shall not be licensed or assigned or transferred to another person except as an ancillary and subordinate part of a good faith assignment or transfer for other reasons of the entire enterprise or line of business to which the defense relates.	TRANSFER OF RIGHT.-- Except for any transfer to the patent owner, the right to assert a defense under this section shall not be licensed or assigned or transferred to another person except as an ancillary and subordinate part of a good-faith assignment or transfer for other reasons of the entire enterprise or line of business to which the defense relates.

The transfer of right clause was included in the statute to prevent a compulsory license-type system. *See* ADVISORY COMM'N ON PATENT LAW REFORM, A REP. TO THE SECRETARY OF COMMERCE, 49, 50-51 (1992) ("Prior user rights should be personal in nature, and should not be transferable, except with that part of the business which exploits the right. This is essential . . . to prevent the personal right from being extended to resemble a compulsory license-like authority . . . ."). Indeed, if prior user rights could be easily transferred from one person to the next, a prior user could transfer his/her prior user rights to a patentee's competitor and the patentee would not be able to sue his/her competitor for patent infringement. This scenario is akin to a compulsory license situation because, in both circumstances, the patentee is unable to choose who is authorized to practice the patented invention. As shown previously, the transfer of right clause did not undergo any changes under the AIA.

3. Limitation on Sites Clause

In addition to limiting the transferability of prior user rights between different people or business entities, pre- and post-AIA 35 U.S.C. § 273 also restrict the geographic sites for which the defense may be used after a transfer of the rights between different persons or business entities has occurred. In particular, the limitation on sites clause only allows a prior user rights defense to be asserted for those geographic sites where the subject matter that would otherwise infringe a claimed invention is in use before the later of: (1) the effective filing date of the claimed invention or patent or (2) the date of the assignment or transfer of such enterprise or line of business. Presumably, this clause was enacted to ensure that the scope of the continued use of the patented invention by the person to whom a transfer of prior user rights has been made is commensurate with the scope of the activity that gave rise to the prior user rights to begin with. *See* THE ADVISORY COMM'N ON PATENT LAW REFORM, A REP. TO THE SECRETARY OF COMMERCE, 49, 50 (1992) ("The right created by the prior use or preparation should be limited to continuation of the particular activity which gives rise to the right."). A side-by-side comparison of the specific clauses governing this restriction is set forth in the following table:

Pre-AIA 35 U.S.C. § 273(b)(7)	Post-AIA 35 U.S.C. § 273(e)(1)(C)
LIMITATION ON SITES.- A defense under this section, when acquired as part of a good faith assignment or transfer of an entire enterprise or line of business to which the defense relates, may only be asserted for uses at sites where the subject matter that would otherwise infringe one or more of the claims is in use before the later of the effective filing date of the patent or the date of the assignment or transfer of such enterprise or line of business.	RESTRICTION ON SITES.-- A defense under this section, when acquired by a person as part of an assignment or transfer described in subparagraph (B), may only be asserted for uses at sites where the subject matter that would otherwise infringe a claimed invention is in use before the later of the effective filing date of the claimed invention or the date of the assignment or transfer of such enterprise or line of business.

As shown previously, the post-AIA statute copies the pre-AIA statute almost verbatim.

4. Premarketing Regulatory Review Clause
The pre-marketing regulatory review clause appears in pre-AIA 35 U.S.C. § 273(a)(1) and post-AIA 35 U.S.C. § 273(c)(1).

Pre-AIA 35 U.S.C. § 273(a)(1)	Post-AIA 35 U.S.C. § 273(c)(1)
[T]he terms "commercially used" and "commercial use" mean use of a method in the United States, so long as such use is in connection with an internal commercial use or an actual arm's-length sale or other arm's-length commercial transfer of a useful end result, whether or not the subject matter at issue is accessible to or otherwise known to the public, except that the subject matter for which commercial marketing or use is subject to a premarketing regulatory review period during which the safety or efficacy of the subject matter is established, including any period specified in section 156(g), shall be deemed "commercially used" and in "commercial use" during such regulatory review period	PREMARKETING REGULATORY REVIEW.-- Subject matter for which commercial marketing or use is subject to a premarketing regulatory review period during which the safety or efficacy of the subject matter is established, including any period specified in section 156(g), shall be deemed to be commercially used for purposes of subsection (a)(1) during such regulatory review period.

As shown previously, the post-AIA statute is a nearly verbatim copy of the pre-AIA statute. The only difference between the two statutes is that the post-AIA statute contains minor non-substantive changes that were necessary to conform the premarketing regulatory review clause to read consistently with some of the other clauses (e.g., 35 U.S.C. § 273(a)(1)) of the post-AIA statute.

The premarketing regulatory review clause adds premarketing regulatory review activities to the list of activities that give rise to prior user rights in post-AIA 35 U.S.C. § 273(a)(1). Premarketing regulatory review activities likely would not fall within the scope of "commercial use" activities mentioned in post-AIA 35 U.S.C. § 273(a)(1) because during premarketing regulatory review, no product is being sold.[p] Rather, a product is undergoing regulatory review so that it can be sold on some future date. Examples of activities that would be subject to the premarketing regulatory review clause are the safety and efficacy studies that a pharmaceutical company conducts for submission to

---

[p]Even if premarketing regulatory review activities would not have been expressly included as "commercial use" activities by virtue of the premarketing regulatory review clause, it is possible that such activities would have fallen within the scope of "internal commercial use[s]" provided for in post-AIA 35 U.S.C. § 273(a)(1). Indeed, Senator Leahy explained that "internally used methods and materials do qualify for the defense of prior user rights when there is evidence of a commitment to put the innovation into use followed by a series of diligent events demonstrating that the innovation has been put into continuous—into a business activity with a purpose of developing new products for the benefit of mankind." 157 Cong. Rec. S5427, 2011.

the U.S. Food and Drug Administration (FDA) to determine whether a new pharmaceutical product can be commercially marketed or used.[q]

5. Nonprofit Laboratory Use Clause

Pre-AIA 35 U.S.C. § 273(a)(2) and post-AIA U.S.C. § 273(c)(2) provide the nonprofit laboratory use clause. A side-by-side comparison of these clauses appears as follows:

Pre-AIA 35 U.S.C. § 273(a)(2)	Post-AIA 35 U.S.C. § 273(c)(2)
[I]n the case of activities performed by a nonprofit research laboratory, or nonprofit entity such as a university, research center, or hospital, a use for which the public is the intended beneficiary shall be considered to be a use described in paragraph (1), except that the use- (A) may be asserted as a defense under this section only for continued use by and in the laboratory or nonprofit entity; and (B) may not be asserted as a defense with respect to any subsequent commercialization or use outside such laboratory or nonprofit entity	NONPROFIT LABORATORY USE.-- A use of subject matter by a nonprofit research laboratory or other nonprofit entity, such as a university or hospital, for which the public is the intended beneficiary, shall be deemed to be a commercial use for purposes of subsection (a)(1), except that a defense under this section may be asserted pursuant to this paragraph only for continued and noncommercial use by and in the laboratory or other nonprofit entity.

The nonprofit laboratory use clause is similar to the premarketing regulatory review clause in that it adds additional activities to the activities that can give rise to prior user rights listed in post-AIA 35 U.S.C. 273(a)(1). The nonprofit laboratory use clause specifically explains that "uses of subject matter by a nonprofit research laboratory or other nonprofit entity" are

[q]As an aside, at least with regard to premarketing regulatory review activities conducted for submission to the FDA, it is not entirely clear why the premarketing regulatory review clause was included in 35 U.S.C. § 273. Indeed, prior user rights are not actually affirmative rights, but are instead only a defense to a claim of patent infringement. Keith M. Kupferschmid, *Prior User Rights: The Inventor's Lottery Ticket*, 21 AIPLA Q.J. 213, 216 (1993). In other words, a prior user does not have prior user rights until he/she is sued for infringement and successfully makes out a prior user rights defense based on his/her prior use of the claimed invention. 35 U.S.C. § 271(e)(1), however, prevents patentees from suing entities for patent infringement based on activities that are undertaken solely for the purpose of developing and submitting information to the FDA. In other words, unless the scope of the activities that fall under 35 U.S.C. § 271(e)(1) differs from the scope of activities that fall within the premarketing regulatory review clause of post-AIA 35 U.S.C. § 273, it seems unnecessary to include the latter clause because a patentee could not bring suit at all against someone for engaging in those activities.

included as "commercial uses" for purposes of the prior user rights defense. The only difference between the pre- and post-AIA statutes is that the latter does not contain subparagraph (B) of the pre-AIA version of the statute. The language of subparagraph (B) was removed by the new patent legislation because it was entirely redundant with subparagraph (A). 157 CONG. REC. S5431, 2011 (statement of Sen. Kyl).

6. Unreasonable Assertion of Defense Clause

A side-by-side comparison of the pre- and post-AIA unreasonable/unsuccessful assertion of defense clauses (hereinafter referred to as the "unreasonable assertion of defense clause") appears as follows:

Pre-AIA 35 U.S.C. § 273(b)(8)	Post-AIA 35 U.S.C. § 273(f)
UNSUCCESSFUL ASSERTION OF DEFENSE.- If the defense under this section is pleaded by a person who is found to infringe the patent and who subsequently fails to demonstrate a reasonable basis for asserting the defense, the court shall find the case exceptional for the purpose of awarding attorney fees under section 285 of this title.	UNREASONABLE ASSERTION OF DEFENSE.-- If the defense under this section is pleaded by a person who is found to infringe the patent and who subsequently fails to demonstrate a reasonable basis for asserting the defense, the court shall find the case exceptional for the purpose of awarding attorney fees under section 285.

The unreasonable assertion of defense clause provides a steep penalty for those defendants who are found to infringe one or more patent claims and who had no reasonable basis for pleading a prior user rights defense but did so anyway. Under the unreasonable assertion of defense clause, such infringers will be liable for both the usual patent infringement damages under 35 U.S.C. § 284, including a reasonable royalty, lost profits, and/or enhanced damages if willful infringement is found, as well as the plaintiff's attorneys fees pursuant to 35 U.S.C. § 285. See 35 U.S.C. § 284 ("Upon finding for the claimant the court shall award the claimant damages adequate to compensate for the infringement but in no event less than a reasonable royalty for the use made of the invention by the infringer, together with interest and costs as fixed by the court."); Takeda Chem. Indus., Ltd. v. Mylan Labs., Inc., 549 F.3d 1381, 1384-85 (Fed. Cir. 2008) (requiring defendants Alphapharm and Mylan to pay $5,400,000 and $11,400,000, respectively for the Plaintiff's attorneys fees where each of the defendants had engaged in litigation misconduct and had changed the focus of their invalidity defenses at trial from the invalidity defenses they had asserted in their paragraph IV certification letters, thereby evincing a lack of good faith basis for their defenses); Yamanouchi Pharm Co., Ltd. v.

Danbury Pharmacal, Inc., 21 F. Supp. 2d 366, 378 (S.D.N.Y. 1998) (explaining that 35 U.S.C. § 285 permits an award of attorneys' fees in "exceptional cases" and requiring a defendant to pay for the Plaintiff's attorneys fees where the defendant "[f]orc[ed the] plaintiff to doubtless incur considerable unjustified expense to defend against an ill-supported claim . . . .").

7. Invalidity Clause

The invalidity clauses of pre-AIA 35 U.S.C. § 273(b)(9) and post-AIA 35 U.S.C. § 273(g) are virtually identical and are shown in the following table:

Pre-AIA 35 U.S.C. § 273(b)(9)	Post-AIA 35 U.S.C. § 273(g)
INVALIDITY.- A patent shall not be deemed to be invalid under section 102 or 103 of this title solely because a defense is raised or established under this section.	INVALIDITY.-- A patent shall not be deemed to be invalid under section 102 or 103 solely because a defense is raised or established under this section.

The invalidity clause of 35 U.S.C. § 273 makes clear that just because an alleged infringer succeeds in asserting a prior user rights defense against one or more claims of a patent does not mean that those claims are invalid under pre- or post-AIA 35 U.S.C. §§ 102 or 103. The invalidity clause was originally enacted to clarify the scope of pre-AIA 35 U.S.C. § 102(g) and make clear that any invalidity defense based on that statute must be established separately from a defense based on pre-AIA 35 U.S.C. § 273.[r] *See* H.R. REP. NO. 106-287, at 49 (1999) ("[Pre-AIA 35 U.S.C. § 273(b)(9)] provides that a party who [sic] uses a process or business method commercially in secrecy before the patent filing date and establishes a § 273 defense is not an earlier inventor for purposes of invalidating the patent [under pre-AIA 35 U.S.C. § 102(g)(2)]."); *id.* ("Under [pre-AIA] law, although the matter has seldom been litigated, a party who commercially used an invention in secrecy before the patent filing date and invented the subject matter before the patent owner's invention may argue that the patent is invalid under [section 102(g)] of the Patent Act. Arguably, commercial use of

[r]One important distinction between the pre-AIA 35 U.S.C. § 102(g) and pre- or post-AIA 35 U.S.C. § 273 defenses can be seen in their effect on a patentee's rights. Indeed, if a defendant successfully makes out a defense under 35 U.S.C. § 102(g), the patent claims for which the defense is established are rendered invalid and cannot be asserted against anyone. In contrast, if an alleged infringer successfully establishes that he/she has prior user rights with regard to the asserted patent claims under pre- or post-AIA 35 U.S.C. § 273, those claims remain valid and can be enforced against anyone except for the person who holds prior user rights to the claimed invention.

an invention in secrecy is not suppression or concealment of the invention within the meaning of § 102(g), and therefore the party's earlier invention will invalidate the patent."); *see also Dunlop Holdings Ltd.*, 524 F.2d at 36-37 (finding that an alleged infringer's nondisclosure of one of the ingredients that made his inventive golf ball durable did not amount to suppression or concealment of his invention for purposes of 35 U.S.C. § 102(g) where "[t]he evidence clearly demonstrate[d] that [the alleged infringer] endeavored to market his [invention] as promptly and effectively as possible" and he publicly used the inventive golf ball before the date of invention of the relevant patent claims). It is not yet clear whether the invalidity clause of post-AIA 35 U.S.C. § 273(g) has any effect on claims that are not subject to pre-AIA 35 U.S.C. § 102(g) (i.e., claims of patents wherein all of the claims have an effective filing date that is after March 15, 2013). Joe Matal, *A Guide to the Legislative History of the America Invents Act: Part II of II*, 21 FED. CIR. B.J. 539, 580 (2012). Thus, post-AIA 35 U.S.C. § 273(g) may have little meaning once pre-AIA 35 U.S.C. § 102(g) no longer applies to patent claims.

8. Abandonment of Use Clause

Pre-AIA 35 U.S.C. § 273(b)(5) and post-AIA 35 U.S.C. § 273(e)(4) provide the abandonment of use clause. As shown in the following table, both the pre- and post-AIA clauses are nearly identical:

Pre-AIA 35 U.S.C. § 273(b)(5)	Post-AIA 35 U.S.C. § 273(e)(4)
ABANDONMENT OF USE.- A person who has abandoned commercial use of subject matter may not rely on activities performed before the date of such abandonment in establishing a defense under this section with respect to actions taken after the date of such abandonment.	ABANDONMENT OF USE.-- A person who has abandoned commercial use (that qualifies under this section) of subject matter may not rely on activities performed before the date of such abandonment in establishing a defense under this section with respect to actions taken on or after the date of such abandonment.

The abandonment of use clause ensures that an alleged infringer who used patented subject matter, then abandoned the subject matter, and then begins using the subject matter again, will not be able to rely on the use of the subject matter that occurred before the abandonment (i.e., by establishing that such use constitutes a prior use under pre- or post-AIA 35 U.S.C. § 273) to avoid liability for patent infringement arising out of the use occurring after the abandonment. The legislative history for this clause makes it clear that the determination as to whether any particular subject matter has been abandoned is fact-intensive. Indeed, Senator Kyl explained that post-AIA 35 U.S.C. § 273(e)(4) "should not be construed to necessarily require continuous use of the subject matter. It is in the nature of some subject matter that it will be used only periodically or seasonally. If such is the case, and the subject [matter] has been so used, its use has not been abandoned." 157 Cong. Rec. S5431, 2011.

9. Burden of Proof Clause

Both pre-AIA 35 U.S.C. § 273(b)(4) and post-AIA 35 U.S.C. § 273(b), which are identical, provide that "[a] person asserting a defense under this section shall have the burden of establishing the defense by clear and convincing evidence." The clear and convincing standard places a higher burden on the defendant to establish prior user rights than the preponderance of the evidence standard places on a patentee to demonstrate that infringement has occurred. *See, e.g., Addington v. Tex.*, 441 U.S. 418, 424-25 (1979) (explaining that the "clear and convincing" standard is an intermediate standard which lies somewhere in between the "beyond a reasonable doubt" and the "preponderance of the evidence" standards of proof).

10. Not a General License Clause

The "not a general license" clause appears in pre-AIA 35 U.S.C. § 273(b)(3)(C) and post-AIA 35 U.S.C. § 273(e)(3). As seen in the following comparison, AIA § 5(a) made only slight changes to the wording of pre-AIA 35 U.S.C. § 273(e) (3) to bring the statute into conformity with post-AIA 35 U.S.C. § 273(a)(1):

Pre-AIA 35 U.S.C. § 273(b)(3)(C)	Post-AIA 35 U.S.C. § 273(e)(3)
NOT A GENERAL LICENSE.- The defense asserted by a person under this section is not a general license under all claims of the patent at issue, but extends only to the specific subject matter claimed in the patent with respect to which the person can assert a defense under this chapter, except that the defense shall also extend to variations in the quantity or volume of use of the claimed subject matter, and to improvements in the claimed subject matter that do not infringe additional specifically claimed subject matter of the patent.	NOT A GENERAL LICENSE.-- The defense asserted by a person under this section is not a general license under all claims of the patent at issue, but extends only to the specific subject matter for which it has been established that a commercial use that qualifies under this section occurred, except that the defense shall also extend to variations in the quantity or volume of use of the claimed subject matter, and to improvements in the claimed subject matter that do not infringe additional specifically claimed subject matter of the patent.

The "not a general license" clause limits the prior user rights defense so that it can only be used to defend against claims of patent infringement for subject matter for which an alleged infringer has established that a prior commercial use (as defined in pre-AIA 35 U.S.C. § 273(a)(1) and (b)(1) and post-AIA 35 U.S.C. § 273(a)(1) and (2)) has occurred. Accordingly, if a patentee asserts a patent that claims two or more embodiments of an invention against an alleged infringer who can only establish a prior commercial use for one of the embodiments, the prior user rights defense does not convey any rights to that alleged infringer to practice the other embodiment for which no prior commercial use can be established. *See* 157 CONG. REC. S5430, 2011 (statement of Sen. Kyl) ("The [prior user rights] defense is similar to the prior-user right that exists in the United Kingdom and Germany. The defense is a relatively narrow one. It does not create a general license with respect to the patented invention, but rather only allows the defendant to keep making the infringing commercial use that he establishes that he made 1 year before the patentee's filing or disclosure.").

While the "not a general license" clause of pre- and post-AIA 35 U.S.C. § 273 restricts the subject matter for which the prior user rights defense can apply to the subject matter that gave rise to the prior user rights to begin with, the clause does allow alleged infringers to vary the quantity or volume of use of the claimed subject matter and also allows alleged infringers to make improvements to the subject matter for which they can establish a prior use so long as the improvements do not infringe additional claims of the patent at-issue.[s] Other countries only permit prior users to continue their activity "to a scope commensurate with the previous activity that triggered the prior user rights." U.S. PATENT & TRADEMARK OFFICE, REPORT TO CONGRESS ON THE PRIOR USER RIGHTS DEFENSE 23 (2012), http://www.uspto.gov/aia_implementation/20120113-pur_report.pdf. Thus, the plain language of the "not a general license" clause of pre- and post-AIA 35 U.S.C. § 273 clearly indicates that the U.S. does not take the same approach as these other countries. Joe Matal, *A Guide to the Legislative History of the America Invents Act: Part II of II*, 21 FED. CIR. B.J. 539, 575 (2012).

11. Derivation Clause

Pre-AIA 35 U.S.C. § 273(b)(3)(B) and post-AIA 35 U.S.C. § 273(e)(2) provide the derivation clause for the prior user rights defense. The following is a side-by-side comparison of the two statutes:

---

[s]The Committee Report for the Moorhead-Schroeder Patent Reform Act that contained nearly identical language to post-AIA 35 U.S.C. § 273(e)(3) explained, "[i]n other words, if the prior user must infringe additional claims of the patent in order to implement an improvement in the claimed subject matter, the prior user would not be able to rely on the [prior user rights] defense provided in this section." Joe Matal, *A Guide to the Legislative History of the America Invents Act: Part II of II*, 21 FED. CIR. B.J. 539, 576 (2012) (quoting H.R. REP. NO. 104-784, at 70 (1996)).

Pre-AIA 35 U.S.C. § 273(b)(3)(B)	Post-AIA 35 U.S.C. § 273(e)(2)
DERIVATION.- A person may not assert the defense under this section if the subject matter on which the defense is based was derived from the patentee or persons in privity with the patentee.	DERIVATION.-- A person may not assert a defense under this section if the subject matter on which the defense is based was derived from the patentee or persons in privity with the patentee.

The wording of the pre- and post-AIA statutes is identical and both act to prohibit a person from using the prior user rights defense if that person obtained the subject matter forming the basis for the defense, either directly or indirectly, from the patentee or someone in privity with the patentee.' Together, the derivation clause of post-AIA 35 U.S.C. § 273(e)(2) and good faith clause of post-AIA 35 U.S.C. § 273(a)(1) combine to define the conduct that is required of an alleged infringer who wishes to rely on the prior user rights defense. *See* U.S. PATENT & TRADEMARK OFFICE, REPORT TO CONGRESS ON THE PRIOR USER RIGHTS DEFENSE 20 (2012), http://www.uspto.gov/aia_implementation/20120113-pur_report.pdf ("The AIA has both a general good faith requirement, and also an articulation of specific conduct that would defeat the prior user rights defense."). It is important to note that the person seeking to rely on the prior user rights defense is not required to have independently invented the subject matter that serves as the basis for that defense. As explained in the Committee Report for the 1996 Moorhead-Schroeder Patent Reform Act:

> The prior user does not have to be a prior inventor in order to assert a defense based on prior use. Prior user rights may be claimed whether the party asserting the right conceived the invention or a third party conceived the invention, so long as the technology that is the basis of the prior use defense was not obtained directly or indirectly from the patentee.
> Joe Matal, *A Guide to the Legislative History of the America Invents Act: Part II of II*, 21 FED. CIR. B.J. 539, 574 (2012) (quoting H.R. REP. NO. 104-784, at 70 (1996)).

'In order for one person to be "in privity" with another person, the two must be engaged in some type of substantive legal relationship, wherein one person has the right to exert control over the other person. *See, e.g., Taylor v. Sturgell*, 553 U.S. 880, 894 (2008) (listing "preceding and succeeding owners of property, bailee and bailor, and assignee and assignor" as examples of substantive legal relationships that give rise to a finding that one person is in privity with another); *Apple Inc. v. Achates Reference Publishing, Inc.*, IPR2013-00080 and IPR2013-00081, at 5-7 (April 3, 2013) and *BAE Systems Information and Electronic Systems Integration, Inc. v. Cheetah Omni, LLC*, No. IPR2013-00175, 2013 WL 5653116, at *1, 2, (Patent Tr. & App. Bd., July 23, 2013) (finding that co-defendants in patent infringement suits were not in privity with one another where they did not have the right to control each other, where they did not fit within one of the substantive legal relationships identified in *Taylor v. Sturgell*, and where each co-defendant had distinct interests from the other co-defendants based on the unique product they were seeking to market).

D. Effective Date of Post-AIA 35 U.S.C. § 273

Only patents that are issued on or after September 16, 2011 are subject to
the prior user rights defense provided by post-AIA 35 U.S.C. § 273. Leahy-
Smith America Invents Act of 2011, Pub. L. No. 112-29, § 5(c), 125 Stat.
284, 299. Patents issued before September 16, 2011 are subject to pre-AIA
35 U.S.C. § 273.

E. Goals That Were Addressed by AIA's Amendments to 35 U.S.C. § 273

The transition from a first-to-invent to a first-inventor-to-file patent system under the AIA was the driving force behind the legislative changes
that were made to pre-AIA 35 U.S.C. § 273. U.S. PATENT & TRADEMARK
OFFICE, REPORT TO CONGRESS ON THE PRIOR USER RIGHTS DEFENSE 48
(2012), http://www.uspto.gov/aia_implementation/20120113-pur_report.
pdf; *see also* 157 CONG. REC. S5429-30, 2011 (statement of Sen. Kyl) (noting
that the expansion of prior user rights was "perhaps the most important"
consequential change in view of the transition to first-inventor-to-file). Indeed, one of the statutory amendments that was necessary to effectuate
the shift to a first-inventor-to-file system was the elimination of pre-AIA
35 U.S.C. § 102(g), which allowed prior inventors of subject matter to invalidate a later inventor's patent claims to that subject matter. Pre-AIA 35
U.S.C. § 102(g); *see also* Leahy-Smith America Invents Act of 2011, Pub.
L. No. 112-29, § 3, 125 Stat. 284, 285-87 (replacing existing Section 102,
including Section 102(g), with new provisions for determining novelty and
non-obviousness based on effective fling dates instead of dates of invention).

Prior invention is not a defense to patent infringement in a first-
inventor-to-file patent system. As explained by the U.S. PTO,

> This raises the possibility, in the absence of a defense mechanism analogous to
> [pre-AIA 35 U.S.C. §] 102(g)(2), that a party that earlier invented and commercially
> used the subject matter without disclosing it could be liable for infringement as
> against another party that later obtained a patent on it . . . In such a situation,
> the earlier inventor may be subject to liability for infringement, which could entail
> a substantial loss of investment, loss of jobs, and erosion of U.S. manufacturing.
> A prior user rights defense addresses this issue by allowing the earlier inventor to
> continue using the invention without liability, subject to certain conditions, while
> the patentee still enjoys exclusive rights as against everyone except the prior user.

**U.S. PATENT & TRADEMARK OFFICE, REPORT TO CONGRESS ON THE PRIOR USER RIGHTS DEFENSE 49-50
(2012), http://www.uspto.gov/aia_implementation/20120113-pur_report.pdf.**

One of the major goals of the AIA's amendments to pre-AIA 35 U.S.C.
§ 273 was to harmonize the U.S. prior user rights defense with those of

other industrialized countries with which the U.S. competes so that businesses would be less inclined to conduct their manufacturing processes overseas. *See* 157 CONG. REC. S5426 (statement of Sen. Blunt) ("The prior user rights provided under section 5 of H.R. 1249 will allow developers of innovative technologies to keep internally used technologies in-house without publication in a patent. This will help U.S. industry to keep jobs at home and provide a basis for restoring and maintaining a technology competitive edge for the U.S. economy."); *see also* Keith M. Kupferschmid, *Prior User Rights: The Inventor's Lottery Ticket*, 21 AIPLA Q.J. 213, 221-22 (1993) (explaining that if the U.S. fails to provide prior user rights that are commensurate in scope with those of other countries that failure may "encourage businesses to practice and exploit their inventions outside the United States, where they can be assured of receiving a prior user right.").[u]

---

[u]Although the goal of amending pre-AIA 35 U.S.C. § 273 was to harmonize the U.S. prior user rights defense with those of other countries, even the U.S. PTO recognizes that U.S. prior user rights are significantly more limited than those provided by other industrialized nations. For instance, in its Report to Congress, the U.S. PTO explained that Denmark attributes prior user rights to those who either (i) exploit the invention commercially, or (ii) make substantial preparations for commercial exploitation of the same. U.S. PATENT & TRADEMARK OFFICE, REPORT TO CONGRESS ON THE PRIOR USER RIGHTS DEFENSE 14 (2012), http://www.uspto.gov/aia_implementation/20120113-pur_report.pdf (citing Law No. 91 of 28 January 2009, Consolidate Patents Act, § 4(1) (Den.)). Similarly, Australia attributes prior user rights to those who either: (i) exploit the product, method, or process; or (ii) take definite steps to do the same. *Id.* at 15. Post-AIA 35 U.S.C. § 273, however, does not attribute prior user rights to those who have made substantial preparations for commercial exploitation of a claimed invention but instead limits the rights to those who have actually commercially used a claimed invention prior to the effective filing date of the patent. Post-AIA 35 U.S.C. § 273(a)(1). Other countries also attribute prior user rights based on uses that occur later in time than those in the U.S. For example, in the United Kingdom, the activity constituting prior use need only take place before the priority date of an invention. U.S. PATENT & TRADEMARK OFFICE, REPORT TO CONGRESS ON THE PRIOR USER RIGHTS DEFENSE 21 (2012), http://www.uspto.gov/aia_implementation/20120113-pur_report.pdf (citing U.K. Patent Act, at § 64)). And in Germany, prior use activity must take place at, or before, the date of a patent filing. *Id.* (citing German Patent Act, at § 12(2)). In contrast, in order to assert the prior user rights defense under the AIA, a commercial use must have occurred at least one year before the earlier of either: "(A) the effective filing date of the claimed invention; or (B) the date on which the claimed invention was disclosed to the public in a manner that qualified for the exemption from the prior art under [post-AIA 35 U.S.C. §] 102(b)." Post-AIA 35 U.S.C. § 273(a)(2). According to the U.S. PTO, "[t]his makes the U.S. temporal approach significantly more restrictive than that for any other prior user rights system." U.S. PATENT & TRADEMARK OFFICE, REPORT TO CONGRESS ON THE PRIOR USER RIGHTS DEFENSE 21 (2012), http://www.uspto.gov/aia_implementation/20120113-pur_report.pdf.

F. Practical Implications of the AIA's Changes to the Prior User Rights Defense

The issue of prior user rights tends to arise in only a very small number of patent cases. *See, e.g.,* Keith M. Kupferschmid, *Prior User Rights: The Inventor's Lottery Ticket*, 21 AIPLA Q.J. 213, 223-24 (1993) ("Even though most countries provide for prior user rights, the amount of litigation involving prior user rights is minimal. Since prior user right litigation is minimal in countries presently having the right, prior user right litigation will also be minimal in the United States."). In fact, even though the U.S. has had a prior user rights statute since 1999, there have been only 6 published opinions that discuss the issue.[v] Given that prior user rights are unlikely to be asserted in the vast majority of cases, the AIA's amendments to pre-AIA 35 U.S.C. § 273 are unlikely to have much effect on the vast majority of patent holders and alleged infringers. *See* U.S. Patent & Trademark Office, Report to Congress on the Prior User Rights Defense 32 (2012), http://www.uspto.gov/aia_implementation/20120113-pur_report.pdf ("There have been prior user rights in the United States for business method patents since the enactment of a 1999 law, but by all accounts this particular defense does not appear to have had a significant impact, either in industrial practice or in patent litigation.").

In the few cases in which an alleged infringer might contemplate raising a prior user rights defense, he/she should evaluate whether there is a reasonable basis to do so, particularly in light of the unreasonable assertion of the defense clause of post-AIA 35 U.S.C. § 273(f). As mentioned earlier, the AIA has placed certain restrictions on the prior user rights defense, including the timing in which a prior commercial use must have occurred in order to give rise to prior user rights and the types of patent claims and patents against which the prior user rights defense can be asserted, such that it may now be available to fewer alleged infringers than the pre-AIA defense was. Thus, any alleged infringer who is thinking about raising the defense should work closely with his/her patent attorney to determine:

[v]The opinions that discuss this issue are: *Vaughan Co. v. Global Bio-Fuels Tech., LLC*, No. 1:12-CV-1292, 2013 U.S. Dist. LEXIS 152068 (N.D.N.Y. Oct. 23, 2013); *Auburn Univ. v. IBM, Corp.*, No. 3:09-CV-694, 2012 U.S. Dist. LEXIS 107759 (E.D. Ala. Aug. 2, 2012); *Sabasta v. Buckaroos, Inc.*, 507 F. Supp. 2d 986 (S.D. Iowa 2007); *Farradyne v. Peterson*, No. C 05-3447, 2006 U.S. Dist. LEXIS 67281 (N.D. Cal. Sept. 5, 2006); *Farradyne v. Peterson*, No. C 05-3447, 2006 U.S. Dist. LEXIS 3408 (N.D. Cal. Jan. 13, 2006); and *Seal-Flex, Inc. v. W.R. Dougherty and Associates, Inc.*, 179 F. Supp. 2d 735 (E.D. Mich. 2002). It is safe to assume that prior user rights have arisen in more than these six cases, however, because it is likely that cases in which an alleged infringer had solid evidence that he/she was entitled to prior user rights would have settled. Given that most settlements are confidential, it is impossible to determine the actual number of cases in which a defense of prior user rights has been raised.

(1) whether his/her alleged prior use of claimed subject matter meets the timing requirement of post-AIA 35 U.S.C. § 273(a)(2), and (2) whether the patent claims, and/or patent, at-issue are even subject to the defense.[w]

Even if an alleged infringer does have a reasonable basis to raise a prior user rights defense, however, he/she may not want to do so. Indeed, by establishing that the subject matter of the patent claims being asserted was used at a time that predates the effective filing date or relevant public disclosure of the claimed invention, an alleged infringer will likely undercut any arguments that he/she does not infringe the patent claims at issue. Alleged infringers should also keep in mind that the patentee's burden for establishing patent infringement (i.e., the preponderance of the evidence standard) is much lower than the evidentiary burden for establishing a prior user rights defense (i.e., a clear and convincing evidence standard). Thus, alleged infringers who rely on the prior user rights defense could end up in the unfortunate position of having put forth enough evidence to establish that they infringe the plaintiff's patent but not enough to establish that they are entitled to prior user rights.

Another important aspect of the prior user rights defense is post-AIA 35 U.S.C. § 273(a)(1)'s requirement that "commercial uses" must occur in the United States in order to give rise to the defense. This requirement is significant because foreign entities that conduct manufacturing processes overseas and then ship the products of those processes into the U.S. can be liable for patent infringement if the processes are patented in the U.S. *See* 35 U.S.C. § 271(g) ("Whoever without authority imports into the United States . . . a product which is made by a process patented in the United States shall be liable as an infringer, if the importation . . . occurs during the term of such process patent."). Thus, although such entities can be liable for patent infringement, they cannot rely on their manufacturing processes practiced abroad to establish that they have prior user rights to defend against claims of patent infringement in the U.S.

Overall, the AIA's amendments to pre-AIA 35 U.S.C. § 273 make it increasingly difficult for an alleged infringer to rely on prior user rights as a defense to patent infringement. Indeed, despite the AIA's efforts to make the defense available against a broader range of patent claims, and for a broader group of alleged infringers, it is likely that the timing requirement for when a prior use must take place in order for prior user rights to vest, as well as the various exceptions to the defense set forth in the statute, will prevent many prior users from being able to rely on the defense.

[w]Alleged infringers should also think about when to raise the defense as well. Indeed, if the meaning of several claim terms is disputed, it might be best to wait until after claim construction to evaluate whether there is a reasonable basis to raise the defense, particularly in light of the unreasonable assertion of defense clause.

III) AIA § 15: ELIMINATION OF THE BEST MODE DEFENSE FROM 35 U.S.C. § 282

A. Background on the Best Mode Defense

Pre-AIA 35 U.S.C. § 112, ¶ 1 and post-AIA 35 U.S.C. § 112(a) require that the specification of a patent set forth the best mode contemplated by the inventor of carrying out his invention. The best mode requirement is "separate and distinct" from the written description and enablement requirements of those statutes and "requires an inventor to disclose the best mode *contemplated by him*, as of the time he executes the application, of carrying out the invention." *In re Gay*, 309 F.2d 769, 772 (C.C.P.A. 1962) (emphasis in original). "The purpose of the best mode requirement is to ensure that the public, in exchange for the rights given the inventor under the patent laws, obtains from the inventor a full disclosure of the preferred embodiment of the invention." *Dana Corp. v. IPC Ltd. P'ship*, 860 F.2d 415, 418 (Fed. Cir. 1988).

To determine whether a patent specification complies with the best mode requirement of pre-AIA 35 U.S.C. § 112, ¶ 1, courts engage in a two-part inquiry. First, the fact-finder determines whether, at the time of filing the application, the inventor possessed a best mode for practicing the invention.[x] *Fonar Corp. v. General Elec. Co.*, 107 F.3d 1543, 1548 (Fed. Cir. 1997). Second, if the inventor possessed a best mode, the fact-finder must determine whether the specification disclosed the best mode such that a person of ordinary skill in the art could practice it.[y] *Id.* The first prong involves a subjective inquiry, focusing on the inventor's state of mind at the time of filing. *U.S. Gypsum Co. v. Nat'l Gypsum Co.*, 74 F.3d 1209, 1212 (Fed. Cir. 1996); *Chemcast Corp. v. Arco Indus. Corp.*, 913 F.2d 923, 928 (Fed. Cir. 1990). The second prong involves an objective inquiry, focusing on the scope of the claimed invention and the level of skill in the art. *Chemcast Corp.*, 913 F.2d at 928. With respect to both the first and second prongs, "[t]he best mode inquiry is directed to what the applicant regards as the invention, which in turn is measured by the claims." *Engel Indus., Inc.*, 946 F.2d at 1531.

---

[x]The first prong focuses on the inventor's state of mind at the time he filed the patent application and asks whether the inventor considered a particular mode of practicing the invention to be superior to all other modes at the time of filing. *N. Telecom Ltd. v. Samsung Elecs. Co., Ltd.*, 215 F.3d 1281, 1286 (Fed. Cir. 2000).

[y]With respect to the second prong of the best mode requirement, the extent of information that an inventor must disclose depends on the scope of the claimed invention. *Engel Indus., Inc. v. Lockformer Co.*, 946 F.2d 1528, 1531 (Fed. Cir. 1991). Accordingly, an inventor need not disclose a mode for obtaining unclaimed subject matter unless the subject matter is novel and essential for carrying out the best mode of the invention. *Applied Med. Resources Corp. v. U.S. Surgical Corp.*, 147 F.3d 1374, 1377 (Fed. Cir. 1998).

While patent claims have not frequently been invalidated based on a specification's failure to disclose the best mode, situations in which they have been invalidated typically involve a failure to disclose a preferred embodiment of the claimed invention or a failure to disclose a preference that materially affects making or using the claimed invention. *Bayer AG v. Schein Pharms., Inc.*, 301 F.3d 1306, 1316 (Fed. Cir. 2002); *see also Great N. Corp. v. Henry Molded Products, Inc.*, 94 F.3d 1569, 1571 (Fed. Cir. 1996) (finding a best mode violation where the specification failed to disclose preferred diamond indentations that were placed on a roll stacker made of papier-mâché because the diamond indentations were crucial to producing a usable version of the invention); *Chemcast Corp.*, 913 F.2d at 928-30 (invalidating a claim to a grommet where the inventor concealed the particular material he preferred to use to make the locking portion of the grommet and skilled practitioners could neither have known what the inventor's contemplated best mode was nor have carried it out); *Spectra-Physics, Inc. v. Coherent, Inc.*, 827 F.2d 1524, 1536-37 (Fed. Cir. 1987) (finding that a failure to disclose a method of a TiCuSil active metal brazing process to attach a copper cup to the inside wall of a ceramic tube structure that was preferred by the inventor rendered the patent claims invalid for failure to disclose the best mode because less than ideal brazing processes affected the efficiency and reliability of the claimed lasers).

Prior to the AIA, best mode challenges could arise in a few different contexts. First, a patent could be challenged in a district court litigation as being invalid for failing to disclose the best mode of carrying out the invention pursuant to pre-AIA 35 U.S.C. § 112, ¶ 1. Intervening prior art[z] could also be used in either a district court litigation or reexamination proceeding to invalidate a patent by arguing that a patent application to which the patent claims priority does not disclose the best mode for carrying out the claimed invention. Alleged infringers could also allege that a patent should be held unenforceable where there is evidence that a patentee committed inequitable conduct by intentionally omitting the best mode from a patent's specification. Lastly, although rare, Patent Examiners could reject claims of a

---

[z]The term "intervening prior art" refers to prior art that has an effective prior art date that is between the priority date of a patent or patent application and the actual filing date of the patent or patent application. When intervening prior art is cited, Patent Examiners or courts must investigate whether the patent or patent application is actually entitled to claim priority to a previously filed application. Under the pre-AIA law, if the previously filed application does not disclose the best mode for carrying out the claimed invention of the later-filed patent or patent application, the later-filed patent application would not be entitled to rely on the previously-filed patent application's filing date to overcome rejections or invalidity arguments based on the intervening prior art.

patent application under pre-AIA 35 U.S.C. § 112, ¶ 1 for failing to comply with the best mode requirement.[aa]

In the years leading up to the AIA's enactment, many criticized the best mode defense for being too subjective, counterproductive, and burdensome for patent litigants. *See, e.g.*, H.R. REP. NO. 112-98, at 52 (2011) ("Many have argued in recent years that the best mode requirement, which is unique to American patent law, is counterproductive. They argue that challenges to patents based on best mode are inherently subjective and not relevant by the time the patent is in litigation, because the best mode contemplated at the time of the invention may not be the best mode for practicing or using the invention years later."); 153 CONG. REC. H10, 303-04, 2007 (statement of Rep. Pence) ("Increasingly in patent litigation defendants have put forth best mode as a defense and a reason to find patents unenforceable. It becomes virtually a satellite piece of litigation in and of itself, detracts from the actual issue of infringement, and literally costs American inventors millions in legal fees."); *Patent Reform Act of 2007: Hearing on H.R. 1908 Before the Subcomm. on Courts, the Internet, and Intellectual Prop.*, 110th Cong. 56 (2007) (statement of Gary L. Griswold, President and Chief IP Counsel of 3M Innovative Properties Companies) ("The 'best mode' requirement is the most *subjective* validity assessment in all of patent law. It requires knowing what the inventor *contemplated* on the day the inventor filed his patent application . . . [it] requires discovery of every mode the inventor knew at the time the patent was sought. This means reviewing every document the inventor wrote—or read—relating to a mode for carrying out the invention."). The problems with the best mode defense caused many to believe that the best mode requirement of pre-AIA 35 U.S.C. § 112, ¶ 1 should be repealed entirely. *See, e.g.*, 157 CONG. REC. E1175, 2011 (statement of Rep. Pence) ("I have maintained since 2007 that best mode should be repealed in full, and I would continue to support a full repeal if possible today.").

B. AIA § 15's Changes to the Best Mode Defense

While the AIA did not go so far as to repeal the best mode requirement from 35 U.S.C. § 112, it did succeed in removing best mode from the list of possible

---

[aa]The reason that a best mode rejection under pre-AIA 35 U.S.C. § 112, ¶ 1 is rare is because Patent Examiners are not usually privy to the type of information that would enable them to reject claims for a failure to disclose the best mode. Indeed, a Patent Examiner would likely need to talk to an inventor, and/or at least see lab notebooks and/or invention records that were kept during the development of a claimed invention, in order to determine whether a best mode existed at the time the patent application was filed and if so, whether that best mode was disclosed in the specification. Patent Examiners, however, typically only examine patent applications in light of information that is publicly available.

defenses provided in 35 U.S.C. § 282 that can be used by defendants in patent litigation. Leahy-Smith America Invents Act of 2011, Pub. L. No. 112-29, § 15(a), 125 Stat. 284, 328; *see also* post-AIA 35 U.S.C. § 112(a) ("The specification . . . shall set forth the best mode contemplated by the inventor or joint inventor of carrying out the invention."). The following table shows a side-by-side comparison of the relevant portions of pre- and post-AIA 35 U.S.C. § 282:

Pre-AIA 35 U.S.C. § 282 Presumption of Validity; Defenses	Post-AIA 35 U.S.C. § 282 Presumption of Validity; Defenses

A patent shall be presumed valid. Each claim of a patent (whether in independent, dependent, or multiple dependent form) shall be presumed valid independently of the validity of other claims; dependent or multiple dependent claims shall be presumed valid even though dependent upon an invalid claim. Notwithstanding the preceding sentence, if a claim to a composition of matter is held invalid and that claim was the basis of a determination of nonobviousness under section 103(b)(1), the process shall no longer be considered nonobvious solely on the basis of section 103(b)(1). The burden of establishing invalidity of a patent or any claim thereof shall rest on the party asserting such invalidity. The following shall be defenses in any action involving the validity or infringement of a patent and shall be pleaded:
(1) Noninfringement, absence of liability for infringement, or unenforceability,
(2) Invalidity of the patent or any claim in suit on any ground specified in part II of this title as a condition for patentability,
(3) Invalidity of the patent or any claim in suit for failure to comply with any requirement of sections 112 or 251 of this title,
(4) Any other fact or act made a defense by this title.
	(a) IN GENERAL.-- A patent shall be presumed valid. Each claim of a patent (whether in independent, dependent, or multiple dependent form) shall be presumed valid independently of the validity of other claims; dependent or multiple dependent claims shall be presumed valid even though dependent upon an invalid claim. The burden of establishing invalidity of a patent or any claim thereof shall rest on the party asserting such invalidity.
	(b) DEFENSES.-- The following shall be defenses in any action involving the validity or infringement of a patent and shall be pleaded:
	(1) Noninfringement, absence of liability for infringement, or unenforceability.
	(2) Invalidity of the patent or any claim in suit on any ground specified in part II as a condition for patentability.
	(3) Invalidity of the patent or any claim in suit for failure to comply with--
	  (A) any requirement of section 112, <u>except that the failure to disclose the best mode shall not be a basis on which any claim of a patent may be canceled or held invalid or otherwise unenforceable;</u> or
	  (B) any requirement of section 251.
	(4) Any other fact or act made a defense by this title.
	(Emphasis added.)

By stating that defenses based on "the failure to disclose the best mode shall not be a basis on which any claim of a patent may be canceled or held invalid or otherwise unenforceable," post-AIA 35 U.S.C. § 282(b)(3)(A) forecloses the possibility of raising the best mode defense in district court litigation, either on the basis of invalidity under pre- or post-AIA 35 U.S.C. § 112 or on the basis that a patent is unenforceable for failure to disclose the best mode requirement under an inequitable conduct theory, or in post-grant proceedings before the Patent Trial and Appeals Board. Indeed, Senator Kyl explained:

> In the new effective-date subsection, the section is made applicable to all "proceedings" commenced after enactment of the Act, in order to make clear that the section's changes to the law will be immediately applicable not just in litigation but also in post-grant reviews of patents under chapter 32.
>
> See 157 CONG. REC. S1378, 2011.

Additionally, AIA § 15(b) amended 35 U.S.C. §§ 119(e)(1) and 120 so that failure to disclose the best mode of carrying out an invention in a provisional or non-provisional application filed in the U.S., or a PCT application meeting the requirements of 35 U.S.C. § 363, could not serve as a basis to deny a claim of priority to such applications. Leahy–Smith America Invents Act of 2011, Pub. L. No. 112-29, 125 Stat. 284, 328. The following tables provide a comparison of the relevant pre- and post-AIA statutory language:

Pre-AIA 35 U.S.C. § 119(e)(1) Benefit of Earlier Filing Date; Right of Priority	Post-AIA 35 U.S.C. § 119(e)(1) Benefit of Earlier Filing Date; Right of Priority
(e)(1) An application for patent filed under section 111(a) or section 363 of this title for an <u>invention disclosed in the manner provided by the first paragraph of section 112 of this title</u> in a provisional application filed under section 111(b) of this title, by an inventor or inventors named in the provisional application, shall have the same effect, as to such invention, as though filed on the date of the provisional application filed under section 111(b) of this title, if the application for patent filed under section 111(a) or section 363 of this title is filed not later than 12 months after the date on which the provisional application was filed and if it contains or is amended to contain a specific reference to the provisional application .... (Emphasis added.)	(e)(1) An application for patent filed under section 111(a) or section 363 <u>for an invention disclosed in the manner provided by section 112(a) (other than the requirement to disclose the best mode)</u> in a provisional application filed under section 111(b), by an inventor or inventors named in the provisional application, shall have the same effect, as to such invention, as though filed on the date of the provisional application filed under section 111(b), if the application for patent filed under section 111(a) or section 363 is filed not later than 12 months after the date on which the provisional application was filed and if it contains or is amended to contain a specific reference to the provisional application .... (Emphasis added.)

Pre-AIA 35 U.S.C. § 120 Benefit of Earlier Filing Date in the United States	Post-AIA 35 U.S.C. § 120 Benefit of Earlier Filing Date in the United States
An application for patent for an invention underlined{disclosed in the manner provided by the first paragraph of section 112 of this title} in an application previously filed in the United States, or as provided by section 363 of this title, which is filed by an inventor or inventors named in the previously filed application shall have the same effect, as to such invention, as though filed on the date of the prior application, if filed before the patenting or abandonment of or termination of proceedings on the first application or on an application similarly entitled to the benefit of the filing date of the first application and if it contains or is amended to contain a specific reference to the earlier filed application . . . . (Emphasis added.)	An application for patent for an invention underlined{disclosed in the manner provided by section 112(a) (other than the requirement to disclose the best mode)} in an application previously filed in the United States, or as provided by section 363, which names an inventor or joint inventor in the previously filed application shall have the same effect, as to such invention, as though filed on the date of the prior application, if filed before the patenting or abandonment of or termination of proceedings on the first application or on an application similarly entitled to the benefit of the filing date of the first application and if it contains or is amended to contain a specific reference to the earlier filed application. (Emphasis added.)

The amendments to 35 U.S.C. §§ 119(e)(1) and 120 remove any possibility that a patent or patent application will be denied a claim of priority to an earlier-filed domestic patent application, or PCT application that complies with 35 U.S.C. § 363, on the basis that the earlier-filed application fails to disclose the best mode for carrying out a claimed invention. Thus, under the post-AIA law, alleged infringers cannot succeed in arguing that a patent is invalid over intervening prior art solely because the application to which the patent claims priority fails to disclose the best mode of practicing the claimed invention.

It is important to note that the AIA did not repeal the best mode requirement from 35 U.S.C. § 119(a), which pertains to priority claims to foreign patent applications. As explained by Joe Matal,

> *AIA section 15 does not repeal the best-mode requirement from § 119(a). This is because there is no best-mode language in § 119(a) to repeal. The courts have sought "to preserve symmetry of treatment between sections 120 and 119," and since pre-AIA § 120 expressly required best-mode disclosure in a domestic parent application, § 119(a) was read to also require such disclosure in a foreign application. Now that the best-mode requirement has been repealed from § 120,*

however, the same "symmetry" rationale requires that best mode be read out of § 119(a) as well.

A Guide to the Legislative History of the America Invents Act: Part II of II, 21 FED. CIR. B.J. 539, 584 (2012) (citations omitted).

Thus, even though the AIA did not amend pre-AIA 35 U.S.C. § 119(a), patent applicants and patentees should not be denied a claim of priority to an earlier-filed foreign patent application on the basis that such application does not describe the best mode for carrying out the claimed invention. Indeed, pre- and post-AIA 35 U.S.C. § 119(a) require that foreign priority applications "shall have the same effect" as U.S.-filed priority applications, and U.S.-filed priority applications need not disclose the best mode for a claimed invention to provide priority for another patent or patent application. *Id.*

The effect of the AIA's amendments to 35 U.S.C. §§ 119(e)(1), 120, and 282 is that the failure to disclose the best mode of a claimed invention can no longer form the basis for invalidating or rendering unenforceable a patent claim in a district court litigation or for canceling a patent claim in a post-grant proceeding before the Patent Trial and Appeal Board. As a result, the only context in which best mode remains relevant is in patent prosecution. *See* 157 CONG. REC. E1175, 2011 (statement of Rep. Pence) (explaining that section 15 "retains the best mode as a specifications requirement for obtaining a patent" but "[o]nce the examiner is satisfied that the best mode has been disclosed, the issue is settled forever"). More specifically, the only context in which a best mode issue may arise in prosecution is during the examination of a patent application that does not claim priority to any other patent application. *See* Robert A. Armitage, *Understanding the America Invents Act*, 40 AIPLA Q.J. 1, 91 (2012) (citing *Transco Prods., Inc. v. Performance Contracting, Inc.*, 38 F.3d 551, 557 (Fed. Cir. 1994)) ("[S]ince the issue of entitlement to priority or benefit of an earlier-filed application for patent does not require that a 'best mode' disclosure be found in the earlier-filed application for patent, there should be no basis during ex parte patent examination to reject a claimed invention on the ground that, as of the effective filing date for the claimed invention, no best mode disclosure was present."). The following table shows a comparison of the contexts in which best mode could arise under the relevant pre- and post-AIA statutes:

Contexts in Which Best Mode Issues Could Arise Prior to the AIA	Contexts in Which Best Mode Issues Could Arise After the AIA
Patent Examination: i) a Patent Examiner could reject a claim in a patent application (that either does or does not claim priority to another patent application) that fails to disclose the best mode under 35 U.S.C. § 112, ¶ 1; and/or ii) a Patent Examiner could reject a claim in a patent application over intervening prior art under 35 U.S.C. §§ 102 and/or 103 where an application to which the patent application claims priority fails to disclose the best mode.	Patent Examination: i) a Patent Examiner could reject a claim in a patent application (that <u>does not</u> claim priority to any other patent application) for failure to disclose the best mode under 35 U.S.C. § 112, ¶ 1.
Post-Grant Proceedings at the U.S. PTO: i) a patent claiming priority to another patent application could be challenged in a reexamination proceeding as being invalid over intervening prior art if a patent challenger could demonstrate that the priority application failed to disclose the best mode of the claimed invention at-issue.	
District Court Litigation: i) a patent could be challenged under 35 U.S.C. § 112, ¶ 1 for failing to disclose the best mode; ii) a patent claiming priority to another patent application could be challenged as being invalid over intervening prior art under 35 U.S.C. §§ 102 and/or 103 if a patent challenger could demonstrate that the priority application failed to disclose the best mode of the claimed invention at-issue; and/or iii) a defendant could allege that a patent is unenforceable if he/she could demonstrate that the patentee had committed inequitable conduct by failing to disclose the best mode.	

C. Effective Date of AIA § 15

AIA § 15(c) states that "[t]he amendments made by this section shall take effect upon the date of the enactment of this Act and shall apply to proceedings commenced on or after that date." Leahy-Smith America Invents Act of 2011, Pub. L. No. 112-29, 125 Stat. 284, 328. The AIA was enacted on September 16, 2011. *Id.* at 284. Accordingly, the best mode defense cannot be asserted in any district court litigation, patent appeal proceeding before the Patent Trial and Appeal Board, or post-grant proceeding before the Patent Trial and Appeal Board that is filed on or after September 16, 2011. Robert A. Armitage, *Understanding the America Invents Act*, 40 AIPLA Q.J. 1, 91 (2012).

D. Legislative Goals Addressed by AIA § 15

Congress' decision to enact § 15 of the AIA appears to have been made to balance the interests of those groups that were lobbying heavily to repeal best mode from pre-AIA 35 U.S.C. § 112, ¶ 1 entirely (e.g., patent-holding companies) and those that were lobbying heavily to maintain the status quo with respect to best mode (e.g., generic drug manufacturers). As explained by Andrew R. Shores[bb]:

> Prior to the AIA, two main camps formed: those arguing to eliminate the best mode requirement completely, and those hoping to retain the requirement as it was ... The compromise struck by Congress, eliminating the best mode defense that led to excessive litigation costs while retaining the best mode requirement to promote the public's benefit of the patent bargain, is the perfect solution.

Change to the Best Mode Requirement in the Leahy-Smith America Invents Act: Why Congress Got it Right, 34 CAMPBELL L. REV. 733, 740 (2012) (citations omitted); *see also Perspective on Patents: Harmonization and Other Matters: Hearing Before the Subcomm. on Intellectual Prop. of the S. Comm. on the Judiciary*, 109th Cong. 77, 116 (2005) (Steven J. Lee, Kenyon & Kenyon, on behalf of Teva North America) ("It is not an overstatement to say that th[e] [best-mode] requirement is a large part of the reason for the United States' technological success.").

[bb]Andrew Shores cites to Wesley D. Markham, *Is Best Mode the Worst? Dueling Arguments, Empirical Analysis, and Recommendations for Reform*, 51 IDEA 129, 132 (2011) (discussing the "pro-big business" group arguing to eliminate the best mode requirement and the "generic [pharmaceutical companies] plus others" group arguing to keep the requirement as it was).

E. Practical Implications of the Elimination of the Best Mode Defense

A question that naturally arises as a result of the AIA's elimination of the best mode defense is whether patent applicants should bother continuing to disclose the best mode for their claimed inventions going forward. While elimination of the defense may tempt many to omit the best mode from future patent applications, concealment of the best mode could still negatively affect the validity of a patent. For example, given that claims are construed to be consistent with what is disclosed in the specification, it is possible that the claims of a patent could be construed to exclude a best mode if it is not disclosed in its specification. *See Merck & Co. v. Teva Pharms. USA, Inc.,* 347 F.3d 1367, 1371 (Fed. Cir. 2003) ("A fundamental rule of claim construction is that terms in a patent document are construed with the meaning with which they are presented in the patent document. Thus claims must be construed so as to be consistent with the specification, of which they are a part.") (citations omitted); *Vitronics Corp. v. Conceptronic, Inc.,* 90 F.3d 1576, 1583 (Fed. Cir. 1996) (explaining that a claim construction that would exclude the preferred embodiment "is rarely, if ever, correct and would require highly persuasive evidentiary support[]"); *see also* Robert A. Armitage, *Understanding the America Invents Act,* 40 AIPLA Q.J. 1, 88 (2012) ("An undisclosed mode cannot be specifically claimed.").

Failing to disclose the best mode for an invention could also impact the validity of a patent's claims under pre- or post-AIA 35 U.S.C. §§ 102, 103 and 112. Indeed, as Robert Armitage points out, "[the best mode] may prove to be the only mode that survives the tests for novelty and non-obviousness" under 35 U.S.C. §§ 102 and 103. *Id.* For instance, it is not difficult to imagine a scenario where an invention that appears to be obvious over certain prior art references is rendered not obvious by virtue of unexpected results that occur when the best mode of the invention is used. And "[i]n many situations, to assure a claimed invention is fully enabled [under post-AIA 35 U.S.C. § 112(a)], it becomes essential that the best mode of carrying out the invention be laid out to bolster that enablement." *Id.*

Last but certainly not least, because the AIA did not repeal the best mode requirement from 35 U.S.C. § 112, patent applicants are not free to ignore the requirement. *See id.* at 87 ("Put another way, so long as the law formally imposes a requirement on patent applicants, it is not up to the applicant to decide which such requirements of law must be observed and which may be ignored."). Indeed, though unlikely, it is possible that a patent practitioner could be disciplined by the U.S. PTO's Office of Enrollment and Discipline if he/she knowingly filed a patent application that failed to disclose the best mode. *See, e.g.*, 37 C.F.R. § 11.303 ("In an *ex parte* proceeding, a practitioner shall inform the tribunal of all material facts known to the practitioner that will enable the tribunal to make an informed decision, whether or not the facts are adverse.").

While it may take several years to determine whether the failure to disclose a best mode will adversely affect the validity of any patents, the most prudent course of action, and the action that is currently required by post-AIA 35 U.S.C. § 112(a), is to continue to disclose the best mode in any patent application that is filed with the U.S. PTO.

CHAPTER EIGHT

AIA's Changes to Patent Prosecution Practices

Contents

America Invents Act Primer. http://dx.doi.org/10.1016/B978-0-12-812096-5.00008-9

I) INTRODUCTION

Several patent prosecution practices have been significantly modified because of the AIA. Indeed, in addition to the changes that were effected as a result of the shift from a first-to-invent to a first-inventor-to-file patent system that are discussed in Chapters 2 and 3, the AIA repealed the statute providing authority for the publication of patent applications as statutory invention registrations, added prioritized examination as an additional option for expediting patent examination, modified the third-party preissuance submission practice in an effort to entice third parties to play a more active role in patent examination, and clarified the statute pertaining to patent term extensions. This chapter will provide a discussion of these changes to patent prosecution practices and provide analyses of the practical implications of the same.

II) AIA § 3(e): REPEAL OF STATUTORY INVENTION REGISTRATIONS

Unless you have been involved in patent law for a long time, it is unlikely you would have ever heard of a statutory invention registration (also known as a "SIR"). A statutory invention registration is a defensive patent application publication that can be used to prevent others from obtaining a patent on the subject matter disclosed therein. While statutory invention registrations benefitted some patent applicants in the past, their disadvantages far outweighed their advantages once the America Inventor's Protection Act ("AIPA") was passed in 1999. It is not surprising, therefore, that Congress repealed the statutory authority for statutory invention registrations through the AIA, especially in light of the fact that very few had been published between 2001 and the time the AIA was passed.

A. Statutory Invention Registrations Under the Pre-AIA Law

Prior to the AIA, a patent applicant could request that the specification and drawings of his/her application be published as a statutory invention registration at any time during the pendency of his/her complete patent application. Pre-AIA 35 U.S.C. § 157(a); pre-AIA 37 C.F.R. § 1.293(a). By publishing a patent application as a statutory invention registration, a patent applicant could ensure that the contents of that application could be used defensively to prevent others from patenting the subject matter described therein. Pre–AIA 35 U.S.C. § 157(c); *see also* pre–AIA 37 C.F.R. § 1.297(b) ("[A statutory invention registration] has the defensive attributes of a patent but does not have the enforceable attributes of a patent."). In particular, as explained in MPEP § 1111 (8th ed. 2012):

> *In accordance with 35 U.S.C. 157(c), a published SIR will be treated the same as a U.S. patent for all defensive purposes, usable as a reference as of its filing date in the same manner as a patent. A SIR is a "constructive reduction to practice" under [pre-AIA] 35 U.S.C. 102(g) and "prior art" under all applicable sections of [pre-AIA] 35 U.S.C. 102 including [pre-AIA] 102(e). SIRS are classified, cross-referenced, and placed in the search files, disseminated to foreign patent offices, stored in U.S. Patent and Trademark Office computer tapes, made available in commercial data bases, and announced in the Official Gazette.*

In 1984, Congress first provided patent applicants with the option to publish their patent applications as statutory invention registrations. Patent Law Amendments Act of 1984, Pub. L. No. 98-622, § 157, 98 Stat. 3383-84. At that time, the U.S. PTO kept patent applications confidential unless and until they issued as patents. *See, e.g.*, Press Release, U.S. PTO, U.S. PTO Will Begin Publishing Patent Applications (Nov. 27, 2000), http://www.uspto.gov/about-us/news-updates/uspto-will-begin-publishing-patent-applications ("The United States Patent and Trademark Office (USPTO) will begin publishing, for the first time, patent applications filed on or after November 29, 2000 eighteen months after the effective filing date of the application."); Michael R. McGurk et al., *Report: The American Inve[n]tors Protection Act of 1999* (1999), http://www.finnegan.com/resources/articles/articlesdetail.aspx?news=36da41a7-178f-4385-b359-6f276f05c303 ("Before enactment [of The Domestic Publication of Foreign Filed Patent Applications Act of 1999], U.S. patent applications were not published until they issued as patents. In contrast, most foreign applications are published eighteen months after the earliest filing date for which benefit is sought."). As a result, patent applications that never issued as patents (i.e., abandoned patent applications) could not be cited as prior art against claims of other patent applications because their contents were not publicly available. *See* MPEP § 2127(I) (9th ed. 2015) ("An abandoned patent application becomes available as prior art only as of the date the public gains access to it."). By publishing patent applications as statutory invention registrations, therefore, patent applicants could ensure that others could not obtain a patent on the subject matter that was disclosed therein.

The major disadvantage of publishing a patent application as a statutory invention registration was that by doing so, a patent applicant gave up all rights to obtain a patent on any inventions that were claimed therein. *See* pre-AIA 35 U.S.C. § 157(a)(3) (requiring that a patent applicant "waive[] the right to receive a patent on the invention" in order to obtain publication of a statutory invention registration); pre-AIA 37 C.F.R. § 1.293(b)(1) (requiring that any request for publication of a statutory invention registration include "[a] waiver of the applicant's right to receive a patent on the invention claimed effective upon the date of publication of the statutory invention registration"). Accordingly, once the U.S. PTO began publishing most non-provisional applications pursuant to the America Inventor's Protection Act of 1999, it made little sense for patent applicants to publish their patent applications as statutory invention registrations, and the number of statutory invention registration publications decreased dramatically. *See, e.g.*, MPEP § 1101 (8th ed. 2012) ("An applicant may find publication of an application to be a desirable alternative to requesting a SIR since publication of the application is achieved without any waiver of patent rights."); Posting of Michael White to http://patentlibrarian.blogspot.com/2013/06/statutory-invention-registration.html (June 2, 2013, 9:06 PM) ("The USPTO has published approximately 2,500 SIRs since the program was established in the mid-1980s as a replacement for the Defensive Publication Program. The number of SIRs per year has steadily decreased since the USPTO began publishing applications in 2001. Only seven were published in 2012.").

B. AIA § 3(e)'s Abolishment of Statutory Invention Registrations

AIA § 3(e)(1) repeals pre-AIA 35 U.S.C. § 157, thereby ending the statutory authorization for statutory invention registrations. Leahy-Smith America Invents Act of 2011, Pub. L. No. 112-29, 125 Stat. 284, 287. As a result, patent applicants no longer have the ability to request publication of their patent applications as statutory invention registrations under the post-AIA law.

C. Effective Date of AIA § 3(e)

AIA § 3(e)(3) provides that AIA § 3(e) "shall take effect upon the expiration of the 18-month period beginning on the date of the enactment of this Act, and shall apply to any request for a statutory invention registration filed on or after that effective date." Leahy-Smith America Invents Act of 2011, Pub. L. No. 112-29, 125 Stat. 284, 288. The "18-month period beginning on the date of the enactment of [the AIA]" expired on March 16, 2013. Accordingly, any request for a statutory invention registration that was filed on or after March 16, 2013 should have been denied, and no further statutory invention registrations should be published.

D. Legislative Goals Addressed by AIA § 3(e)

The legislative history on AIA § 3(e) is sparse. What little was said about statutory invention registrations, however, indicated that they would not be needed once certain provisions of the AIA took effect. Indeed, during his March 8, 2011 comments on the AIA, Senator Kyl explained that:

> SIRs are needed only so long as interferences exist. The [AIA] repeals the authority to initiate interferences 18 months after the date of enactment. The added effective-date language also repeals SIRs 18 months after enactment, making clear that preexisting SIRs will remain effective for purposes of pending interferences, which may continue under this bill.
>
> **157 Cong. Rec. S1371, 2011.**

Congress recognized that statutory invention registrations were unnecessary as early as 1992. *See The Patent System Harmonization Act of 1992: Joint Hearing Before the Subcomm. on Patents, Copyrights and Trademarks of the Senate Committee on the Judiciary*, 102nd Cong. 228-29 (1992) (showing that Congress sought to repeal pre-AIA 35 U.S.C. § 157 in the Patent System Harmonization Act of 1992). Thus, it comes as no surprise that the statutory authority for these publications was repealed through the AIA.

E. Practical Effects of AIA § 3(e)

The repeal of statutory invention registration publications by AIA § 3(e) will have virtually no effect on U.S. patent law because the practice of publishing patent applications as statutory invention registrations had become obsolete long before the AIA was passed. Indeed, as discussed in Section II.A., patent applicants would have been ill advised to request publication of their patent applications as statutory invention registrations once the U.S. PTO began publishing most non-provisional patent applications in 2001. *See* 35 U.S.C. § 122(b)(1)(A) (explaining that subject to certain exceptions, "each application for a patent shall be published, in accordance with procedures determined by the Director, promptly after the expiration of a period of 18 months from the earliest filing date for which a benefit is sought under this title."); MPEP § 1120(I) (9th ed. 2015) (explaining that non-provisional utility and plant applications that are filed on or after November 29, 2000 are published promptly after the expiration of a period of eighteen months from the earliest filing date for which a benefit is sought). Moreover, even though some patent applications are excluded from the mandatory publication requirements of 35 U.S.C. § 122(b)(1)(A), such patent applications could still be published upon request by the patent applicant. *See* 37 C.F.R. § 1.221(a) (permitting patent applicants to request publication of an application filed before, but pending on, November 29, 2000).

III) AIA § 8—THIRD-PARTY PREISSUANCE SUBMISSIONS

Third parties (e.g., competitors of patent applicants) often monitor the prosecution of published patent applications to ensure that any technology they are utilizing is not likely to infringe another's patent rights. While monitoring the prosecution of particular patent applications, these third parties may realize that some or all of the claims that are pending in an application read on prior art (e.g., a published patent application, a patent, or a journal article that qualifies as prior art under post-AIA 35 U.S.C. § 102(a)) that has not yet been cited by either the patent applicant or the Patent Examiner. In other cases, a third party might recognize that a reference that has not yet been cited by either the patent applicant or the Patent Examiner is relevant to the determination of whether sufficient written description and/or enablement exists for certain claims in the patent application under 35 U.S.C. § 112. To ensure that claims encompassing such prior art, or claims that do not meet the requirements of 35 U.S.C. § 112, are not ultimately allowed, the U.S. PTO has provided third parties with a mechanism for submitting references in patent applications to which such references are relevant. This mechanism is referred to as "third-party preissuance submission practice." Once a third party properly submits a reference in a pending patent application, that reference is included in the file wrapper of the patent application and the Patent Examiner, as well as the general public, may review the reference.

Third-party preissuance submission practice was rarely utilized prior to the AIA. One reason the practice was not used much was because the U.S. PTO's regulations imposed a very early deadline on third parties who wished to submit references in a patent application. Another reason third parties scarcely used the preissuance submission practice was because they were prohibited from providing any explanation as to why the reference(s) they submitted had any relevance to a patent application. Many third parties therefore felt it was not worth the effort of submitting a reference because, without any explanation as to why the reference was relevant, a Patent Examiner likely would ignore it. In an effort to bolster third-party preissuance submission practice, therefore, Congress enacted post-AIA 35 U.S.C. § 122(e), which eliminates some of the problems third parties encountered with preissuance submission practice under the pre-AIA law.

A. AIA § 8(a)'s Modifications to Third-Party Preissuance Submission Practices

AIA § 8(a) amends pre-AIA 35 U.S.C. § 122 by adding subsection (e), which delineates the requirements for filing third-party submissions[a] in the record of a patent application. Leahy-Smith America Invents Act of 2011, Pub. L. No. 112-29, 125 Stat. 284, 315-16. Post-AIA 35 U.S.C. § 122(e) reads as follows:

> (e) PREISSUANCE SUBMISSIONS BY THIRD PARTIES.—
>> (1) IN GENERAL.— Any third party may submit for consideration and inclusion in the record of a patent application, any patent, published patent application, or other printed publication of potential relevance to the examination of the application, if such submission is made in writing before the earlier of—
>>> (A) the date a notice of allowance under section 151 is given or mailed in the application for patent; or
>>> (B) the later of—
>>>> (i) 6 months after the date on which the application for patent is first published under section 122 by the Office, or
>>>> (ii) the date of the first rejection under section 132 of any claim by the examiner during the examination of the application for patent.
>> (2) OTHER REQUIREMENTS.— Any submission under paragraph (1) shall—
>>> (A) set forth a concise description of the asserted relevance of each submitted document;
>>> (B) be accompanied by such fee as the Director may prescribe; and
>>> (C) include a statement by the person making such submission affirming that the submission was made in compliance with this section.

---

[a] The provision that was passed as AIA § 8(a) initially read that "any person may submit for consideration . . . ." Congress changed the wording before the AIA was enacted, however, to state that "[a]ny third party may submit for consideration . . . ." to ensure that patent applicants themselves did not submit references under post-AIA 35 U.S.C. § 122(e). *See* 157 CONG. REC. S1377, 2011 ("In paragraph (1) of new section 122(e) of title 35, the word 'person' has been replaced with 'third party,' so that submissions may only be submitted by third parties. This addresses the Office's concern that applicants might otherwise use section 122(e) to submit prior art and thereby evade other examination disclosure requirements.").

Third-party preissuance submission practice was not governed by a statute prior to the AIA. Instead, it was governed by pre-AIA 37 C.F.R. § 1.99. A comparison of the pre- and post-AIA U.S. PTO regulations governing third-party preissuance submissions appears in the following table:

Pre-AIA 37 C.F.R. § 1.99	Post-AIA 37 C.F.R. § 1.290
(a) A submission by a member of the public of patents or publications relevant to a pending published application may be entered in the application file if the submission complies with the requirements of this section and the application is still pending when the submission and application file are brought before the examiner.	(a) A third party may submit, for consideration and entry in the record of a patent application, any patents, published patent applications, or other printed publications of potential relevance to the examination of the application if the submission is made in accordance with 35 U.S.C. 122(e) and this section. A third-party submission may not be entered or considered by the Office if any part of the submission is not in compliance with 35 U.S.C. 122(e) and this section.
(b) A submission under this section must identify the application to which it is directed by application number and include:	(b) Any third-party submission under this section must be filed prior to the earlier of:
(1) The fee set forth in § 1.17(p);	(1) The date a notice of allowance under § 1.311 is given or mailed in the application; or
(2) A list of the patents or publications submitted for consideration by the Office, including the date of publication of each patent or publication;	(2) The later of:
(3) A copy of each listed patent or publication in written form or at least the pertinent portions; and	(i) Six months after the date on which the application is first published by the Office under 35 U.S.C. 122(b) and § 1.211, or
(4) An English language translation of all the necessary and pertinent parts of any non-English language patent or publication in written form relied upon.	(ii) The date the first rejection under § 1.104 of any claim by the examiner is given or mailed during the examination of the application.
(c) The submission under this section must be served upon the applicant in accordance with § 1.248.	(c) Any third-party submission under this section must be made in writing.

(Continued on next page)

Pre-AIA 37 C.F.R. § 1.99	Post-AIA 37 C.F.R. § 1.290
(d) A submission under this section shall not include any explanation of the patents or publications, or any other information. The Office will not enter such explanation or information if included in a submission under this section. A submission under this section is also limited to ten total patents or publications.	(d) Any third-party submission under this section must include:
	(1) A document list identifying the documents, or portions of documents, being submitted in accordance with paragraph (e) of this section;
	(2) A concise description of the asserted relevance of each item identified in the document list;
(e) A submission under this section must be filed within two months from the date of publication of the application (§ 1.215(a)) or prior to the mailing of a notice of allowance (§ 1.311), whichever is earlier. Any submission under this section not filed within this period is permitted only when the patents or publications could not have been submitted to the Office earlier, and must also be accompanied by the processing fee set forth in § 1.17(i). A submission by a member of the public to a pending published application that does not comply with the requirements of this section will not be entered.	(3) A legible copy of each item identified in the document list, other than U.S. patents and U.S. patent application publications;
	(4) An English language translation of any non-English language item identified in the document list; and
	(5) A statement by the party making the submission that:
	(i) The party is not an individual who has a duty to disclose information with respect to the application under § 1.56; and
	(ii) The submission complies with the requirements of 35 U.S.C. 122(e) and this section.
(f) A member of the public may include a self-addressed postcard with a submission to receive an acknowledgment by the Office that the submission has been received. A member of the public filing a submission under this section will not receive any communications from the Office relating to the submission other than the return of a self-addressed postcard. In the absence of a request by the Office, an applicant has no duty to, and need not, reply to a submission under this section.	(e) The document list required by paragraph (d)(1) of this section must include a heading that identifies the list as a third-party submission under § 1.290, identify on each page of the list the application number of the application in which the submission is being filed, list U.S. patents and U.S. patent application publications in a separate section from other items, and identify each:
	(1) U.S. patent by patent number, first named inventor, and issue date;
	(2) U.S. patent application publication by patent application publication number, first named inventor, and publication date;

(*Continued on next page*)

Pre-AIA 37 C.F.R. § 1.99	Post-AIA 37 C.F.R. § 1.290
	(3) Foreign patent or published foreign patent application by the country or patent office that issued the patent or published the application; the applicant, patentee, or first named inventor; an appropriate document number; and the publication date indicated on the patent or published application; and (4) Non-patent publication by author (if any), title, pages being submitted, publication date, and, where available, publisher and place of publication. If no publication date is known, the third party must provide evidence of publication. (f) Any third-party submission under this section must be accompanied by the fee set forth in § 1.17(o) for every ten items or fraction thereof identified in the document list. (g) The fee otherwise required by paragraph (f) of this section is not required for a submission listing three or fewer total items that is accompanied by a statement by the party making the submission that, to the knowledge of the person signing the statement after making reasonable inquiry, the submission is the first and only submission under 35 U.S.C. 122(e) filed in the application by the party or a party in privity with the party. (h) In the absence of a request by the Office, an applicant need not reply to a submission under this section. (i) The provisions of § 1.8 do not apply to the time periods set forth in this section.

In an effort to revive the public's interest in employing third-party pre-issuance submission practice, Congress made three significant changes, as well as some minor changes, to the pre-AIA practice.[b] The first significant change that the AIA made to third-party preissuance submission practice was to extend the deadline by which a third party must make a preissuance submission in order for such submission to be entered into the record (i.e., the file wrapper) of a patent application. In particular, pre-AIA 37 C.F.R. § 1.99(e) requires that a preissuance submission "must be filed within two months from the date of publication of the application . . . or prior to the mailing of a notice of allowance, whichever is earlier."[c] In contrast, post-AIA 37 C.F.R. § 1.290(b) requires that third-party submissions be filed "prior to the earlier of:

(1) The date a notice of allowance under [37 C.F.R. §] 1.311 is given or mailed in the application; or

(2) the later of:

    (i) Six months after the date on which the application is first published by the Office under 35 U.S.C. 122(b) and [37 C.F.R.] 1.211, or

    (ii) The date the first rejection under [37 C.F.R.] 1.104 of any claim by the examiner is given or mailed during the examination of the application."

---

[b]Despite the changes Congress made, there are several similarities in the content of what must be provided in pre- and post-AIA third party preissuance submissions. For instance, under both the pre- and post-AIA laws, a third party may submit patents, published patent applications, or other printed publications that are relevant to a patent application. *Compare* pre-AIA 37 C.F.R. § 1.99(a) (allowing entry of relevant patents or publications in the file wrapper of a pending published application) *with* post-AIA 37 C.F.R. § 1.290(a) (permitting entry of any patents, published patent applications, or other printed publications of potential relevance to the examination of a patent application). Both pre- and post-AIA practices also require third parties to provide a list of the documents they are submitting as well as a copy of the non-patent publication documents identified in the list. Pre-AIA 37 C.F.R. § 1.99(b) (2) and (3); post-AIA 37 C.F.R. § 1.290(d)(1) and (3). In addition, both the pre- and post-AIA regulations require that third parties submit an English language translation of any non-English item identified in the document list. Pre-AIA 37 C.F.R. § 1.99(b)(4); post-AIA 37 C.F.R. § 1.290(d)(4).

[c]Note that pre-AIA 37 C.F.R. § 1.99(e) does allow patents or publications to be submitted after this deadline where such documents "could not have been submitted to the Office earlier." For instance, if a relevant patent application was published after the deadline set forth in 37 C.F.R. § 1.99(e), a third party could still submit that document under the pre-AIA law. Post-AIA 37 C.F.R. § 1.290 surprisingly does not include an analogous clause.

Most patent applications are: (1) filed, (2) published several months after filing,[d] (3) rejected several months after publication, and (4) either allowed or abandoned approximately 12 months after the first rejection of one or more claims is made.[e] Under the pre-AIA law, therefore, the deadline for a third-party preissuance submission in most patent applications would be "within two months from the date of publication of the application" because that date would be earlier than the date on which a notice of allowance is mailed. Pre-AIA 37 C.F.R. § 1.99(e). Under the post-AIA law, however, the deadline for a third-party preissuance submission in most patent applications would likely be prior to "the date the first rejection under [37 C.F.R. §] 1.104 of any claim by the examiner is given or mailed during the examination of the application."[f] This date typically occurs several months after the publication of a patent application. Indeed, "the date the first rejection under [37 C.F.R. §] 1.104 of any claim by the examiner is given or mailed" typically occurs later than the date that is "six months after the date on which the application is first published" but is earlier than "the date a notice of allowance under [37 C.F.R. §] 1.311 is given or mailed in the application." 37 C.F.R. § 1.290(b)(1) and (2). Accordingly, the post-AIA deadline by which a third-party preissuance submission must be filed should be considerably later than the pre-AIA deadline in most instances.

The second significant change the AIA made to the pre-AIA third-party preissuance submission practice lies in the content of what may be included

---

[d] New patent applications must be published 18 months after filing. Post-AIA 35 U.S.C. § 122(b)(1)(A). Continuing applications (i.e., continuation, divisional, and continuation-in-part applications) are supposed to be published within 18 months of their earliest filing date, i.e., 18 months after the filing date of the earliest filed application to which they claim priority. Id. Continuing applications, however, are often filed after the 18-month deadline has expired. Thus, the U.S. PTO usually publishes continuing applications as soon as possible after they are filed.

[e] Some patent applications are allowed on a first office action called a "Quayle" action. In these cases, none of the claims in the patent application is rejected. Accordingly, a third-party preissuance submission would likely be due in such patent applications "within two months from the date of publication of the application" under the pre-AIA law and "six months after the date on which the application is first published by the Office" under the post-AIA law.

[f] It is important to note that the deadline for filing a preissuance submission under the post-AIA law is "prior to" any of the dates that are mentioned in post-AIA 37 C.F.R. § 1.290(b). Thus, a third-party preissuance submission would be untimely if it was filed on the date on which the first rejection under 37 C.F.R. § 1.104 of any claim by the examiner was given or mailed during the examination of the application because that date is not "prior to" the imposed deadline.

in such submissions. Indeed, the pre-AIA law explicitly forbade parties from including any statements that would indicate why a particular document included in the preissuance submission had any relevance to the patent application in which it was being filed. *See* pre-AIA 37 C.F.R. § 1.99(d) ("A submission under this section shall not include any explanation of the patents or publications, or any other information. The Office will not enter such explanation or information if included in a submission under this section."). The post-AIA law, however, allows third parties to submit a concise description of the asserted relevance of each document that is being submitted in his/her pre-issuance submission. Post-AIA 35 U.S.C. § 122(e)(2) (A); post-AIA 37 C.F.R. § 1.290(d)(2). Allowing third parties to include a description of why a particular document is relevant to a patent application is significant because it minimizes the chance that the Patent Examiner will disregard the document being submitted.

The third significant change that AIA § 8 made to third-party preissuance submission practice is that it removed the limitation on the number of documents that could be submitted under the pre-AIA practice. In particular, pre-AIA 37 C.F.R. § 1.99(d) provides that third-party preissuance submissions are "limited to ten total patents or publications." Thus, if a third party wanted to submit more than ten patents or publications for consideration in a patent application, he/she could not do so. In contrast, post-AIA 37 C.F.R. § 1.290(f) simply requires that a fee be paid "for every ten items or fraction thereof" submitted in a third-party preissuance submission. Moreover, if a third party submits three or fewer documents, and states that to the knowledge of the person signing the statement, the submission is the first and only submission under post-AIA 35 U.S.C. § 122(e) filed in the application by the party or a party in privity with the party, the fee required by post-AIA 37 C.F.R. § 1.290(f) is waived. Post-AIA 37 C.F.R. § 1.290(g). Accordingly, the post-AIA law no longer places any limit on the number of documents that may be submitted in a third-party preissuance submission.

Besides the major changes discussed previously, AIA § 8 made some minor changes to the pre-AIA third-party preissuance submission practice. For instance, AIA § 8 eliminated pre-AIA 37 C.F.R. § 1.99(c)'s requirement that third-party preissuance submissions be served upon the applicant in accordance with 37 C.F.R. § 1.248. In addition, post-AIA 37 C.F.R. § 1.290(d)(5) requires that the party making the preissuance submission state that: "(i) [t]he party is not an individual who has a duty to disclose information with respect to the application under [37 C.F.R. §] 1.56; and (ii) [t]he submission complies with the requirements of 35 U.S.C. 122(e) and this section." Pre-AIA 37 C.F.R. § 1.99

did not require third parties to make such statements. The post-AIA law also eliminated pre-AIA 37 C.F.R. § 1.99(a)'s requirement that a submission be made in a "pending published application." *See* FAQ AIA Preissuance Submissions, http://www.uspto.gov/sites/default/files/documents/FAQ%20AIA%20 Preissuance%20Submissions.pdf (last visited March 16, 2016) ("A third party may file a submission in any non-provisional utility, design, or plant application, as well as in any continuing application, even if the application to which the submission is directed has been abandoned or has not been published.").

B. Effective Date of AIA § 8(a)

AIA § 8(a)'s amendments to pre-AIA 35 U.S.C. § 122 took effect on September 16, 2012. Leahy-Smith America Invents Act of 2011, Pub. L. No. 112-29, § 8(b), 125 Stat. 284, 316. Post-AIA 35 U.S.C. § 122(e) applies to any patent application filed before, on, or after September 16, 2012. *Id.* Thus, all third-party preissuance submissions filed on or after September 16, 2012 are governed by post-AIA 35 U.S.C. § 122(e) and post-AIA 37 C.F.R. § 1.290.

C. Legislative Goals That Were Addressed by AIA § 8

The final Committee Report on the AIA explains the reasoning behind the changes that were made to the pre-AIA third-party preissuance submission practice:

> *After an application is published, members of the public—most likely, a competitor or someone else familiar with the patented invention's field—may realize they have information relevant to a pending application. The relevant information may include prior art that would prohibit the pending application from issuing as a patent. Current USPTO rules permit the submission of such prior art by third parties only if it is in the form of a patent or publication, but the submitter is precluded from explaining why the prior art was submitted or what its relevancy to the application might be. Such restrictions decrease the value of the information to the examiner and may, as a result, deter such submissions.*
>
> *The Act improves the process by which third parties submit relevant information to the UPSTO [sic] by permitting those third parties to make statements concerning the relevance of the patents, patent applications, and other printed publications that they bring to the USPTO's attention.*
>
> **H.R. Rep. No. 112-98, at 48-49 (2011) (citations omitted).**

The legislative history for AIA § 8 further explains that the pre-AIA third-party preissuance submission practice was modified to improve the quality of patent examination and issued patents. *See Driving American Innovation: Creating Jobs and Boosting the Economy: Hearing Before the Subcomm. on Intellectual Property, Competition, and the Internet of the H. Comm. on the*

Judiciary, 112th Cong. 27 (2011) (testimony of Scott Smith, Ph.D., Professor and Chair, Department of Mechanical Engineering and Engineering Science, University of North Carolina at Charlotte) (explaining that "[b]etter quality patent reviews could be achieved by allowing third parties to submit printed references to the patent office for a pending patent."); 157 CONG. REC. S1326, 2011 (statement of Sen. Sessions) (explaining that third party submission of prior art will help the PTO determine if the invention is already in the public domain and should not be patented); 157 CONG. REC. S1350, 2011 (statement of Sen. Leahy) ("At the outset, [AIA § 8(a)] makes the common-sense change that third parties who see a patent application and know that it is not novel and nonobvious, can assist the PTO examiners by providing relevant information and explaining its relevance."); 157 CONG. REC. S1097, 2011 (statement of Sen. Hatch) ("The America Invents Act also creates a mechanism for third parties to submit relevant information during the patent examination process. This provision would provide the USPTO with better information about the technology and claimed invention by leveraging the knowledge of the public. This will also help the agency increase the efficiency of examination and the quality of patents.")

D. Practical Implications of AIA § 8's Changes to Third-Party Preissuance Submission Practice

The changes to third-party preissuance submission practice that were implemented through the AIA have made the practice more enticing. Indeed, by giving third parties more time to file preissuance submissions, and allowing them to explain the relevance of such submissions, it appears that many third parties are taking advantage of the practice. *See* U.S. PTO Preissuance Statistics, http://www.uspto.gov/sites/default/files/documents/Preissuance%20Statistics%20as%20of%2011-09-15.pdf, (last visited March 16, 2016) (indicating that 9878 total documents have been submitted in third-party preissuance submissions between September 16, 2012 and November 6, 2015).

The fact that many third parties are taking advantage of the preissuance submission practice is not surprising given that it costs far less to file such a submission than it would to challenge a patent in a post-grant proceeding or in a district court. Moreover, even though many third-party preissuance submissions may not prevent a patent from issuing, the submissions could force a patent applicant to distinguish his/her claims over a cited reference and/or amend his/her claims to avoid reading on a reference, thereby making it easier for others to design around his/her patent claims. Furthermore, a third party who makes a preissuance submission is not estopped from using the references included in that submission in a later challenge to any patent that may issue from the patent application in which the submission was made. Still, third parties should be strategic about submitting references in a preissuance submission. Indeed, if a reference is cited, and the Patent Examiner ultimately allows the patent application over that reference, the third party might have a more difficult time thereafter proving that the patent is invalid over that reference. *See Am. Hoist & Derrick Co. v. Sowa & Sons, Inc.,* 725 F.2d 1350, 1360 (Fed. Cir. 1984) ("When new evidence touching validity of the patent not considered by the PTO is relied on [in patent litigation], the tribunal considering it is not faced with having to *disagree* with the PTO or with *deferring* to its judgment or with taking its expertise into account. The evidence may, therefore, carry more weight and go further toward sustaining the attacker's unchanging burden.") (emphasis in original).

IV) AIA § 11(h)—PRIORITIZED EXAMINATION

When a non-provisional patent application is filed with the U.S. PTO, the application is typically taken up for examination by a Patent Examiner on a first-come, first-serve basis. MPEP § 708 (9th ed. 2015). In other words, Patent Examiners usually examine patent applications with earlier U.S. filing dates before examining those with later filing dates. *Id.* A patent application can be advanced ahead of others that were filed before it, however, if the application undergoes examination under one of the expedited examination procedures provided by 37 C.F.R. § 1.102. *Id.* Before the AIA was enacted, 37 C.F.R. § 1.102 provided three permanent procedures[g] for expediting the

---

[g]In addition to the three permanent procedures for expediting examination, MPEP § 708.01 (9th ed. 2015) lists a number of special cases in which an application will be examined out of turn. For instance, applications that have been pending more than five years, reissue applications, and applications that have been remanded by an appellate tribunal will be examined out of turn. *Id.* From time to time, the U.S. PTO also implements certain temporary programs to expedite the examination of certain types of patent applications. For instance, the Office ran an Enhanced First Action Interview Pilot Program from October 2009 through March 2011 to allow applicants to conduct an interview with a Patent Examiner before receiving a first Office Action on the merits. Enhanced First Action Interview Pilot Program, http://www.uspto.gov/patent/initiatives/first-action-interview/enhanced-first-action-interview-pilot-program (last visited April 6, 2016). The U.S. PTO still offers a First Action Interview Pilot Program today. First Action Interview Pilot Program, http://www.uspto.gov/patents-application-process/applying-online/full-first-action-interview-pilot-program (last visited April 6, 2016). Because the procedures for expediting examination are constantly changing, it is important to confer with patent counsel before making a decision as to what program will best expedite the examination of a particular patent application.

examination of a patent application. In particular, patent applicants could file a petition to make special under the accelerated examination program, a petition to make special under the Patent Prosecution Highway ("PPH") program, or a petition to make special based on an applicant's age or health.[h] Pre-AIA 37 C.F.R. § 1.102(c)(1); MPEP § 708.02(a) and (c) (9th ed. 2015). As a result of the enactment of AIA § 11(h), a prioritized examination procedure[i] was added to 37 C.F.R. § 1.102.[j] Leahy-Smith America Invents Act of 2011, Pub. L. No. 112-29, 125 Stat. 284, 324-25. This section will explore the similarities and differences between prioritized examination and the other expedited examination procedures that were available prior to the AIA.

---

[h]Prior to August 25, 2006, patent applicants could file petitions to make special based on several reasons other than an applicant's age or health. Once the accelerated examination program was implemented on August 25, 2006, however, petitions to make special that were based on anything other than an applicant's age or health, or that were filed under the Patent Prosecution Highway, were required to be filed under the accelerated examination program. See 71 Fed. Reg. 36,324 (June 26, 2006) (explaining that starting on August 25, 2006, petitions to make special that are based on manufacture, infringement, environmental quality, energy, recombinant DNA, superconductivity materials, HIV/AIDS and cancer, countering terrorism, and biotechnology applications filed by small entities will be processed and examined using the revised procedure for accelerated examination); MPEP § 708.02 (9th ed. 2015) ("Any petition to make special, other than those based on applicant's health or age or participation in the Patent Prosecution Highway (PPH) pilot program, filed on or after August 25, 2006 must meet the requirements for the revised accelerated examination program set forth in MPEP § 708.02(a).").

[i]In addition to the prioritized examination procedure that was added by AIA § 11(h), AIA § 25 added subsection (G) to 35 U.S.C. § 2(b)(2) to provide priority examination for patent applications directed to "products, processes, or technologies that are important to the national economy or national competitiveness without recovering the aggregate extra cost of providing such prioritization." Leahy-Smith America Invents Act of 2011, Pub. L. No. 112-29, 125 Stat. 284, 337-38. AIA § 25 was enacted to "create jobs, incentivize investment, and support innovation" by fast-tracking patent applications that are important to the national economy. 157 Cong. Rec. S1052, 2011 (statement of Sen. Menendez). It does not appear, however, that the U.S. PTO has enacted any guidelines for this particular type of priority examination. Thus, it is unclear which technologies are eligible for prioritized examination under post-AIA 35 U.S.C. § 2(b)(2)(G).

[j]The U.S. PTO actually proposed the prioritized examination procedure (which the Office also refers to as "Track One" prioritized examination) that was ultimately enacted by AIA § 11(h) in June 2010. 76 Fed. Reg. 59,050 (Sept. 23, 2011). Unfortunately, the Office had to delay implementation of the procedure due to funding limitations. Id. at 59,051. Accordingly, the procedure only became available to patent applicants after the AIA was enacted.

A. Background on Expedited Examination Procedures That Were Available Prior to the AIA

Each of the three permanent expedited examination procedures that was available prior to the AIA was developed at different times to fast track patent examination for specific types of patent applications or patent applicants. For instance, the practice of submitting a petition to make special based on an applicant's age or health has been available since December 1959, making it the oldest of the three permanent expedited examination procedures. USPTO Patent Examination Acceleration Programs and Proposals, http://www.uspto. gov/sites/default/files/patents/process/file/accelerated/comp_chart_dom_ accel.pdf (last visited April 6, 2016). As the name suggests, petitions to make special based on an applicant's age or health expedite the examination of patent applications filed by patent applicants who are advanced in age (i.e., who are 65 years of age or older) or whose health could prevent them from being available to assist in the prosecution of an application if it were to run its normal course. 37 C.F.R. § 1.102(c)(1); MPEP § 708.02(I) and (II) (9th ed. 2015).

More than forty years passed between the time that petitions to make special based on an applicant's age or health were first made available and the date on which expedited examination under the Patent Prosecution Highway was offered. In fact, patent applicants were not able to file petitions to make special under the Patent Prosecution Highway until 2006. See Michelle K. Lee, *Implementation of the Global and IP5 Patent Prosecution Highway (PPH) Pilot Programs with Participating Offices*, Feb. 10, 2014, http://www.uspto.gov/patents/law/notices/global-ip5.pdf ("Beginning in July 2006, the United States Patent and Trademark Office (USPTO) has partnered with several other offices in Patent Prosecution Highway (PPH) programs."). The Patent Prosecution Highway expedites the examination process for non–reissue utility applications, and international applications entering the national stage under 35 U.S.C. § 371, for which related patent applications have been filed in other countries which have partnering Patent Prosecution Highway offices. FAQs, General PPH, http://www.uspto.gov/sites/default/files/documents/Global%20PPH%20FAQs%20-%20082815_updated.pdf. The Patent Prosecution Highway serves two main purposes. First, it "enables an applicant who receives a positive ruling on patent claims from one participating office to request accelerated prosecution of corresponding claims in another participating office, which allows the applicant to obtain a patentability decision in the second office more quickly." Michelle K. Lee, *Implementation of the Global and IP5 Patent Prosecution Highway (PPH) Pilot*

Programs with Participating Offices, Feb. 10, 2014, http://www.uspto.gov/patents/law/notices/global-ip5.pdf. Second, it "promotes . . . efficiency by allowing [a patent] examiner in the office of later examination (OLE) to reuse the search and examination results from the office of earlier examination (OEE), thereby reducing workload and duplication of effort." *Id.*

Like the Patent Prosecution Highway, accelerated examination did not become available until 2006. *See* 71 Fed. Reg. 36,323-24 (June 26, 2006) (explaining that the accelerated examination program will commence on August 25, 2006). Petitions to make special under the accelerated examination program can be filed by any patent applicant filing a non-reissue utility application under 35 U.S.C. § 111(a), including a continuation application, a continuation-in-part application, or a divisional application, or a design application filed under 35 U.S.C. § 111(a). MPEP § 708.02(a)(I)(B) (9th ed. 2015). Unlike the Patent Prosecution Highway, which requires that patent applications meet three specific criteria[k] in order to qualify for expedited examination under the program, no such criteria are required to be met by patent applications for which expedited examination under the accelerated examination procedure is sought. MPEP § 708.02(a)(I) (9th ed. 2015). Thus, the accelerated examination procedure targets a different category of patent applications than are targeted by the Patent Prosecution Highway and a different group of patent applicants than are targeted by petitions to make special based on an applicant's age or health.

B. AIA § 11(h)—Prioritized Examination

AIA § 11(h) changes the pre-AIA law by adding prioritized examination as a fourth option for expediting the prosecution of a patent application. In particular, AIA § 11(h) modifies pre-AIA 35 U.S.C. § 41 to establish a

---

[k]In particular, in order to be eligible for examination under the Patent Prosecution Highway, a patent application must meet the following three conditions: (1) a partnering Patent Prosecution Highway office must have indicated that there is allowable subject matter in a related patent application, (2) the related application for which the partnering Patent Prosecution Highway office indicated there is allowable subject matter shares a common earlier priority date with the application for which expedited examination under the Patent Prosecution Highway is being sought in the U.S. PTO, and (3) examination must not have begun on the patent application for which expedited examination under the Patent Prosecution Highway is being sought in the U.S. PTO. FAQ, General PPH, http://www.uspto.gov/sites/default/files/documents/Global%20PPH%20FAQs%20-%20082815_updated.pdf.

fee for prioritized examination. Leahy–Smith America Invents Act of 2011, Pub. L. No. 112-29, 125 Stat. 284, 324-25. AIA § 11(h) reads as follows:

(h) PRIORITIZED EXAMINATION FEE.—

(1) IN GENERAL.—

 (A) FEE.—

  (i) PRIORITIZED EXAMINATION FEE.— A fee of $4,800 shall be established for filing a request, pursuant to section 2(b)(2)(G) of title 35, United States Code, for prioritized examination of a nonprovisional application for an original utility or plant patent.

  (ii) ADDITIONAL FEES.— In addition to the prioritized examination fee under clause (i), the fees due on an application for which prioritized examination is being sought are the filing, search, and examination fees (including any applicable excess claims and application size fees), processing fee, and publication fee for that application.

 (B) REGULATIONS; LIMITATIONS.—

  (i) REGULATIONS.— The Director may by regulation prescribe conditions for acceptance of a request under subparagraph (A) and a limit on the number of filings for prioritized examination that may be accepted.

  (ii) LIMITATION ON CLAIMS.— Until regulations are prescribed under clause (i), no application for which prioritized examination is requested may contain or be amended to contain more than 4 independent claims or more than 30 total claims.

  (iii) LIMITATION ON TOTAL NUMBER OF REQUESTS.— The Director may not accept in any fiscal year more than 10,000 requests for prioritization until regulations are prescribed under this subparagraph setting another limit.

(2) REDUCTION IN FEES FOR SMALL ENTITIES.— The Director shall reduce fees for providing prioritized examination of non-provisional applications for original utility and plant patents by 50 percent for small entities that qualify for reduced fees under section 41(h)(1) of title 35, United States Code.

(3) DEPOSIT OF FEES.— All fees paid under this subsection shall be credited to the United States Patent and Trademark Office Appropriation Account, shall remain available until expended, and may be used only for the purposes specified in section 42(c)(3)(A) of title 35, United States Code.

(4) EFFECTIVE DATE AND TERMINATION.—

 (A) EFFECTIVE DATE.— This subsection shall take effect on the date that is 10 days after the date of the enactment of this Act.

 (B) TERMINATION.— The fee imposed under paragraph (1)(A)(i), and the reduced fee under paragraph (2), shall terminate on the effective date of the setting or adjustment of the fee under paragraph (1)(A)(i) pursuant to the exercise of the authority under section 10 for the first time with respect to that fee.

In addition, AIA § 11(h)(1)(B)(i) authorizes the U.S. PTO to enact regulations that prescribe conditions for acceptance of requests for prioritized examination. Leahy-Smith America Invents Act of 2011, Pub. L. No. 112-29, 125 Stat. 284, 324. In accordance with AIA § 11(h)(1)(B)(i), the U.S. PTO enacted 37 C.F.R. § 1.102(e). 37 C.F.R. § 1.102(e) reads as follows:

(e) A request for prioritized examination under this paragraph must comply with the requirements of this paragraph and be accompanied by the prioritized examination fee set forth in § 1.17(c), the processing fee set forth in § 1.17(i), and if not already paid, the publication fee set forth in § 1.18(d). An application for which prioritized examination has been requested may not contain or be amended to contain more than four independent claims, more than thirty total claims, or any multiple dependent claim. Prioritized examination under this paragraph will not be accorded to international applications that have not entered the national stage under 35 U.S.C. 371, design applications, reissue applications, provisional applications, or reexamination proceedings. A request for prioritized examination must also comply with the requirements of paragraph (e)(1) or paragraph (e)(2) of this section.

(1) A request for prioritized examination may be filed with an original utility or plant nonprovisional application under 35 U.S.C. 111(a). The application must include a specification as prescribed by 35 U.S.C. 112 including at least one claim, a drawing when necessary, and the inventor's oath or declaration on filing, except that the filing of an inventor's oath or declaration may be postponed in accordance with § 1.53(f)(3) if an application data sheet meeting the conditions specified in § 1.53(f)(3)(i) is present upon filing. If the application is a utility application, it must be filed via the Office's electronic filing system and include the filing fee under § 1.16(a), search fee under § 1.16(k), and examination fee under § 1.16(o) upon filing. If the application is a plant application, it must include the filing fee under § 1.16(c), search fee under § 1.16(m), and examination fee under § 1.16(q) upon filing. The request for prioritized examination in compliance with this paragraph must be present upon filing of the application, except that the applicant may file an amendment to cancel any independent claims in excess of four, any total claims in excess of thirty, and any multiple dependent claim not later than one month from a first decision on the request for prioritized examination. This one-month time period is not extendable.

(2) A request for prioritized examination may be filed with or after a request for continued examination in compliance with § 1.114. If the application is a utility application, the request must be filed via the Office's electronic filing system. The request must be filed

before the mailing of the first Office action after the filing of the request for continued examination under § 1.114. Only a single such request for prioritized examination under this paragraph may be granted in an application.

C. Differences Between Prioritized Examination and the Pre-AIA Expedited Examination Procedures

Several differences exist between prioritized examination and the three expedited examination procedures that existed before the AIA was enacted, including: (1) the types of patent applications that are eligible for the various types of expedited examination procedures, (2) the filing requirements, including required fees, to file requests for the different types of expedited examination procedures, (3) the manner in which patent applications will be processed by the U.S. PTO under each of the four expedited examination procedures, (4) restrictions on time extensions that may be taken by applicants whose applications are undergoing examination under each of the four expedited examination procedures, and (5) the U.S. PTO's goals for the time it takes to complete examination under the different procedures (i.e., the U.S. PTO's "final disposition" goals). Before deciding whether to request expedited examination under any of the available procedures, therefore, it is important to understand these differences and assess the advantages and disadvantages of each expedited examination procedure that may be available for a given patent application.

i. Types of Patent Applications That Are Eligible for Each Expedited Examination Procedure

Not all patent applications and not all patent applicants are eligible to undergo expedited examination. Accordingly, it is important to determine whether a particular patent application and whether a particular patent applicant is eligible for expedited examination before assessing the advantages and disadvantages of any particular expedited examination procedure. The table below compares the types of patent applications and the types of patent applicants that are eligible and are not eligible to undergo expedited examination under each of the four expedited examination procedures:

Type of Expedited Examination Procedure	Types of Patent Applications that Are Eligible for the Expedited Examination Procedure	Types of Patent Applications that are NOT Eligible for the Expedited Examination Procedure
Petitions to Make Special Based on Age or Health	There are no restrictions on the types of patent applications for which a petition to make special based on an applicant's age or health may be filed. The U.S. PTO only places restrictions on the types of patent applicants who may file such petitions. (37 C.F.R. § 1.102(c)(1); MPEP § 708.02(I) and (II) (9th ed. 2015))	There are no restrictions on the types of patent applications for which a petition to make special based on an applicant's age or health may be filed. The U.S. PTO only places restrictions on the types of patent applicants who may file such petitions. (37 C.F.R. § 1.102(c)(1); MPEP § 708.02(I) and (II) (9th ed. 2015))
Accelerated Examination	1) Non-reissue utility applications (including continuation, continuation-in-part, or divisional applications) filed under 35 U.S.C. § 111(a); and 2) Design applications filed under 35 U.S.C. § 111(a). (MPEP § 708.02(a)(I)(B) (9th ed. 2015))	1) Plant applications; 2) Reissue applications; 3) Applications entering the national stage from an international application after compliance with 35 U.S.C. § 371; 4) Applications involved in reexamination proceedings; 5) RCEs under 37 C.F.R. § 1.114 (unless the application was previously granted special status under the program); 6) Applications for which a petition to make special based on applicant's health or age was filed; 7) Applications for which a petition to make special based on participation in the Patent Prosecution Highway pilot program was filed; and 8) Provisional applications. (MPEP § 708.02(a)(VIII) (9th ed. 2015))

(*Continued on next page*)

Type of Expedited Examination Procedure	Types of Patent Applications that Are Eligible for the Expedited Examination Procedure	Types of Patent Applications that are NOT Eligible for the Expedited Examination Procedure
Patent Prosecution Highway (PPH)	Applications that meet the following three conditions are eligible to participate in the Patent Prosecution Highway (PPH): 1) a partnering PPH office has indicated that there is allowable subject matter in a related application; 2) the related application for which the partnering PPH office indicated there is allowable subject matter shares a common earlier priority date with the application that has been filed with the U.S. PTO; and 3) examination has not yet begun on the patent application that has been filed with the U.S. PTO. (FAQ, General PPH, http://www.uspto. gov/sites/default/files/ documents/Global%20 PPH%20FAQs%20-%20 082815_updated.pdf)	Any applications that do not meet the criteria listed on the column to the left are not eligible to participate in the Patent Prosecution Highway. For example, patent applications that do not have any foreign counterparts could not take advantage of the Patent Prosecution Highway because those applications would not have any related applications for which a partnering PPH office could indicate claim allowable subject matter. In addition, the following applications are not eligible to participate in the Patent Prosecution Highway: 1) Provisional applications; 2) Plant applications; 3) Design applications; 4) Reissue applications; 5) Patent applications that are undergoing reexamination proceedings; and 6) Patent applications that are subject to a secrecy order. (FAQ, General PPH, http://www. uspto.gov/sites/default/files/docu- ments/Global%20PPH%20FAQs%20 -%20082815_updated.pdf)
Prioritized Examination (a.k.a. "Track One" prioritized examination)	1) Utility non-provisional applications (including continuations, continuation-in-parts, or divisional applications) filed under 35 U.S.C. § 111(a)); 2) Plant non-provisional applications filed under 35 U.S.C. § 111(a);	1) International applications that have entered the national stage under 35 U.S.C. § 371, except that such application may undergo prioritized examination if an RCE is filed; 2) Design patent applications; 3) Reissue applications; 4) Provisional applications; and 5) Applications involved in reexamination proceedings.

(*Continued on next page*)

Type of Expedited Examination Procedure	Types of Patent Applications that Are Eligible for the Expedited Examination Procedure	Types of Patent Applications that are NOT Eligible for the Expedited Examination Procedure
	3) Applications listed in #1 and 2 above in which a request for continued examination (RCE) in compliance with 37 C.F.R. § 1.114 has been filed (note, however, that the request for prioritized examination must be filed before a first Office Action is mailed after the filing the RCE. and only a single request for prioritized examination may be granted for any given application); and 4) "By-pass continuation applications" (i.e., a new patent application that has been filed in the U.S. under 35 U.S.C. § 111(a) that claims the benefit of an earlier international application under 35 U.S.C. § 365(c)). (37 C.F.R. § 1.102(e)(1) and (2); MPEP §§ 708.02(b)(I) (A)(1) and 708.02(b)(I)(C) (9th ed. 2015))	(37 C.F.R. § 1.102(e); MPEP §§ 708.02(b)(I)(A)(1) and 708.02(b)(I)(C) (9th ed. 2015))

As shown in the previous table, the U.S. PTO's guidelines are fairly restrictive with regard to the types of patent applications that are eligible for expedited examination under the accelerated examination, prioritized examination and Patent Prosecution Highway procedures. For instance, applications involved in reexamination procedures, reissue applications, and provisional applications[l] are not eligible to participate in any of the

---

[l]It is somewhat misleading to list provisional applications as a type of application that is not eligible to undergo expedited examination because provisional applications are not examined by the U.S. PTO. See MPEP § 201.01(I) (9th ed. 2015) (explaining that one major difference between non-provisional applications filed under 35 U.S.C. § 111(a) and provisional applications filed under 35 U.S.C. § 111(b) are that the latter will not be examined for patentability).

three expedited examination procedures. 37 C.F.R. § 1.102(e); MPEP §§ 708.02(a)(VIII) and 708.02(b)(I)(A)(1) (9th ed. 2015); FAQ, General PPH, http://www.uspto.gov/sites/default/files/documents/Global%20 PPH%20FAQs%20-%20082815_updated.pdf. Additionally, request for continued examination (RCE) applications are not eligible for accelerated examination but are eligible for both prioritized examination[m] and examination under the Patent Prosecution Highway. 37 C.F.R. § 1.102(e) (2); see also FAQ, General PPH, http://www.uspto.gov/sites/default/files/ documents/Global%20PPH%20FAQs%20-%20082815_updated.pdf (listing Provisional applications, plant applications, design applications, reissue applications, patent applications involved in reexamination proceedings, and applications subject to a secrecy order as the only applications that are excluded from the Patent Prosecution Highway program). Furthermore, design applications are only eligible for accelerated examination and are not eligible for either prioritized examination or examination under the Patent Prosecution Highway program. 37 C.F.R. § 1.102(e); MPEP § 708.02(a) (I)(B) (9th ed. 2015); FAQ, General PPH, http://www.uspto.gov/sites/ default/files/documents/Global%20PPH%20FAQs%20-%20082815_up-dated.pdf. In addition, plant applications are only eligible for prioritized examination and are not eligible for accelerated examination or examination under the Patent Prosecution Highway. 37 C.F.R. § 1.102(e) and (e)(1); MPEP §§ 708.02(a)(VIII) (9th ed. 2015); FAQ, General PPH, http://www. uspto.gov/sites/default/files/documents/Global%20PPH%20FAQs%20 -%20082815_updated.pdf. International applications that have entered the national stage under 35 U.S.C. § 371 are eligible for examination under the Patent Prosecution Highway program but are not eligible for accelerated or prioritized examination.[n] Id. Lastly, applications for which a petition to make special based on the applicant's health or age, or for which a petition

---

[m]Note that only a single prioritized examination request may be granted for an RCE under 37 C.F.R. § 1.114 in a patent application. MPEP § 708.02(b)(I)(C)(4) (9th ed. 2015). Thus, if prioritized examination has been granted for one RCE, the patent applicant may not request prioritized examination if a second RCE is later filed in the application. Id.

[n]Although applications that have entered the national stage under 35 U.S.C. § 371 are not eligible for prioritized examination, non-provisional applications that are filed under 35 U.S.C. § 111(a) and claim the benefit of an international application under 35 U.S.C. § 365(c) (i.e., by-pass continuations) are eligible for prioritized examination. MPEP § 708.02(b)(I)(B)(1) (9th ed. 2015). Furthermore, prioritized examination can be requested for a national stage application in which a proper request for continued examination (RCE) has been filed. MPEP § 708.02(b)(I)(C)(1) (9th ed. 2015).

to make special based on participation in the Patent Prosecution Highway program has been filed, are not eligible for accelerated examination. MPEP §§ 708.02(a)(VIII) (9th ed. 2015).

The only applications that are eligible for prioritized examination are original (i.e., non-reissue) utility non-provisional applications filed under 35 U.S.C. § 111(a), including continuation applications, divisional applications, and continuation-in-part applications, plant non-provisional applications filed under 35 U.S.C. § 111(a), and RCE applications filed pursuant to 37 C.F.R. § 1.114.° 37 C.F.R. § 1.102(e)(1) and (2). Furthermore, the only applications that are eligible for accelerated examination are non-reissue utility applications filed under 35 U.S.C. § 111(a), including continuation applications, divisional applications, and continuation-in-part applications, and design applications filed under 35 U.S.C. § 111(a). MPEP § 708.02(a)(I)(B) (9th ed. 2015)). Moreover, only patent applications that meet the three conditions listed in the previous table are eligible to participate in the Patent Prosecution Highway. FAQ, General PPH, http://www.uspto.gov/sites/default/files/documents/Global%20PPH%20FAQs%20-%20082815_updated.pdf. A patent application that does not have any foreign counterpart applications could not meet these conditions because no "related application for which the partnering PPH office indicated there is allowable subject matter" would exist.ᴾ

Although the U.S. PTO does not place restrictions on the types of patent applications that may undergo expedited examination pursuant to a petition to make special based on an applicant's age or health, it does place restrictions on the types of patent applicants who may seek such examination. Indeed, only patent applicants whose state of health is such that they might not be available to assist in the prosecution of the application if it were to run its normal course are eligible to expedite the examination of the patent application by filing a petition to make special based on an applicant's health. MPEP § 708.02(I) (9th ed. 2015). Additionally, only patent applicants who are 65 years of age or older are eligible to expedite examination of a patent application by filing a petition to make special based on an applicant's age. MPEP § 708.02(II) (9th ed. 2015).

°A request for prioritized examination may only be filed for an RCE application if an Office Action in response to the RCE has not yet been mailed. 37 C.F.R. § 1.102(e)(2).

ᴾThe U.S. PTO provides a list of partnering PPH offices at: http://www.uspto.gov/patents/init_events/pph/index.jsp.

For those patent applications that are eligible for prioritized examination and at least one other type of expedited examination procedure, it is important to evaluate the advantages and disadvantages of each before deciding which procedure to pursue. Section IV.C.ii. and IV.C.iii. discuss the advantages and disadvantages of prioritized examination as compared to the other expedited examination procedures that are currently available.

ii. Major Advantages of Prioritized Examination

Prioritized examination offers three major advantages over accelerated examination, examination under the Patent Prosecution Highway, and examination that is conducted pursuant to a petition to make special based on an applicant's age or health. The first, and possibly most significant advantage of prioritized examination is that the filing requirements for requesting the expedited examination procedure, especially as compared to accelerated examination, are minimal. The tables below provide an overview of the filing requirements for each type of expedited examination procedure:

Type of Expedited Examination Procedure	Requirements for Filing a Request for the Expedited Examination Procedure
Prioritized Examination (a.k.a. "Track One" prioritized examination)	Requests for prioritized examination that are filed with non-provisional utility or non-provisional plant applications filed under 35 U.S.C. § 111(a) must include the following at the time of filing: 1) a specification as prescribed by 35 U.S.C. § 112, including at least one claim, and a drawing when necessary; 2) an inventor's oath or declaration, except that the filing of an inventor's oath or declaration may be postponed in accordance with 37 C.F.R. § 1.53(f)(3) if an application data sheet meeting the conditions of 37 C.F.R. § 1.53(f)(3)(i) is present upon filing; 3) the publication fee set forth in 37 C.F.R. § 1.18(d); 4) if the application is a utility application, the fees required by 37 C.F.R. § 1.16(a), (k), and (o) must be filed; and 5) if the application is a plant application, the fees required by 37 C.F.R. § 1.16(c), (m), and (q) must be filed. In addition, requests for prioritized examination of utility applications must be filed via the U.S. PTO's electronic filing system. (37 C.F.R. § 1.102(e)(1))

(Continued on next page)

Type of Expedited Examination Procedure	Requirements for Filing a Request for the Expedited Examination Procedure
Accelerated Examination	A petition to make special under the accelerated examination program must include the following at the time of filing: 1) the basic filing fee, search fee, examination fee, and application size fee (if applicable) for the application for which the petition is being filed; 2) an executed inventor's oath or declaration for each inventor; 3) a statement that the applicant will agree not to separately argue the patentability of any dependent claim during any appeal in the application; 4) a statement that the applicant will agree to make an election without traverse in a telephonic interview if the U.S. PTO determines that all the claims presented are not directed to a single invention; 5) a statement that the applicant will agree to have an interview before a first Office Action to discuss the prior art and any potential rejections or objections with the intention of clarifying and possibly resolving all issues with respect to patentability at that time; 6) a statement that a preexamiantion search was conducted, including an identification of the field of search by U.S. class and subclass and the date of the search, where applicable, and for database searches, the search logic or chemical structure or sequence used as a query, the name of the file or files searched and the database service, and the date of the search; and 7) an accelerated examination support document. In addition, petitions to make special under the accelerated examination program must be filed electronically using the USPTO's electronic filing system (EFS) or EFS-Web. (MPEP § 708.02(a)(I) (9th ed. 2015))
Petition to Make Special Based on Applicant's Age or Health	A petition to make special based on an applicant's health or age must include the following at the time of filing: 1) evidence that the state of health of the applicant is such that he or she might not be available to assist in the prosecution of the application if it were to run its normal course, such as a doctor's certificate or other medical certificate; or 2) evidence that the applicant is 65 years of age or more, such as a statement by the applicant or by a registered practitioner that he or she has evidence that the application is 65 years of age or older. (MPEP § 708.02(I) and (II) (9th ed. 2015))

(*Continued on next page*)

Type of Expedited Examination Procedure	Requirements for Filing a Request for the Expedited Examination Procedure

Patent Prosecution Highway (PPH)	In order to participate in the Patent Prosecution Highway, an applicant must submit the following before examination begins in his/her U.S. PTO application:
	1) a PPH Request Form (see http://www.uspto.gov/patents/ init_events/pph/index.jsp for a list of forms);
	2) either:
	  a) a copy of the office action issued by the office of earlier examination (OEE) that indicates there is allowable subject matter in a related application if the office action is not available via the Dossier Access System, or
	  b) if the office action is available via the Dossier Access System, a request that the U.S. PTO obtain the copy via the Dossier Access System; and
	3) if the office action issued by the OEE is in a language other than English, an English translation of the office action.
	(FAQ, General PPH, http://www.uspto.gov/sites/default/files/documents/Global%20PPH%20FAQs%20-%20082815_updated.pdf)

As shown in the first of the two previous tables, unlike requests for accelerated examination, requests for prioritized examination need not include a statement that a preexamination search was conducted or an examination support document.[q] *Compare* MPEP § 708.02(a)(I)(H) and (I) (9th ed. 2015) *with* 37 C.F.R. § 1.102(e)(1). The fact that the preexamination search statement and examination support document are not required for prioritized examination requests is significant because conducting the preexamination search required to make such a statement, and preparing the examination support document, are burdensome tasks. Indeed, in most cases, the preexamination search required for accelerated examination requests must include a search of U.S. patents and patent application publications, foreign patent documents, and non-patent literature.[r] MPEP § 708.02(a)(I)(H)(1) (9th ed. 2015).

---

[q]Like requests for prioritized examination, petitions to make special based on an applicant's health or age and petitions to make special under the Patent Prosecution Highway are not required to include a preexamination search statement or an examination support document. MPEP § 708.02(I) and (II) (9th ed. 2015); FAQ, General PPH, http://www.uspto.gov/sites/default/files/documents/Global%20PPH%20FAQs%20-%20082815_updated.pdf.

[r]An applicant can exclude one of the aforementioned sources in the preexamination search ▶ if he/she can justify with reasonable certainty that no references more pertinent than those already identified are likely to be found in the eliminated source and include such a justification with the preexamination search statement that is required for filing a request for

The search must also be directed to the claimed invention and encompass all of the features of the claims, giving the claims their broadest reasonable interpretation. MPEP § 708.02(a)(I)(H)(2) (9th ed. 2015). Most patent attorneys and patent agents utilize patent searchers (e.g., Landon IP, Global Prior Art, Inc., Science IP) to conduct patentability searches such as the one required to make a preexamination search statement for the purpose of requesting accelerated examination. Such searches can easily cost $1500 or more.

The examination support document required for requesting accelerated examination can also be burdensome to prepare. Among other things, the examination support document must cite each reference that is most closely related to the subject matter of each of the claims and provide a detailed explanation of how each of the claims is patentable over the reference(s) cited. MPEP § 708.02(a)(I)(I)(1) and (3) (9th ed. 2015). The examination support document must also include a showing of where each claim limitation finds support in the specification under pre-AIA 35 U.S.C. § 112, ¶ 1 or post-AIA 35 U.S.C. § 112(a) and must identify any cited references that may be disqualified as prior art under post-AIA 35 U.S.C. § 102(b)(2)(C) or pre-AIA 35 U.S.C. § 103(c). MPEP § 708.02(a)(I)(I)(5) and (6) (9th ed. 2015). The preparation of an accelerated examination support document involves a great deal of analysis and could therefore take a significant amount of time. Accordingly, an applicant would likely pay more in attorneys' fees or patent agent fees to prepare an accelerated examination support document than he/she would pay to commission the preexamination search that is required to request accelerated examination.

Aside from the burdens applicants undertake to provide a preexamination search statement and an examination support document, the inclusion of these items in the file history of a patent application may significantly diminish the strength of any patent that may be granted from such an application. Indeed, a patentee can virtually be guaranteed that any defendant against whom a patent is enforced will carefully scrutinize the preexamination search statement and examination support document. And a defendant would likely cite any failure to conduct a thorough preexamination search and/or provide a comprehensive and accurate examination support document as evidence of inequitable conduct. Furthermore, even assuming that the preexamination search and accelerated support document are thorough and accurate, both could be used to provide support for a narrower interpretation of any issued patent claims. For example, a defendant could argue

◄ accelerated examination. MPEP § 708.02(a)(I)(H)(1) (9th ed. 2015). Realistically, it would be difficult to justify with reasonable certainty that no references more pertinent than those already identified are likely to be found in a source without first searching that source, so the best practice is to conduct a preexamination search of all of the aforementioned sources.

that the claims could not be interpreted to cover any subject matter that is outside of the field of search that was used for the preexamination search. In addition, a defendant could rely on the examination support document's explanations of how each of the claims is patentable over the cited references to argue that limitations beyond those that are expressly included in the claims should be read into the claims. By omitting the requirements for a preexamination search statement and examination support document, prioritized examination avoids many of the problems that patent applicants may experience with accelerated examination. For this reason alone, prioritized examination is a far more attractive option for expediting examination of a patent application than is accelerated examination.

The filing requirements for prioritized examination also offer advantages over petitions to make special based on an applicant's age or health and over petitions to make special under the Patent Prosecution Highway. Indeed, unlike petitions to make special based on an applicant's age or health, requests for prioritized examination do not need to include evidence of the applicant's age or state of health. 37 C.F.R. § 1.102(e)(1); MPEP § 708.02(I) and (II) (9th ed. 2015). Furthermore, unlike petitions to make special under the Patent Prosecution Highway, requests for prioritized examination are not required to include a copy of the office action issued by the office of earlier examination (OEE) that indicates there is allowable subject matter in a related application. 37 C.F.R. § 1.102(e)(1); FAQ, General PPH, http://www.uspto.gov/sites/default/files/documents/Global%20PPH%20 FAQs%20-%20082815_updated.pdf. Overall, aside from the requirement that utility applications seeking prioritized examination be filed electronically, requests for prioritized examination are not required to include anything more than what is required to file a patent application under 35 U.S.C. § 111(a).

The second major advantage of prioritized examination is that, with one exception, it follows normal prosecution procedures. This is not the case for examination that is conducted under the accelerated examination procedure.[5] The following table provides an overview of the

[5]Examination of patent applications under the Patent Prosecution Highway, and pursuant to a petition to make special based on an applicant's age or health, follows normal prosecution procedures. MPEP § 708.02 (9th ed. 2015); FAQ, General PPH, http://www.uspto.gov/ sites/default/files/documents/Global%20PPH%20FAQs%20-%20082815_updated.pdf. Thus, prioritized examination does not offer any particular advantage over these expedited examination procedures.

special prosecution procedures that apply to prioritized and accelerated examination:

Type of Expedited Examination Procedure	Modified Prosecution Procedure(s) for the Expedited Examination Procedure
Prioritized Examination (a.k.a. "Track One" prioritized examination)	1) Applicants may not file any replies to Office Actions that adds claims that would result in more than four independent claims, more than thirty total claims, or any multiple dependent claims. If these claim limits are exceeded in a reply by an applicant, prioritized examination will be terminated. (MPEP § 708.02(b)(I)(A)(3) (9th ed. 2015))
Accelerated Examination	1) The patent applicant may not file any preliminary amendments with his/her application; 2) The patent application must be "in condition for examination" at the time of filing, meaning it must include all of the required filing fees, an executed inventor's oath or declaration for each inventor, a specification, title, abstract, drawings (if necessary), sequence listings (if necessary), any domestic benefit claim (if applicable), any foreign priority claim (if applicable), and any required English translations; 3) If the application is filed before September 16, 2012, it cannot include a petition under 37 C.F.R. § 1.47 for a non-signing inventor; 4) If the examiner determines that all of the claims presented are not directed to a single invention, the examiner will request that the applicant make a telephonic election without traverse. If the applicant refuses to make such election, or the examiner cannot reach the applicant after a reasonable effort, the examiner will treat the first claimed invention as constructively elected without traverse for examination; 5) If the examiner determines that a possible rejection or other issue must be addressed, the examiner will call the applicant to discuss the issue and any possible amendment or submission to resolve such issue. The U.S. PTO also will not issue an Office Action (other than a notice of allowance) unless either: (A) an interview was conducted but did not result in the application being placed in condition for allowance; or (B) there is a determination that an interview is unlikely to result in the application being placed in condition for allowance. Furthermore, prior to mailing any Office Action rejecting the claims, the U.S. PTO will conduct a conference to review the rejections set forth in the Office Action;

(Continued on next page)

Type of Expedited Examination Procedure	Modified Prosecution Procedure(s) for the Expedited Examination Procedure
	6) If an Office Action other than a notice of allowance is mailed, the applicant will only be given a shortened statutory period of two months to reply;
	7) If the applicant appeals an examiner's rejection, the applicant must agree not to argue the patentability of dependent claims separately from independent claims;
	8) Applicants may not file any replies to Office Actions that: A) add claims that would result in more than three independent claims or more than twenty dependent claims in the application; B) present claims that are not encompassed by the preexamination search or an updated examination support document; or (C) present claims that are directed to a nonelected invention; and
	9) Prior to the mailing of a final Office Action, the U.S. PTO will conduct a conference to review the rejections set forth in the final Office Action. (MPEP §§ 708.02(a)(I), 708.02(a)(VIII)(C), and 708.02(a)(III) and (IV) (9th ed. 2015))

As shown in the previous table, applicants whose patent applications are undergoing accelerated examination must, among other things, agree to participate in an interview with a Patent Examiner before an Office Action on the merits is mailed. MPEP § 708.02(a)(III) (9th ed. 2015). Furthermore, applicants whose patent applications are undergoing accelerated examination are fairly restricted in replying to Office Actions. Indeed, such replies cannot: (1) add claims that would result in more than three independent claims or more than twenty dependent claims, (2) present claims that are not encompassed by the preexamination search or an updated examination support document, or (3) present claims that are directed to a nonelected invention. MPEP § 708.02(a)(IV) (9th ed. 2015). Given that the examination support document must include a "detailed explanation of how each of the claims are patentable over the references cited," it is likely that an applicant would have to submit an updated examination support document if he/she made any substantive claim amendments in a reply to an Office Action. MPEP § 708.02(a)(I)(I)(3) (9th ed. 2015). Applicants whose applications are undergoing accelerated examination must also submit to shortened time periods for responding to certain actions by the U.S. PTO. For example, if an Office Action other than a notice of allowance is mailed, applicants are given a shortened statutory period of two months to reply. MPEP § 708.02(a)(III) (9th ed. 2015). Under normal prosecution procedures, applicants are given a shortened statutory period of three months to reply to Office Actions on the merits. MPEP § 710.02(b) (9th ed. 2015). Other than prohibiting applicants from filing a reply to an Office Action that adds claims that would result in more than four independent claims, thirty total claims, or any multiple dependent claims, prioritized examination follows normal prosecution procedures. MPEP § 708.02(b)(II) (9th ed. 2015).

The third and last major advantage of prioritized examination, especially as compared to examination under the Patent Prosecution Highway and examination that is conducted pursuant to a petition to make special based on an applicant's age or health, is that the U.S. PTO has set a specific goal for completing examination (i.e., a "final disposition goal") under the procedure. *Compare* MPEP § 708.02(b)(II) (9th ed. 2015) ("The goal of the Office [for prioritized examination] is to provide a final disposition within twelve months, on average, of the date that prioritized status was granted.") *with* MPEP § 708.02(VI) (9th ed. 2015) (omitting any reference to a final disposition goal for petitions to make special) and FAQ, General PPH, http://www. uspto.gov/sites/default/files/documents/Global%20PPH%20FAQs%20 -%20082815_updated.pdf (explaining that while the PPH program does not have a final disposition goal, PPH requests are generally decided within four months from the filing of the PPH request and examination usually begins within two and three months from the grant of a PPH request provided that the application is ready for examination). The U.S. PTO has similarly set a final disposition goal for the accelerated examination procedure. *See* MPEP § 708.02(a)(VIII)(F) (9th ed. 2015) ("The objective of the accelerated examination program is to complete the examination of an application within twelve months from the filing date of the application."). Even though the U.S. PTO does not guarantee that the final disposition goal will be met, having such a goal is advantageous as it should provide applicants with more certainty and predictability with regard to how quickly their applications will be examined.

iii. Major Disadvantages of Prioritized Examination

Apart from the many advantages offered by prioritized examination, there are at least four major disadvantages to the expedited examination procedure. The first disadvantage is that the U.S. PTO's fee for requesting prioritized examination is significantly higher than the fees required for requesting the other three types of expedited examination. *Compare* MPEP § 708.02(a)(I)(A) (9th ed. 2015) (requiring that a fee of only $140, which is discounted for micro and small entities, be included with a request for accelerated examination and that no fee be submitted for applications claiming subject matter that is directed to environmental quality, the development or conservation of energy resources, or countering terrorism) and 37 C.F.R. § 1.102(d) (requiring that petitions to make special under the Patent Prosecution Highway be accompanied by a fee of $140 but requiring no fee for petitions to make special filed based on an applicant's age or health) *with* MPEP § 708.02(b)(I)(A)(2)(A) (9th ed. 2015) (requiring that a fee of $4000, which is discounted for micro and small entities, be included with a request for prioritized examination). Importantly the fee for prioritized examination will be refunded to a patent applicant if the U.S. PTO dismisses an original request for prioritized examination. MPEP § 708.02(b)(III) (9th ed. 2015). The fee will not be refunded, however, in instances where a petition for prioritized examination is granted but the prioritized examination is later terminated.[t] *Id.*

When comparing the costs of prioritized examination and accelerated examination, the high filing fee for the former is somewhat misleading. Indeed, as discussed in Section IV.C.ii., applicants who request accelerated examination can expect to spend a great deal in order to satisfy the requirements for the preexamination search statement and examination support document. Thus, in many instances, the cost for filing a request for prioritized examination may not be as high as that for accelerated examination even though the U.S. PTO's fee for requesting the former type of expedited examination is much higher.

Apart from the high fee required to request prioritized examination, a second major disadvantage of the procedure is that the U.S. PTO limits the

---

[t]Prioritized examination will be terminated if an applicant files any one of the following: (1) a petition for extension of time to extend the time period for filing a reply, (2) an amendment to amend the application to contain more than four independent claims, more than thirty total claims, or a multiple dependent claim, (3) a request for continued examination (RCE), (4) a notice of appeal, or (5) a request for suspension of action. USPTO's Prioritized Patent Examination Program FAQs, http://www.uspto.gov/patents/init_events/track1_FAQS.jsp (last visited April 6, 2016). Prioritized examination will also terminate if a notice of allowance is mailed, a final Office Action is mailed, the patent application is abandoned, or examination is completed as defined in 37 C.F.R. § 41.102. *Id.*

number of patent applications that may be granted prioritized examination to 10,000 per fiscal year.[u] MPEP § 708.02(b)(I)(A)(4) (9th ed. 2015). No such limit exists for accelerated examination, petitions to make special based on an applicant's age or health, or petitions to make special under the Patent Prosecution Highway. *See* MPEP § 708.02 (9th ed. 2015) (omitting any reference to a limit on the number of patent applications that may be examined pursuant to a petition to make special, including a petition to make special filed under the Patent Prosecution Highway program); MPEP § 708.02(a) (9th ed. 2015) (omitting any reference to a limit on the number of patent applications that may be examined using the accelerated examination procedure).

A third disadvantage of prioritized examination is that patent applications that are examined under the procedure may not contain or be amended to contain more than four (4) independent claims, more than thirty (30) total claims, or any multiple dependent claims.[v] 37 C.F.R. § 1.102(e). While the limit for the number of claims in applications undergoing prioritized examination is not as low as for applications undergoing accelerated examination, applications for which examination is expedited due to a petition to make special based on an applicant's age or health or under the Patent Prosecution Highway are not subject to any claim limitations. *See* MPEP § 708.02(a)(I)(E) (9th ed. 2015) (imposing a limit of three (3) or fewer independent claims, twenty (20) or fewer total claims, and zero (0) multiple dependent claims on patent applications undergoing accelerated examination); MPEP § 708.02 (9th ed. 2015) (omitting any reference to claim limitations for applications examined pursuant to a petition to make special, including a petition to make special under the Patent Prosecution Highway).

A fourth major disadvantage of prioritized examination is that applicants are prohibited from seeking extensions of time under 37 C.F.R. § 1.136, or from requesting a suspension of action, under the expedited procedure. *See* MPEP § 708.02(b)(II) (9th ed. 2015) (explaining that if an applicant files a petition for an extension of time to file a reply or files a request for suspension of action, the petition or request will be acted upon, but the prioritized

---

[u]Patent applicants can determine whether the limit of 10,000 applications has been met by visiting: http://www.uspto.gov/corda/dashboards/patents/main.dashxml?CTNAVID=1007.
[v]A multiple dependent claim is a claim that depends on more than one other claim. 37 C.F.R. § 1.75(c).

examination of the application will be terminated). Applicants whose patent applications are undergoing accelerated examination are similarly prohibited from taking time extensions for responding to Office Actions. *See* MPEP § 708.02(a)(III) (9th ed. 2015) ("Extensions of this shortened statutory period under 37 [C.F.R. §] 1.136(a) will be permitted. However, filing a petition for extension of time will result in the application being taken out of the accelerated examination program."). No restrictions on taking time extensions are imposed on applicants whose examination has been expedited pursuant to a petition to make special based on an applicant's age or health or pursuant to a petition to make special under the Patent Prosecution Highway. *See* MPEP § 708.02 (9th ed. 2015) (omitting any reference to restrictions on taking time extensions); FAQ, General PPH, http://www.uspto.gov/sites/default/files/documents/Global%20PPH%20 FAQs%20-%20082815_updated.pdf (explaining that extensions of time under 37 C.F.R. § 1.136(a) are available for patent applications that have been accepted into the Patent Prosecution Highway program).

Applicants are usually given a period of three months to respond to Office Actions on the merits and a period of two months to respond to Restriction Requirements.[w] MPEP § 710.02(b) (9th ed. 2015). Under normal, non-expedited examination procedures, it is not uncommon for applicants to request extensions of time to obtain up to six months to respond to an Office Action or Restriction Requirement. Indeed, doing so can give applicants time to collect and submit data that may be helpful for arguing for the patentability of one or more claims and/or prepare and submit any declarations (e.g., a declaration under 37 C.F.R. § 1.132) that may be needed to support patentability. Thus, applicants who are contemplating expedited examination should be sure they will be ready to respond to Office Actions in a relatively short period of time before requesting prioritized examination. Indeed, the U.S. PTO will not provide a refund of the $4000 fee for a request for prioritized examination if prioritized examination is terminated

---

[w]Another difference between accelerated and prioritized examination is that under the former procedure, applicants are given period of only two months to reply to an Office Action on the merits, whereas applicants are given a period of three months to reply to an Office Action on the merits under the latter procedure. *See* MPEP § 708.02(a)(III) (9th ed. 2015) ("If an Office action other than a notice of allowance or a final Office action is mailed [in an application undergoing accelerated examination], the Office action will set a shortened statutory period of two months [for reply]."); MPEP § 708.02(b)(II) (9th ed. 2015) ("The time periods set for reply in Office actions for applications undergoing prioritized examination will be the same as set forth in MPEP § 710.02(b).").

due to an applicant's petition for an extension of time or request for suspension of action. MPEP § 708.02(b)(III) (9th ed. 2015).

D. Effective Date of AIA § 11(h)

AIA § 11(h)(4)(A) provides that AIA § 11(h) "shall take effect on the date that is 10 days after the date of the enactment of this Act." Leahy-Smith America Invents Act of 2011, Pub. L. No. 112-29, 125 Stat. 284, 324-25. Furthermore, a request for prioritized examination must be filed on the filing date of the patent application for which the request is being made. USPTO's Prioritized Patent Examination Program FAQs, http://www. uspto.gov/patents/init_events/track1_FAQS.jsp (last visited April 6, 2016). The AIA was enacted on September 16, 2011. Leahy-Smith America Invents Act of 2011, Pub. L. No. 112-29, 125 Stat. 284. Accordingly, a request for prioritized examination can be filed for any non-provisional utility or plant application filed under 35 U.S.C. § 111(a) that is filed on or after September 26, 2011.[x]

E. Legislative History of AIA § 11(h)

Aside from some comments supporting the reduction of fees for small entities seeking prioritized examination, there appears to be no discussion of prioritized examination in the legislative history of the AIA. *See* 157 Cong. Rec. S956, 2011 (statement of Sen. Bennet) ("My first amendment . . . can help small businesses utilize the Patent Office's Track I program by reducing their fees for participating. Track I allows applicants to get their patent processed more quickly, but the cost can be burdensome for small entities. This amendment would reduce small business costs by 50 percent."). Congress likely had little reason to discuss prioritized examination given that the U.S. PTO had proposed implementing the expedited examination procedure as far back as June 2010. 76 Fed. Reg. 59,050 (Sept. 23, 2011). Indeed, the U.S. PTO wanted to implement prioritized examination in a three-track examination program that was designed to provide patent applicants with greater control over when their original utility or plant applications are examined. *Id.*

---

[x]Unlike non-provisional utility applications and plant applications, which are eligible for prioritized examination if they were filed on or after September 26, 2011, only RCE applications that are filed on or after December 19, 2011 are eligible for the expedited examination procedure. USPTO's Prioritized Patent Examination Program FAQs, http://www.uspto. gov/patents/init_events/track1_FAQS.jsp (last visited April 6, 2016).

F. Practical Effects of AIA § 11(h)

Prioritized examination has already proven to be far more popular than accelerated examination. Indeed, according to the U.S. PTO's own data, approximately 27,300 patent applications received a final disposition under the prioritized examination procedure in the first four and a half years of the procedure's existence (i.e., during the period starting on September 26, 2011 and ending in February 2016). USPTO's Data Visualization Center, https://www.uspto.gov/corda/dashboards/patents/main. dashxml?CTNAVID=1007 (last visited April 6, 2016). By comparison, approximately 2523 patent applications received a final disposition under the accelerated examination procedure in the first four and a half years of its existence (i.e., during the period starting in August 2006 and ending in January 2011). Kathleen Fonda, *Moving Patent Applications Through the USPTO: Options for Applicants* (2011).

The fact that prioritized examination has been far more popular than accelerated examination is not surprising given that the latter procedure requires applicants to conduct a preexamination search and provide an examination support document. As discussed in Section IV.C.ii., the preexamination search statement and examination support document are burdensome to complete and may significantly weaken any patent that issues from an application in which they are submitted. Furthermore, even though the fee to request prioritized examination is quite high (i.e., $4000 unless the patent applicant qualifies for small or micro entity status), patent applicants are likely to spend at least that amount to prepare the preexamination search statement and the examination support document that are required for requesting accelerated examination. Thus, even though the fee for requesting accelerated examination is much lower than that for requesting prioritized examination (i.e., $140 unless the patent applicant qualifies for small or micro entity status), it is likely that a patent applicant requesting accelerated examination will end up spending more than a patent applicant who requests prioritized examination as of the time his/her patent application is filed.

Prioritized examination also appears to proceed much faster than accelerated examination. The following table provides an overview of some of the statistics that have been collected by the U.S. PTO:

Statistic Measured	Accelerated Examination	Prioritized Examination
Average Number of Months from the Date a Patent Application was Filed until the Date a Petition for Expedited Examination Was Granted	4.9[y]	1.4[z]
Average Number of Months from the Date a Petition for Expedited Examination Was Granted until a First Office Action was Mailed	1.8	2.1
Average Number of Months from the Date a Petition for Expedited Examination Was Granted until Final Disposition Occurred	11.2[aa]	6.5

The speedier examination provided by prioritized examination can provide many benefits to patent applications. For instance, faster examination can allow patent applicants to avoid third-party preissuance submissions. Indeed, third parties would have no knowledge of a patent application until it is published, and most patent applications are not published until 18 months from the date they are filed. 35 U.S.C. § 122(b)(1). Because a patent application undergoing prioritized examination is likely to receive a final disposition less than a year from the date the application is filed, it is unlikely a third party would have a chance to file a preissuance submission for such application. Faster examination is also beneficial for those patent applicants who are trying to build a patent portfolio to attract investors. Indeed, many investors want to see that a company owns issued patents as opposed to pending patent applications. Prioritized examination may also benefit patent owners who are looking to enforce a patent against a particular competitor. For instance, a patent owner may discover that the claim scope of a patent he/she owns is not ideal for proving infringement and/ or for avoiding certain invalidity defenses. If the patentee still has a related

[y]Unless otherwise indicated, all of the statistics given for accelerated examination were measured during the time period of 2011–2015. Accelerated Examination, http://www.uspto.gov/patent/initiatives/accelerated-examination#heading-1 (last visited April 6, 2016).

[z]Unless otherwise indicated, all of the statistics given for prioritized examination were measured for the time period beginning in September 2011 and ending in February 2016. USPTO's Data Visualization Center, http://www.uspto.gov/corda/dashboards/patents/main.dashxml?CTNAVID=1007 (last visited April 6, 2016).

[aa]Kathleen Fonda, *Moving Patent Applications Through the USPTO: Options for Applicants* (2011).

application pending at the U.S. PTO, he/she may opt to file a continuing application and seek prioritized examination so that a patent with a more ideal claim scope could be issued before he/she files suit.

V) AIA § 37—CALCULATION OF THE 60-DAY PERIOD FOR PATENT TERM EXTENSION

A. Background on U.S. Patent Terms

In general, as long as a patent owner pays the required maintenance fees for a patent, the patent will remain in force for a period of time that extends 20 years from the date on which the application for the patent was filed.[bb] 35 U.S.C. § 154(a)(2). For example, if an applicant filed patent application B on January 1, 2000, and the application did not claim the benefit of an earlier application, the patent term would extend from the date a patent issued from patent application B until January 1, 2020. Several exceptions to the 20-year patent term exist, however, such that the term of some patents ends on a date that is earlier than the date that is 20 years from the actual filing date of the patent application and the term of other patents ends on a date that is later than the date that is 20 years from the actual filing date of the patent application. For instance, if a patent application references an earlier-filed application under 35 U.S.C. §§ 120, 121, 365(c) or 386(c),[cc] the patent arising from such patent application would extend from the date the patent issues until the date that is 20 years from the date on which the

---

[bb]Note that this is the term for utility and plant patents. Design patents have a term of fourteen years from the date of patent grant, except that any design patent issued from applications filed on or after May 13, 2015 has a term of fifteen years from the date of patent grant. MPEP § 2701 (9th ed. 2015).

[cc]35 U.S.C. § 120 pertains to claims for the benefit of an earlier filing date in the U.S., e.g., a claim for the benefit of a filing date of a provisional or non-provisional application. 35 U.S.C. § 121 provides that an invention that is made the subject of a divisional application which complies with the requirements of 35 U.S.C. § 120 shall be entitled to the benefit of the filing date of the original application. 35 U.S.C. § 365(c) permits international applications that designate the United States to claim the benefit of a prior national application or the benefit of a prior international application that designates the United States, and permits national applications to claim the benefit of the filing date of a prior international application that designates the United States. Furthermore, 35 U.S.C. § 386(c) permits international design applications designating the U.S. to claim the benefit of a filing date of certain prior national and international applications and also permits a national application to claim the benefit of the filing date of a prior international design application designating the United States.

earliest such application was filed.[dd] Furthermore, if a patent applicant experiences a certain amount of delay during the examination of his/her patent application, he/she may be entitled to a patent term adjustment ("PTA") under 35 U.S.C. § 154(b). Patent term adjustments are calculated on a case-by-case basis and are intended to make up for unreasonable delays caused by the U.S. PTO that occur between the time a patent application is filed and the time a patent from such patent application issues. If an applicant obtains a patent term adjustment, the term of his/her patent would extend from the date on which the patent issued until a date that is later than 20 years from the filing date of the application for the patent.

B. Background on Patent Term Extensions

Similar to a patent term adjustment ("PTA"), a patent term extension ("PTE") causes the term of a patent to extend from the date on which the patent issued until a date that is later than 20 years from the filing date of the application for the patent. Patent term extensions are distinct from patent term adjustments, however, in that they do not apply to all types of utility patents. Indeed, patent term extensions may only be obtained for patents on certain human drugs, food or color additives, medical devices, animal drugs, and veterinary biological products. MPEP § 2750 (9th ed. 2015). Furthermore, unlike patent term adjustments which make up for delays caused by the U.S. PTO, patent term extensions enable patent owners to recover time lost from their patent term that was spent waiting for premarket approval from a regulatory agency, such as the FDA.[ee] *Id.* In other words, if a patent owner is unable to market his/her product upon issuance of the patent covering that product because the product has not yet received regulatory

---

[dd]For instance, if patent application B mentioned previously was a continuation of nonprovisional patent application A that was filed on February 1, 1999, the patent term would extend from the date the patent issued from patent application B until February 1, 2019 (i.e., the date that is 20 years from the date on which patent application A was filed).

[ee]The statute that provides patent term extensions, 35 U.S.C. § 156, was enacted in the Drug Price Competition and Patent Term Restoration Act of 1984, Public Law 98-417, 98 Stat. 1585, which sought to eliminate the distortion to the normal "patent term produced by the requirement that certain products must receive premarket regulatory approval." *Eli Lilly & Co. v. Medtronic Inc.,* 496 U.S. 661, 669 (1990). By allowing patent applicants to recapture the portion of their patent term during which they are unable to market a product because they are awaiting approval of their product from a regulatory agency, 35 U.S.C. § 156 incentivizes companies to engage in research and development on such products. MPEP § 2750 (9th ed. 2015).

approval, the patent owner can obtain a patent term extension, which will restore the period of the patent term that was spent waiting for the regulatory approval.[ff]

The U.S. PTO's role in awarding patent term extensions is ministerial.[gg] Indeed, as long as a patent is eligible for a patent term extension, and the patent owner or patent owner's agent submits an application for a patent term extension "within the sixty-day period beginning on the date the product received permission for commercial marketing or use under the provision of law under which the applicable regulatory review period occurred for commercial marketing or use," the Office will grant the patent term extension. 35 U.S.C. § 156(d)(1); *see also* Joe Matal, *A Guide to the Legislative History of the America Invents Act: Part II of II*, 21 FED. CIR. B.J. 539, 647 (2012) ("The USPTO's grant of [a patent term] extension is ministerial—the patentee is entitled to the extension if his term ran during the review period. The only additional requirement, per [35 U.S.C. § 156](d)(1), is that the patentee must seek the extension 'within the sixty-day period beginning on the date the product received permission . . . for commercial marketing or use.'") (internal citations omitted).

C. Context in Which AIA § 37 Arose

While the process for granting patent term extensions seems straightforward, the U.S. PTO created an enormous amount of controversy when it denied The Medicines Company's ("MDCO's") patent term extension for Angiomax, an anticoagulant drug, on the basis that the application for the extension was filed outside of the sixty-day statutory deadline. The

---

[ff]Sales of many blockbuster drugs (e.g., Lipitor, Abilify, Crestor) total in the billions of dollars each year, and others total in the hundreds of millions per year. Not surprisingly, generic manufacturers want to try to cut into the sales of these drugs as soon as possible. Generics cannot begin marketing their own copies of the blockbuster drugs, however, until patent issues are resolved (i.e., until the patent covering the drug expires, its claims are found invalid, or it is rendered unenforceable). In cases where a company is making hundreds of millions of dollars per year or more from the sale of a drug, every single day of patent coverage is extremely valuable. For this reason, patent term extensions are of utmost importance to innovative pharmaceutical companies.

[gg]Unlike the U.S. PTO's role in awarding patent term extensions, the FDA's role is more complicated. Indeed, the FDA is tasked with determining the length of the regulatory review period so that the U.S. PTO can determine the length of the patent term extension to which a patent applicant is entitled. *Medicines Co. v. Kappos*, 731 F. Supp. 2d 470, 472 (E.D. Va. 2010); MPEP § 2750 (9th ed. 2015).

pertinent facts, as described in *Medicines Co. v. Kappos*, 731 F. Supp. 2d 470 (E.D.Va. 2010), are as follows:

> MDCO filed a new drug application (NDA) for ANGIOMAX on December 23, 1997. The FDA approved that application in December 2000. The FDA's approval was set forth in a letter faxed to MDCO at 6:17 p.m. on Friday, December 15, 2000. The FDA then published the approval date for ANGIOMAX as December 19, 2000 on one page of its website...
>
> MDCO filed its patent term extension application on February 14, 2001 [i.e., 61 days after December 15, 2000 and 57 days after December 19, 2000 if December 15, 2000 is not included in the total count] under the Hatch-Waxman Act. Such an extension would [have] change[d] the expiration date of [MDCO's] patent from March 23, 2010 to December 2014...
>
> On September 6, 2001, in response to a request from the PTO, the FDA asserted that ANGIOMAX was approved on December 15, 2000 and that MDCO's application was untimely within the meaning of 35 U.S.C. § 156(d)(1). The FDA did not address the fact that a page on its website listed December 19 as the approval date for ANGIOMAX.
>
> On December 18, 2001, MDCO received an undated Notice of Final Determination from the PTO denying MDCO's application. The Notice accepted the FDA's view that ANGIOMAX was approved on December 15, 2000 and that the extension application was untimely because it was filed one day late...
>
> On October 2, 2002, MDCO filed a timely Request for Reconsideration with the PTO. Among other things, MDCO pointed out that the FDA approval letter for ANGIOMAX was faxed after business hours on a Friday evening, and that under [the] FDA's practices,[hh] facsimiles submitted to FDA after the close of business are considered received by the Agency on the next business day. For that reason, MDCO asked the PTO to treat December 18, 2000 as the effective approval date of ANGIOMAX®, and would have made MDCO's February 14, 2001 application timely.
>
> On April 26, 2007, the PTO denied MDCO's Request for Reconsideration. The PTO offered no explanation of what it acknowledged was the FDA's seemingly inconsistent approach to determining the effective date of submissions to the

---

[hh]MDCO was referring to the FDA's practices for receiving and sending out communications. See *Medicines Co.*, 731 F. Supp. 2d at 473 ("The FDA treats submissions to the FDA received after its normal business hours differently than it treats communications from the agency after normal business hours. The agency considers the date of submission of a new drug application received after 4:30 p.m. EST to be the next business day. If an applicant submits an electronic application or sends a fax to the FDA at 6:17 p.m. on a Friday night, the FDA will deem that application to be submitted on the following Monday (or Tuesday, if the Monday is a federal holiday). This FDA practice has the consequence of making the regulatory review period defined in § 156(g) commence days later than if the application was considered submitted on Friday and can operate to reduce the overall length of the patent term extension granted.").

agency and communications from the agency . . . Using the FDA's December 15, 2000 approval date, the PTO also determined that MDCO's application was filed two days after the 60-day period expired. The change in the PTO's calculation was due to a change in the agency's interpretation of § 156(d)(1) that was apparently announced for the first time in its reconsideration decision in this case.[ii]

Medicines Co., 731 F. Supp. 2d at 472-73.

In January 2010, MDCO filed suit against the U.S. PTO to challenge its interpretation of 35 U.S.C. § 156(d)(1)[iii] in the Eastern District of Virginia. That court vacated the U.S. PTO's decision to deny MDCO's patent term extension application as untimely and remanded the matter back to the U.S. PTO to allow the agency to "reconsider [the] matter free from its erroneous conception of the bounds of the law." *Medicines Co. v. Kappos*, 699 F. Supp. 2d 804, 811-12 (E.D.Va. 2010). The Court specifically found that the U.S. PTO erred in interpreting the term "date" in 35 U.S.C. § 156(d)(1) as spanning the course of 24 hours instead of ending with the close of business. *See id.* at 810 (explaining that "[t]he PTO provided no analysis of the putative plain meaning it assigned to the term ["date"], the ways in which the term is used throughout *§ 156*, the different contexts of the usage, or the regulatory impact of competing alternative interpretations. It merely asserted that the term date in *§ 156(d)(1)* must be a calendar day and could not be a business day.") (emphasis in original).

[ii] *See Medicines Co.*, 731 F. Supp. 2d at 473-74 ("For years, in applying § 156(d)(1)'s 60-day deadline, the PTO followed the general rule of starting the count on the first day after the triggering event. [In MDCO's case, therefore, the PTO would have started the count on December 16, 2000.] In its 2007 Decision, however, the PTO concluded that it had been misreading § 156(d)(1). It then changed course and announced that it would count the date of FDA approval as one of the sixty days included in the time period for filing a PTE application. *In re Patent Term Extension Application for U.S. Patent No. 5,817,338*, 2008 WL 5477176 (Comm'r Pat. Dec. 16, 2008) ("Prilosec Decision") . . . The PTO takes the view that the date of FDA approval counts as the first day of the 60-day-period even where the application is not approved until after the close of business, even as late as 11:59 p.m. This interpretation can mean that an applicant is afforded only 59 days rather than 60 days [to file their patent term extension application].").

[iii] 35 U.S.C. § 156(d)(1) states: "[t]o obtain an extension of the term of a patent under this section, the owner of record of the patent or its agent shall submit an application to the Director. Except as provided in paragraph (5), such an application may only be submitted within the sixty-day period beginning on the date the product received permission under the provision of law under which the applicable regulatory review period occurred for commercial marketing or use, or in the case of a drug product described in subsection (i), within the sixty-day period beginning on the covered date (as defined in subsection (i))."

On March 19, 2010, a mere 72 hours after Judge Hilton of the Eastern District of Virginia handed down his opinion, the U.S. PTO issued a 15-page opinion that again denied MDCO's application for a patent term extension. Kurt R. Karst, *PTO Once Again Denies PTE for ANGIOMAX Patent . . . But Not Before Issuing an Interim Extension; MDCO is Outraged*, March 22, 2010, http://www.fdalawblog.net/fda_law_blog_hyman_phelps/2010/03/page/2/. Outraged at the U.S. PTO's quick denial of its patent term extension application, MDCO filed yet another suit against the U.S. PTO in the Eastern District of Virginia on March 26, 2010 asking the court to vacate the U.S. PTO's March 19th denial of its patent term extension, declare that the company timely filed its patent term extension application, order the U.S. PTO to "immediately" grant an interim patent term extension for the '404 patent, "and to take any additional present or future actions as are necessary to enable MDCO to protect its rights and to ensure that [the '404 patent] does not expire prior to issuance of a certificate of extension." Kurt R. Karst, *Here We Go Again! MDCO Launches Another Lawsuit Against the PTO Over ANGIO-MAX PTE*, March 26, 2010, http://www.fdalawblog.net/fda_law_blog_hyman_phelps/2010/03/here-we-go-again-mdco-launches-another-lawsuit-against-the-pto-over-angiomax-pte.html.

On August 3, 2010, Judge Hilton granted summary judgment for MDCO and remanded the case to the U.S. PTO to consider MDCO's patent term extension application for the '404 patent "timely filed and to adopt an interpretation of § 156(d)(l) that includes a next business day construction for filing of a [PTE] application." Medicines Co. v. Kappos, No. 01:10-cv-286 (E.D. Va. Aug. 3, 2010) (order granting summary judgment for MDCO); *see also Medicines Co. v. Kappos*, 731 F. Supp. 2d 470, 482 (E.D. Va. 2010) ("The Court finds the proper interpretation of § 156(d)(1) is a business day construction of the phrase 'beginning on the date.' Of the parties' competing interpretations the business day construction is consistent with the statute's text, structure, and purpose.").

On September 9, 2010, the government notified the Eastern District of Virginia that it would not appeal the August 3, 2010 decision. Kurt R. Karst, *All Eyes are on APP's Intervention Motion as the Government Bows Out of ANGIOMAX PTE Litigation*, September 9, 2010, http://www.fdalawblog.net/fda_law_blog_hyman_phelps/2010/09/all-eyes-are-on-apps-intervention-motion-as-the-government-bows-out-of-angiomax-pte-litigation.html. Although it would seem that the government's decision not to appeal would have ended the case, APP Pharmaceuticals, LLC moved to inter-vene[kk] and appealed the district court's decision to the Federal Circuit.

Many might wonder why MDCO spent so much time and money litigating the issue of whether the U.S. PTO properly interpreted and applied 35 U.S.C. § 156(d)(1) to its application for a patent term extension instead of simply suing its attorneys for malpractice for missing the patent term extension application deadline. The reason is fairly simple. The value of the patent term extension to which MDCO was entitled was approximately one and a half billion dollars. Joe Matal, *A Guide to the Legislative History of the America Invents Act: Part II of II*, 21 FED. CIR. B.J. 539, 648 (2012). Given that no law firm has malpractice insurance that can cover such an enormous liability, nor do even the most prominent firms have assets from which such a sum can be extracted, MDCO could not possibly recoup the money it was owed from a malpractice suit. *Id.* In its desperate quest to obtain its patent term extension, therefore, MDCO tirelessly lobbied Congress to change the law on the patent term extension application deadline while its litigation regarding the proper interpretation of 35 U.S.C. § 156(d)(1) was ongoing. Andrew Pollack, *Patent Bill Viewed as Bail Out for a Law Firm*, September 7, 2011. Indeed, between 2005 and 2011, MDCO spent more than $17 million on prominent lobbyists to try to get the law changed. *Id.* It was against this backdrop that AIA § 37 arose.

[kk]Fed. R. Civ. P. 24(a)(2) provides third parties with a right to intervene in a case between other parties where the third party "claims an interest relating to the property or transaction that is the subject of the action, and is so situated that disposing of the action may as a practical matter impair or impede the movant's ability to protect its interest, unless existing parties adequately represent that interest." APP Pharmaceuticals, LLC ("APP") argued that it had a right to intervene in MDCO's case against the U.S. PTO because APP had made substantial investments to develop a generic version of ANGIOMAX and had relied on the March 23, 2010 expiration date of MDCO's patent when filing its Abbreviated New Drug Application ("ANDA") with the FDA. Mem. in Supp. of Motion of APP Pharmaceuticals LLC's Motion for Leave to Intervene in *Medicines Co. v. Kappos*, No. 1:10-cv-286 (E.D.Va. 2010). According to APP, the August 3, 2010 decision by the Eastern District of Virginia "caused a retrospective change of PTO rules and procedures that the pharmaceutical industry (other than Plaintiff) has followed and relied upon for many years." *Id.* at 4. As a result of the retrospective change in the PTO's rules and procedures, the expiration of MDCO's patent could have been delayed until late 2014. *Id.* APP argued that it would be prejudiced by such delay because the FDA would not approve APP's ANDA application until MDCO's patent expired. *Id.* at 4-5.

D. AIA § 37's Effective Date and Its Impact on MDCO's Dispute

AIA § 37(a) amends pre-AIA 35 U.S.C. § 156(d)(1) as follows:

(1) To obtain an extension of the term of a patent under this section, the owner of record of the patent or its agent shall submit an application to the Director. Except as provided in paragraph (5), such an application may only be submitted within the sixty-day period beginning on the date the product received permission under the provision of law under which the applicable regulatory review period occurred for commercial marketing or use. The application shall contain -

(A) the identity of the approved product and the Federal statute under which regulatory review occurred;

(B) the identity of the patent for which an extension is being sought and the identity of each claim of such patent;

(C) information to enable the Director to determine under subsections (a) and (b) the eligibility of a patent for extension and the rights that will be derived from the extension and information to enable the Director and the Secretary of Health and Human Services or the Secretary of Agriculture to determine the period of the extension under subsection (g);

(D) a brief description of the activities undertaken by the applicant during the applicable regulatory review period with respect to the approved product and the significant dates applicable to such activities; and

(E) such patent or other information as the Director may require.

For purposes of determining the date on which a product receives permission under the second sentence of this paragraph, if such permission is transmitted after 4:30 P.M., Eastern Time, on a business day, or is transmitted on a day that is not a business day, the product shall be deemed to receive such permission on the next business day. For purposes of the preceding sentence, the term 'business day' means any Monday, Tuesday, Wednesday, Thursday, or Friday, excluding any legal holiday under section 6103 of title 5.

Leahy-Smith America Invents Act of 2011, Pub. L. No. 112-29, 125 Stat. 284, 341
(emphasis added).

AIA § 37(a), therefore, codifies Judge Hilton's August 3, 2010 opinion in the long-running dispute between MDCO and the U.S. PTO. Moreover, AIA § 37(b) states that "[t]he amendment made by subsection (a) shall apply to any application for extension of a patent term under section 156 . . . that is pending on, that is filed after, or as to which a decision regarding the application is subject to judicial review on, the date of the enactment of this Act." *Id.* The AIA was enacted on September 16, 2011, and APP's appeal against MDCO was still pending in the Federal Circuit as of that date. Joe Matal, *A Guide to the Legislative History of the America Invents Act: Part II of II*, 21 FED. CIR. B.J. 539, 648 (2012). Accordingly, AIA § 37(b) works to apply the provision of AIA § 37(a) to MDCO's patent term extension application that was filed over ten years before the AIA was enacted.

E. Practical Implications of AIA § 37

Overall, AIA § 37 will extend the deadline for patent term extension applications in cases where the FDA transmits its permission under 35 U.S.C. § 156(d)(1) after 4:30 p.m. on a business day or on any day that is not a business day. Indeed, the language added to pre-AIA 35 U.S.C. § 156(d)(1) makes clear that if the FDA transmits permission for a product's marketing after 4:30 p.m. Eastern Time on a business day, or transmits such permission on a day that is not a business day, the product "shall be deemed to receive such permission on the next business day" for purposes of calculating the 60-day deadline for filing a patent term extension application with the U.S. PTO.

F. Legislative Goals Addressed by AIA § 37

It is safe to say that if the patent term extension application for Angiomax had not been denied, the provision that was ultimately passed as AIA § 37 would never have been proposed. Unlike all of the other AIA provisions that were enacted to address extensive issues with the patent system that affected the general public, many believe that AIA § 37 was only enacted to bail out MDCO, its law firm, and its law firm's malpractice insurance company. *See* 157 CONG. REC. H4489, 2011 (statement of Rep. Lamar Smith) (arguing that AIA § 37 would interfere with pending litigation and explaining that "as a practical matter, [AIA § 37] is a special fix for one company."); 157 CONG. REC. S5442, 2011 (statement of Sen. McCain) ("As I noted earlier today when I spoke in support of the amendment offered by my colleague from Alabama, Senator Sessions, needed reform of our patent laws should not be diminished nor impaired by inclusion of the shameless special interest provision, dubbed 'The Dog Ate My Homework Act' that benefits a single drug manufacturer, Medicines & Company, to excuse their failure to follow the drug patent laws on the books for over 20 years."); 157 CONG. REC. S5403, 2011 (statement of Sen. Sessions) ("Average people have to suffer when they miss the statute of limitations. Poor people suffer when they miss the statute of limitations. But we are undertaking, at great expense to the taxpayers, to move a special interest piece of legislation that I don't believe can be justified as a matter of principle. I agree with the Wall Street

Journal that the adoption of it corrupts the system. We ought not be a part of that.").[ll]

Proponents of AIA § 37 characterized the provision as a "technical revision" that would eliminate "confusion regarding the deadline for patent term extension applications." 157 Cong. Rec. H4489, 2011 (statement of Rep. Conyers); *see also* 157 Cong. Rec. S5425, 2011 (statement of Sen. Kerry) ("Section 37 is, in fact, a very important clarification of a currently confusing deadline for filing patent term extension applications under the Hatch-Waxman Act. Frankly, this is a clarification, I would say to the Senator from Alabama, that benefits everybody in the country."). However, the context in which AIA § 37 arose, as well as the retroactive effect of the provision that enabled its application to the Angiomax patent term extension application, undercuts the notion that the provision was enacted simply to clarify the existing patent term extension statute.

[ll]For a powerful account of AIA § 37's highly controversial legislative history, see Joe Matal, *A Guide to the Legislative History of the America Invents Act: Part II of II*, 21 Fed. Cir. B.J. 539, 647-53 (2012).

CHAPTER NINE

The AIA's Changes to Inventor Oaths and Declarations and Patent Application Filings by Non-Inventors

Contents

I) INTRODUCTION

The AIA made significant amendments to pre-AIA 35 U.S.C. §§ 115 and 118, which pertain to the practices associated with inventor oaths and declarations and patent application filings by non-inventors. Pre-AIA 35 U.S.C. § 116 was also amended by the AIA, which prompted the U.S. PTO to make significant changes to the procedures for correcting inventorship in a patent or patent application. This chapter will discuss how the AIA's statutory amendments affect the inventor oath and declaration requirements for patent applications and the ability of non-inventors to file and prosecute patent applications before the U.S. PTO.

America Invents Act Primer. http://dx.doi.org/10.1016/B978-0-12-812096-5.00009-0
Copyright © 2017 Sarah Hasford. Published by Elsevier Inc. All rights reserved.

II) INVENTOR OATHS AND DECLARATIONS

Both the pre- and post-AIA laws require that the inventor(s) of any non-provisional patent application, as well as any application that enters the national stage from an international application under 35 U.S.C. § 371, provide an oath or declaration as prescribed by 35 U.S.C. § 115. Pre- and post-AIA 35 U.S.C. § 111(a)(2)(C); pre- and post-AIA 35 U.S.C. § 371(c) (4). The pre-AIA law assumed that patent applications were generally filed by their inventor; i.e., that the "patent applicant" and the "inventor" were the same person. H.R. REP. No. 112-98, at 43 (2011). As a result, the statutes and regulations governing the inventor oath or declaration requirements made it difficult for anyone other than the inventor (e.g., an assignee or other person with sufficient proprietary interest in the patent application) to file a patent application. *See* 77 Fed. Reg. 48,778 (Aug. 14, 2012) ("Traditionally, being the applicant (or the person who may 'make the application') under 35 U.S.C. chapter 11 has been synonymous with being the person who must execute the oath or declaration under 35 U.S.C. 115."). In reality, inventors are rarely the ones who file patent applications for their inventions. Indeed, the assignee or entity to which an obligation of assignment exists typically carries out the task of filing a patent application for any given invention. In recognition of the fact that the patent applicant is typically distinct from the inventor of a patent application, Congress amended pre-AIA 35 U.S.C. §§ 115 and 118 to streamline the inventor oath and declaration requirements and to make it easier for persons other than the inventor to file a patent application. The U.S. PTO also separated its regulations pertaining to patent applicants from the regulations that pertain to inventor oaths and declarations. *See* 77 Fed. Reg. 48,778-79 (Aug. 14, 2012) (explaining that post-AIA 37 C.F.R. §§ 1.41, 1.42, 1.43, 1.45, and 1.46 now pertain to patent applicants while post-AIA 37 C.F.R. §§ 1.63, and 1.64 now pertain to inventors).

A. AIA § 4—Changes to the Pre-AIA Inventor Oath and Declaration Practices

Through its amendments to pre-AIA 35 U.S.C. §§ 115 and 118, AIA § 4 modifies many of the pre-AIA requirements for inventor oaths and declarations, as well as some of the pre-AIA procedures for filing and prosecuting patent applications. More specifically, AIA § 4(a)(1) amends pre-AIA 35 U.S.C. § 115, which governs what types of documents may be submitted to the U.S. PTO to fulfill the inventor oath/declaration requirements, the content of these documents, and the timing of when such documents must be filed with the U.S. PTO. Furthermore, AIA § 4(b)(1) amends pre-AIA 35 U.S.C. § 118, which sets forth the specific guidelines for patent application filings by non-inventors, such as assignees, obligated assignees, and other people with sufficient proprietary interests in the patent application. The amendments to pre-AIA 35 U.S.C. §§ 115 and 118 are discussed in Sections II.A.i. and II.A.ii.

i. AIA § 4(a)(1)'s Amendments to Pre-AIA 35 U.S.C. § 115

The following table provides a comparison of pre- and post-AIA 35 U.S.C. § 115:

Pre-AIA 35 U.S.C. § 115 Oath of Applicant	Post-AIA 35 U.S.C. § 115 Inventor's Oath or Declaration
The applicant shall make oath that he believes himself to be the original and first inventor of the process, machine, manufacture, or composition of matter, or improvement thereof, for which he solicits a patent; and shall state of what country he is a citizen. Such oath may be made before any person within the United States authorized by law to administer oaths, or, when made in a foreign country, before any diplomatic or consular officer of the United States authorized to administer oaths, or before any officer having an official seal and authorized to administer oaths in the foreign country in which the applicant may be, whose authority is proved by certificate of a diplomatic or consular officer of the United States, or apostille of an official designated by a foreign country which, by treaty or convention, accords like effect to apostilles of designated officials in the United States.	(a) NAMING THE INVENTOR; INVENTOR'S OATH OR DECLARATION.-- An application for patent that is filed under section 111(a) or commences the national stage under section 371 shall include, or be amended to include, the name of the inventor for any invention claimed in the application. Except as otherwise provided in this section, each individual who is the inventor or a joint inventor of a claimed invention in an application for patent shall execute an oath or declaration in connection with the application. (b) REQUIRED STATEMENTS.-- An oath or declaration under subsection (a) shall contain statements that-- (1) the application was made or was authorized to be made by the affiant or declarant; and (2) such individual believes himself or herself to be the original inventor or an original joint inventor of a claimed invention in the application. (c) ADDITIONAL REQUIREMENTS.-- The Director may specify additional information relating to the inventor and the invention that is required to be included in an oath or declaration under subsection (a).

(*Continued on next page*)

Pre-AIA 35 U.S.C. § 115 Oath of Applicant	Post-AIA 35 U.S.C. § 115 Inventor's Oath or Declaration
Such oath is valid if it complies with the laws of the state or country where made. When the application is made as provided in this title by a person other than the inventor, the oath may be so varied in form that it can be made by him. For purposes of this section, a consular officer shall include any United States citizen serving overseas, authorized to perform notarial functions pursuant to section 1750 of the Revised Statutes, as amended (22 U.S.C. 4221).	(d) SUBSTITUTE STATEMENT.— (1) IN GENERAL.-- In lieu of executing an oath or declaration under subsection (a), the applicant for patent may provide a substitute statement under the circumstances described in paragraph (2) and such additional circumstances that the Director may specify by regulation. (2) PERMITTED CIRCUMSTANCES.-- A substitute statement under paragraph (1) is permitted with respect to any individual who-- (A) is unable to file the oath or declaration under subsection (a) because the individual-- (i) is deceased; (ii) is under legal incapacity; or (iii) cannot be found or reached after diligent effort; or (B) is under an obligation to assign the invention but has refused to make the oath or declaration required under subsection (a). (3) CONTENTS.-- A substitute statement under this subsection shall-- (A) identify the individual with respect to whom the statement applies; (B) set forth the circumstances representing the permitted basis for the filing of the substitute statement in lieu of the oath or declaration under subsection (a); and (C) contain any additional information, including any showing, required by the Director. (e) MAKING REQUIRED STATEMENTS IN ASSIGNMENT OF RECORD.-- An individual who is under an obligation of assignment of an application for patent may include the required statements under subsections (b) and (c) in the assignment executed by the individual, in lieu of filing such statements separately.

(Continued on next page)

Pre-AIA 35 U.S.C. § 115 Oath of Applicant	Post-AIA 35 U.S.C. § 115 Inventor's Oath or Declaration
	(f) TIME FOR FILING.-- The applicant for patent shall provide each required oath or declaration under subsection (a), substitute statement under subsection (d), or recorded assignment meeting the requirements of subsection (e) no later than the date on which the issue fee for the patent is paid. (g) EARLIER–FILED APPLICATION CONTAINING REQUIRED STATEMENTS OR SUBSTITUTE STATEMENT.-- (1) EXCEPTION.-- The requirements under this section shall not apply to an individual with respect to an application for patent in which the individual is named as the inventor or a joint inventor and that claims the benefit under section 120, 121, or 365(c) of the filing of an earlier-filed application, if-- (A) an oath or declaration meeting the requirements of subsection (a) was executed by the individual and was filed in connection with the earlier-filed application; (B) a substitute statement meeting the requirements of subsection (d) was filed in connection with the earlier filed application with respect to the individual; or (C) an assignment meeting the requirements of subsection (e) was executed with respect to the earlier-filed application by the individual and was recorded in connection with the earlier-filed application. (2) COPIES OF OATHS, DECLARATIONS, STATEMENTS, OR ASSIGNMENTS.-- Notwithstanding paragraph (1), the Director may require that a copy of the executed oath or declaration, the substitute statement, or the assignment filed in connection with the earlier-filed application be included in the later-filed application.

(Continued on next page)

Pre-AIA 35 U.S.C. § 115 Oath of Applicant	Post-AIA 35 U.S.C. § 115 Inventor's Oath or Declaration
	(h) SUPPLEMENTAL AND CORRECTED STATEMENTS; FILING ADDITIONAL STATEMENTS.-- (1) IN GENERAL.-- Any person making a statement required under this section may withdraw, replace, or otherwise correct the statement at any time. If a change is made in the naming of the inventor requiring the filing of 1 or more additional statements under this section, the Director shall establish regulations under which such additional statements may be filed. (2) SUPPLEMENTAL STATEMENTS NOT REQUIRED.-- If an individual has executed an oath or declaration meeting the requirements of subsection (a) or an assignment meeting the requirements of subsection (e) with respect to an application for patent, the Director may not thereafter require that individual to make any additional oath, declaration, or other statement equivalent to those required by this section in connection with the application for patent or any patent issuing thereon. (3) SAVINGS CLAUSE.-- A patent shall not be invalid or unenforceable based upon the failure to comply with a requirement under this section if the failure is remedied as provided under paragraph (1). (i) ACKNOWLEDGMENT OF PENALTIES.-- Any declaration or statement filed pursuant to this section shall contain an acknowledgment that any willful false statement made in such declaration or statement is punishable under section 1001 of title 18 by fine or imprisonment of not more than 5 years, or both.

As shown in the previous table, AIA § 4(a)(1) made significant changes to pre-AIA 35 U.S.C. § 115. These changes are most easily understood by analyzing each subsection of post-AIA 35 U.S.C. § 115.[a] A discussion of each subsection is set forth in Sections II.A.i.1–II.A.i.7.

1. Post-AIA 35 U.S.C. § 115(a)—Naming the Inventor; Inventor's Oath or Declaration

Post-AIA 35 U.S.C. § 115(a) does not itself alter the pre-AIA inventor oath and declaration requirements, but it does provide the basis for the changes that are made through the other subsections of post-AIA 35 U.S.C. § 115. Post-AIA 35 U.S.C. § 115(a) states that:

> *An application for patent that is filed under section 111(a) or commences the national stage under section 371 shall include, or be amended to include, the name of the inventor for any invention claimed in the application. Except as otherwise provided in this section, each individual who is the inventor or a joint inventor of a claimed invention in an application for patent shall execute an oath or declaration in connection with the application.*

The first sentence of post-AIA 35 U.S.C. § 115(a) simply requires that non-provisional patent applications (i.e., applications filed under section 111(a) and applications entering the national stage under section 371) include, or be amended to include, the name of the inventor for any invention claimed in the application. The pre-AIA law imposed the same requirements on non-provisional and national stage applications. *See* pre-AIA 35 U.S.C. § 111(a)(2) (requiring that a non-provisional application include an oath or declaration in accordance with pre-AIA 35 U.S.C. § 115 in order to be considered complete); pre-AIA 37 C.F.R. § 1.63(a)(2) (requiring that oaths or declarations filed in non-provisional applications identify each inventor by full name); pre-AIA 35 U.S.C. § 371(c)(4) (requiring that national stage applications filed under pre-AIA 35 U.S.C. § 371 include an oath or declaration of the inventor that complies with the requirements of 35 U.S.C. § 115); pre-AIA 37 C.F.R. § 1.497(a)(3) (requiring that an oath or declaration that is filed under pre-AIA 35 U.S.C. § 371(c)(4) identify each inventor for any invention claimed in the application). Thus,

---

[a] Note that subsections (c) and (i) of post-AIA 35 U.S.C. § 115 are discussed in the context of other subsections.

the requirement to identify the name(s) of the inventor(s) for any claimed invention in a non-provisional or national stage application was not altered by the AIA.

Furthermore, even though the second sentence of post-AIA 35 U.S.C. § 115(a) does not itself change the pre-AIA inventor oath and declaration requirements, it provides the basis for the changes that are made through the remaining subsections of post-AIA 35 U.S.C. § 115. In particular, the second sentence of post-AIA 35 U.S.C. § 115(a) states that: "[e]xcept as otherwise provided in this section, each individual who is the inventor or a joint inventor of a claimed invention in an application for patent shall execute an oath or declaration in connection with the application." As discussed in detail later on in this chapter, subsections (d), (e) and (g) of post-AIA 35 U.S.C. § 115 set forth various exceptions to the requirement to file an inventor oath or declaration set forth in post-AIA 35 U.S.C. § 115(a). Many of these exceptions did not exist under the pre-AIA law.

2. Post-AIA 35 U.S.C. § 115(b)—Streamlining the Oath or Declaration Requirements

Both pre- and post-AIA 35 U.S.C. § 115 require that only two statements be included in an inventor oath or declaration. In particular, with regard to the content of inventor oaths and declarations, pre-AIA 35 U.S.C. § 115 states "[t]he applicant shall make [an] oath that he believes himself to be the original and first inventor of the process, machine, manufacture, or composition of matter, or improvement thereof, for which he solicits a patent; and shall state of what country he is a citizen." Pre-AIA 35 U.S.C. § 115 therefore requires that inventors state: (1) "that he believes himself to be the original and first inventor of the process, machine, manufacture, or composition of matter, or improvement thereof, for which he solicits a patent," and (2) "of what country he is a citizen." Similar to pre-AIA 35 U.S.C. § 115, post-AIA 35 U.S.C. § 115(b) requires inventors to make two statements in their oaths or declarations. In particular, post-AIA 35 U.S.C. § 115(b) states:

> REQUIRED STATEMENTS.—— An oath or declaration under subsection (a) shall contain statements that——
> (1) the application was made or was authorized to be made by the affiant or declarant; and
> (2) such individual believes himself or herself to be the original inventor or an original joint inventor of a claimed invention in the application.

At first glance, therefore, it would appear that the post-AIA statute did little to change the pre-AIA inventor oath and declaration requirements. The inventor oath and declaration requirements of pre-AIA 35 U.S.C. § 115 and post-AIA 35 U.S.C. § 115(b), however, are implemented through pre- and post-AIA 37 C.F.R. § 1.63.[b] And as can be seen in the following table, pre-AIA 37 C.F.R. § 1.63 underwent significant changes as a result of the AIA.[c]

Required Statements in an Inventor's Oath or Declaration Under Pre-AIA 37 C.F.R. § 1.63	Required Statements in an Inventor's Oath or Declaration Under Post-AIA 37 C.F.R. § 1.63
Statements required under pre-AIA 37 C.F.R. § 1.63(a): (1) An identification of each inventor by full name, including the family name, and at least one given name without abbreviation together with any other given name or initial; (2) An identification of the country of citizenship of each inventor; and	*Statements required under post-AIA 37 C.F.R. § 1.63(a):* (1) An identification of the inventor or joint inventor executing the oath or declaration by his or her legal name; (2) An identification of the application to which it is directed;

(*Continued on next page*)

[b]When Congress enacts a new statute or amends an existing statute that affects procedures and practices before the U.S. PTO, the statute will often provide the U.S. PTO with the authority to enact regulations that further specify how one can comply with the statute. For example, when Congress amended pre-AIA 35 U.S.C. § 115 through the AIA, it provided the U.S. PTO with authority to enact further regulations that would govern the content of inventor's oaths and declarations. *See* post-AIA 35 U.S.C. § 115(c) ("The Director [of the U.S. PTO] may specify additional information relating to the inventor and the invention that is required to be included in an oath or declaration under subsection (a)."). The U.S. PTO's regulations are codified in the Code of Federal Regulations, which is abbreviated as "C.F.R." These regulations are usually helpful to patent practitioners and applicants because they set forth specific guidelines for meeting the requirements of a statute, which, by their nature, are more general.

[c]Despite the many differences between pre- and post-AIA 37 C.F.R. § 1.63, both require that an inventor oath and declaration: (1) identify the application to which it is directed; and (2) identify the inventor's mailing address and residence (if an inventor lives at a location that is different from where the inventor customarily receives mail) if such information is not included in the application data sheet ("ADS"). Pre-AIA 37 C.F.R. §§ 1.63(b)(1) and (c)(1); post-AIA 37 C.F.R. §§ 1.63(a)(2) and (b)(2).

Required Statements in an Inventor's Oath or Declaration Under Pre-AIA 37 C.F.R. § 1.63	Required Statements in an Inventor's Oath or Declaration Under Post-AIA 37 C.F.R. § 1.63
(3) A statement that the person making the oath or declaration believes the named inventor or inventors to be the original and first inventor or inventors of the subject matter which is claimed and for which a patent is sought. *Statements required under pre-AIA 37 C.F.R. § 1.63(b):* (1) An identification of the application to which it is directed; (2) A statement that the person making the oath or declaration has reviewed and understands the contents of the application, including the claims, as amended by any amendment specifically referred to in the oath or declaration; and (3) A statement that the person making the oath or declaration acknowledges the duty to disclose to the Office all information known to the person to be material to patentability as defined in 37 C.F.R. § 1.56. *Statements required under pre-AIA 37 C.F.R. § 1.63(c):* (1) An identification of the mailing address, and the residence if an inventor lives at a location which is different from where the inventor customarily receives mail, of each inventor[a]; and (2) An identification of any foreign application for patent (or inventor's certificate) for which a claim for priority is made pursuant to 37 C.F.R. § 1.55, and any foreign application having a filing date before that of the application on which priority is claimed, by specifying the application number, country, day, month, and year of its filing.[a]	(3) A statement that the person executing the oath or declaration believes the named inventor or joint inventor to be the original inventor or an original joint inventor of a claimed invention in the application for which the oath or declaration is being submitted; and (4) A statement that the application was made or was authorized to be made by the person executing the oath or declaration. *Statements required under post-AIA 37 C.F.R. § 1.63(b):* (1) An identification of each inventor by his or her legal name[b]; and (2) An identification of the mailing address where the inventor customarily receives mail, and residence, if an inventor lives at a location which is different from where the inventor customarily receives mail, for each inventor.[b]

[a]Indicates statements that only need to be included in the oath or declaration if the statement has not been included in the application data sheet ("ADS") accompanying the patent application pursuant to 37 C.F.R. § 1.76.
[b]Indicates statements that only need to be included in the oath or declaration if the statement has not been supplied in the application data sheet ("ADS") accompanying the patent application pursuant to 37 C.F.R. § 1.76.

Overall, as compared to pre-AIA 37 C.F.R. § 1.63, post-AIA 37 C.F.R. § 1.63 reduces the total number of statements that must be included in an inventor's oath or declaration. Post-AIA 37 C.F.R. § 1.63 also modifies the content of one of the statements that was required under pre-AIA 37 C.F.R. § 1.63 and adds one new statement that was not required under the U.S. PTO's pre-AIA regulations.

More specifically, post-AIA 37 C.F.R. § 1.63 no longer requires that the following statements, which were required by pre-AIA 37 C.F.R. § 1.63, be included in post-AIA inventor oaths and declarations:

(1) a statement that the person executing the oath or declaration has reviewed and understands the contents of the application, including the claims, as amended by any amendment specifically referred to in the oath or declaration (pre-AIA 37 C.F.R. § 1.63(b)(2));

(2) a statement that the person executing the oath or declaration acknowledges the duty to disclose to the Office all information known to the person to be material to patentability as defined in 37 C.F.R. § 1.56 (pre-AIA 37 C.F.R. § 1.63(b)(3))[d];

(3) an acknowledgement that willful false statements may jeopardize the validity of the application or any patent issuing thereon, and that all statements made of the declarant's own knowledge are true and that all statements made on information and belief are believed to be true (77 Fed. Reg. 48,779 (Aug. 14, 2012))[e];

(4) an identification of any foreign application for patent (or inventor's certificate) for which a claim for priority is made pursuant to § 1.55, and any foreign application having a filing date before that of the

[d]Note that although the statements required by pre-AIA 37 C.F.R. § 1.63(b)(2) and (3) are no longer required by the post-AIA law, post-AIA 37 C.F.R. § 1.63(c) states that "[a] person may not execute an oath or declaration for an application unless that person has reviewed and understands the contents of the application, including the claims, and is aware of the duty to disclose to the Office all information known to the person to be material to patentability as defined in [37 C.F.R.] § 1.56." Thus, the best practice is to maintain these statements in a post-AIA inventor oath or declaration so that it is clear that the person executing the document has complied with post-AIA 37 C.F.R. § 1.63(c).

[e]Post-AIA 35 U.S.C. § 115(i) still requires that any oath or declaration, substitute statement, or assignment-statement that is filed pursuant to post-AIA 35 U.S.C. § 115 include an acknowledgment of penalties that any willful false statement made in such declaration or statement is punishable under section 1001 of title 18 by fine or imprisonment of not more than 5 years, or both. The statement required by post-AIA 35 U.S.C. § 115(i) has the effect of making an inventor oath or declaration, substitute statement, or assignment-statement properly denominated as a "declaration" for legal purposes. 77 Fed. Reg. 48,779 (Aug. 14, 2012).

application on which priority is claimed, by specifying the application number, country, day, month, and year of its filing (pre-AIA 37 C.F.R. § 1.63(c)(2))[f]; and

(5) an identification of the inventor's country of citizenship (*see* pre-AIA 37 C.F.R. § 1.63(a)(3)).

Furthermore, unlike under pre-AIA 37 C.F.R. § 1.63(a)(2), post-AIA 37 C.F.R. § 1.63(b)(1) does not require an oath or declaration to indicate the name of each inventor[g] if the applicant provides an application data sheet ("ADS") that indicates the legal name, residence, and mailing address of each inventor.

Besides eliminating several of the statements that were required under pre-AIA 37 C.F.R. § 1.63, post-AIA 37 C.F.R. § 1.63 modifies one of the statements that was required under the pre-AIA law and requires one additional statement that was not required under pre-AIA 37 C.F.R. § 1.63.

---

[f]The post-AIA law's elimination of the requirement that patent applicants identify any foreign application for patent (or inventor's certificate) for which a claim for priority is made in either the application data sheet or the inventor oath or declaration harmonizes the practice regarding foreign priority claims with the practice regarding domestic priority claims. Indeed, under the pre-AIA law, foreign applications to which a patent applicant wanted to claim priority could be identified in either the application data sheet ("ADS") or the inventor oath or declaration, whereas domestic applications to which a patent applicant wanted to claim priority could be identified in either the first sentence of the specification or the ADS. Pre-AIA 37 C.F.R. § 1.63(c); pre-AIA 37 C.F.R. § 1.78(a)(2)(iii). The ability to place different priority claims in different places caused confusion for Patent Examiners and applicants. 77 Fed. Reg. 48,789 (Aug. 14, 2012). Indeed, to determine what priority claims were being made under the pre-AIA law, Patent Examiners had to look at the specification, any amendments to the specification, the inventor oath(s) or declaration(s), and the ADS. 77 Fed. Reg. at page 48,788 (Aug. 14, 2012). To add to Examiners' confusion, the information provided in those documents was often inconsistent. Under the post-AIA law, both foreign and domestic priority claims must be presented in only one location; i.e., the ADS. The harmonization of the practices for foreign and domestic priority claims should eliminate confusion for Examiners and promote efficiency during patent examination.

[g]By eliminating the requirement to identify each inventor, the post-AIA law makes it somewhat easier for inventors to execute an oath or declaration under post-AIA 35 U.S.C. § 115. Indeed, under the pre-AIA law, the inventor oath or declaration had to list all inventors for a patent application. Thus, patent practitioners would often prepare one oath or declaration and send it to all inventors. Once one inventor signed the document, he/she would then send the signed document to the other inventors for their signatures so that all of the signatures appeared in one document. Not surprisingly, the task of collecting inventors' signatures in one document became more onerous as the number of inventors increased. Under the post-AIA law, each inventor executes his/her own document (i.e., an oath, declaration, or substitute statement). Accordingly, there should no longer be a need to pass the required documentation from one inventor to the next.

Specifically, consistent with the shift from a first-to-invent to a first-inventor-to-file patent system, post-AIA 37 C.F.R. § 1.63 no longer requires that an inventor's oath or declaration state that the person making the oath or declaration believes himself/herself to be the original and <u>first</u> inventor of the subject matter which is claimed and for which a patent is sought. *Compare* pre-AIA 37 C.F.R. § 1.63(a)(4) (requiring that an inventor's oath or declaration state that the person making the oath or declaration "believes the named inventor or inventors to be the original and first inventor or inventors of the subject matter which is claimed and for which a patent is sought") *with* post-AIA 37 C.F.R. § 1.63(a)(3) (requiring only that the person executing the oath or declaration state that he/she "believes the named inventor or joint inventor to be the original inventor or an original joint inventor of a claimed invention in the application for which the oath or declaration is being submitted."). Post-AIA 37 C.F.R. § 1.63(a)(4) further requires that an inventor's oath or declaration "[s]tate that the application was made or was authorized to be made by the person executing the oath or declaration." This statement was not required under pre-AIA 37 C.F.R. § 1.63.

3. Post-AIA 35 U.S.C. § 115(d)—Substitute Statements

Post-AIA 35 U.S.C. § 115(d) reads as follows:

> *(d) SUBSTITUTE STATEMENT.--*
> > *(1) IN GENERAL.-- In lieu of executing an oath or declaration under subsection (a), the applicant for patent may provide a substitute statement under the circumstances described in paragraph (2) and such additional circumstances that the Director may specify by regulation.*
> > *(2) PERMITTED CIRCUMSTANCES.-- A substitute statement under paragraph (1) is permitted with respect to any individual who--*
> > > *(A) is unable to file the oath or declaration under subsection (a) because the individual--*
> > > > *(i) is deceased;*
> > > > *(ii) is under legal incapacity; or*
> > > > *(iii) cannot be found or reached after diligent effort;*
> > > > *or*
> > > *(B) is under an obligation to assign the invention but has refused to make the oath or declaration required under subsection (a).*
> > *(3) CONTENTS.-- A substitute statement under this subsection shall--*
> > > *(A) identify the individual with respect to whom the statement applies;*
> > > *(B) set forth the circumstances representing the permitted basis for the filing of the substitute statement in lieu of the oath or declaration under subsection (a); and*
> > > *(C) contain any additional information, including any showing, required by the Director.*

Post-AIA 35 U.S.C. § 115(d) alters the U.S. PTO's inventor oath and declaration practices by permitting a patent applicant to file a substitute statement in lieu of an inventor's oath or declaration in circumstances where:

(1) an individual is unable to file an oath or declaration because he/she is deceased;

(2) an individual is unable to file an oath or declaration because he/she is under legal incapacity;

(3) an individual is unable to file an oath or declaration because he/she cannot be found or reached after diligent effort; or

(4) an individual is under an obligation to assign the invention but has refused to make the oath or declaration that is required under post-AIA 35 U.S.C. § 115(a).

Post-AIA 35 U.S.C. § 115(d)(2)(A) and (B); *see also* post-AIA 37 C.F.R. § 1.64(a) ("An applicant under [37 C.F.R.] § 1.43, 1.45 or 1.46 may execute a substitute statement in lieu of an oath or declaration under § 1.63 if the inventor is deceased, is under a legal incapacity, has refused to execute the oath or declaration under § 1.63, or cannot be found or reached after diligent effort.").

Substitute statements were not available prior to the AIA. Indeed, before the new legislation was enacted, the U.S. PTO required that an oath or declaration be filed for every inventor listed on a patent application even if an inventor was unavailable due to the existence of one of the aforementioned circumstances, i.e., even if the inventor was deceased, legally incapacitated, could not be found or reached after diligent effort, or refused to make the oath or declaration. *See* pre-AIA 35 U.S.C. § 117 ("Legal representatives of deceased inventors and of those under legal incapacity may make application for patent upon compliance with the requirements and on the same terms and conditions applicable to the inventor."); pre-AIA 37 C.F.R. § 1.42 ("In case of the death of the inventor, the legal representative (executor, administrator, etc.) of the deceased inventor may make the necessary oath or declaration, and apply for and obtain the patent."); pre-AIA 37 C.F.R. § 1.43 ("In case an inventor is insane or otherwise legally incapacitated, the legal representative (guardian, conservator, etc.) of such inventor may make the necessary oath or declaration, and apply for and obtain the patent."); MPEP § 409.03(a) (9th ed. 2015) (explaining that if a joint inventor "cannot be found or reached after diligent effort" or refuses to "join in an application," all of the available joint inventors must make an oath or declaration on behalf of the nonsigning joint inventor as required by pre-AIA 37 C.F.R. § 1.64). Furthermore, in circumstances where an inventor could not be found or reached after diligent effort, or refused to make an oath or declaration, the pre-AIA law required that the oath or declaration filed on behalf of the inventor be accompanied by a petition including proof of the pertinent facts, and an identification of the last known address of the nonsigning inventor. Pre-AIA 37 C.F.R. § 1.47(a); MPEP § 409.03(d) (9th ed. 2015).[h]

The shift from requiring an oath or declaration for every inventor, regardless of whether one or more inventors is unavailable, to permitting a substitute statement in certain situations is significant in part because the content of a substitute statement differs from that of a pre-AIA oath or declaration. The following table compares the statements that must be included in a pre-AIA inventor oath or declaration with those that must be included in a substitute statement:

Statements Required in an Oath or Declaration Filed on Behalf of an Unavailable Inventor Under the Pre-AIA Law	Statements Required in a Substitute Statement
Statements required under pre-AIA 37 C.F.R. § 1.63(a): (1) An identification of each inventor by full name, including the family name, and at least one given name without abbreviation together with any other given name or initial; (2) An identification of the country of citizenship of each inventor; and (3) A statement that the person making the oath or declaration believes the named inventor or inventors to be the original and first inventor or inventors of the subject matter which is claimed and for which a patent is sought.	*Statements required under post-AIA 35 U.S.C. § 115(d)(3), post-AIA 37 C.F.R. § 1.63(a), and post-AIA 37 C.F.R. § 1.64:* (1) An identification of the application to which it is directed (post-AIA 37 C.F.R. §§ 1.63(a)(2) and 1.64(b)(1)); (2) A statement that the person executing the oath or declaration believes the named inventor or joint inventor to be the original inventor or an original joint inventor of a claimed invention in the application for which the oath or declaration is being submitted (post-AIA 37 C.F.R. §§ 1.63(a)(3) and 1.64(b)(1));

(Continued on next page)

---

[h] The requirement to file a petition with the oath or declaration is significant because prior to the AIA, a patent application could not be examined until the Office of Petitions granted the petition. MPEP § 409.03(j) (9th ed. 2015) (explaining that examination of a patent application with which a petition under pre-AIA 37 C.F.R. § 1.47(a) is submitted will only be acted on by a Patent Examiner once the Office has granted such petition).

Statements Required in an Oath or Declaration Filed on Behalf of an Unavailable Inventor Under the Pre-AIA Law	Statements Required in a Substitute Statement
Statements required under pre-AIA 37 C.F.R. § 1.63(b): (1) An identification of the application to which the oath or declaration is directed; (2) A statement that the person making the oath or declaration has reviewed and understands the contents of the application, including the claims, as amended by any amendment specifically referred to in the oath or declaration; and (3) A statement that the person making the oath or declaration acknowledges the duty to disclose to the Office all information known to the person to be material to patentability as defined in § 1.56. *Statements required under pre-AIA 37 C.F.R. § 1.63(c):* (1) The mailing address, and the residence if an inventor lives at a location which is different from where the inventor customarily receives mail, of each inventor[a]; and (2) An identification of any foreign application for patent (or inventor's certificate) for which a claim for priority is made pursuant to § 1.55, and any foreign application having a filing date before that of the application on which priority is claimed, by specifying the application number, country, day, month, and year of its filing.[a] *Statements required under pre-AIA 37 C.F.R. § 1.64(b):* (1) A statement describing the relationship between the person executing the oath on behalf of the inventor and the inventor, and, upon information and belief, the facts which the inventor is required to state.	(3) A statement that the application was made or was authorized to be made by the person executing the oath or declaration (post-AIA 37 C.F.R. §§ 1.63(a)(4) and 1.64(b)(1)); (4) An identification of the inventor or joint inventor with respect to whom a substitute statement in lieu of an oath or declaration is executed, and a statement upon information and belief of the facts which such inventor is required to state (post-AIA 35 U.S.C. § 115(d)(3)(A); post-AIA 37 C.F.R. §§ 1.63(a)(1) and 1.64(b)(1)); (5) An identification of the person executing the substitute statement and the relationship of such person to the inventor or joint inventor with respect to whom the substitute statement is executed, and unless such information is supplied in an application data sheet in accordance with 37 C.F.R. § 1.76, the residence and mailing address of the person signing the substitute statement (post-AIA 37 C.F.R. § 1.64(b)(2)); (6) An identification of the circumstances permitting the person to execute the substitute statement in lieu of an oath or declaration under post-AIA 37 C.F.R. § 1.63, namely whether the inventor is deceased, is under a legal incapacity, cannot be found or reached after a diligent effort was made, or has refused to execute the oath or declaration under 37 C.F.R. § 1.63 (post-AIA 35 U.S.C. § 115(d)(3)(B); post-AIA 37 C.F.R. § 1.64(b)(3));

(Continued on next page)

Statements Required in an Oath or Declaration Filed on Behalf of an Unavailable Inventor Under the Pre-AIA Law	Statements Required in a Substitute Statement
(2) If the person signing the oath or declaration is the legal representative of a deceased inventor, the oath or declaration shall also state that the person is a legal representative and the citizenship, residence, and mailing address of the legal representative.	(7) Unless the following information is supplied in an application data sheet ("ADS") in accordance with § 1.76, an identification of: (i) Each inventor by his or her legal name; and (ii) The last known mailing address where the inventor customarily receives mail, and last known residence, if an inventor lives at a location which is different from where the inventor customarily receives mail, for each inventor who is not deceased or under a legal incapacity (post-AIA 37 C.F.R. § 1.64(b)(4)(i) and (ii)); and (8) An acknowledgment that any willful false statement made in such statement is punishable under section 1001 of title 18 by fine or imprisonment of not more than 5 years, or both (post-AIA 37 C.F.R. § 1.64(e)).

aIndicates statements that only need to be included in the oath or declaration if the statement has not been included in the application data sheet ("ADS") accompanying the patent application pursuant to 37 C.F.R. § 1.76.

Substitute statements are similar to pre-AIA inventor oaths and declarations in that both are required to: (1) identify the application to which they are directed, and (2) provide the inventor's mailing address and residence if the inventor lives at a location that is different from where the inventor customarily receives mail (assuming such information is not supplied in an ADS). Pre-AIA 37 C.F.R. § 1.63(b)(1) and (c)(1); post-AIA 37 C.F.R. §§ 1.63(a)(2), 1.64(b)(1) and 1.64(b)(4)(ii). Despite these similarities, there are several differences between the statements that are required in a pre-AIA inventor oath or declaration and those required in a post-AIA substitute statement. For example, unlike the oath or declaration required under the pre-AIA law, the following statements are not required in a substitute statement under the post-AIA law:

1) an identification of the country of citizenship of the inventor with respect to whom a substitute statement in lieu of an oath or declaration is being executed (pre-AIA 37 C.F.R. § 1.63(a)(3));

2) a statement that the person making the oath or declaration has reviewed and understands the contents of the application, including the
claims, as amended by any amendment specifically referred to in the
oath or declaration (pre-AIA 37 C.F.R. § 1.63(b)(2));

3) a statement that the person making the oath or declaration acknowledges the duty to disclose to the Office all information known to the
person to be material to patentability as defined in 37 C.F.R. § 1.56
(pre-AIA 37 C.F.R. § 1.63(b)(3));

4) an identification of any foreign application for patent or inventor's certificate for which a claim for priority is made pursuant to 37 C.F.R.
§ 1.55, and any foreign application having a filing date before that of the
application on which priority is claimed, by specifying the application
number, country, day, month, and year of its filing (pre-AIA 37 C.F.R.
§ 1.63(c)(2)).[i]

In addition, a substitute statement that is filed under post-AIA 35 U.S.C.
§ 115(d) and post-AIA 37 C.F.R. § 1.64 requires the following statements
that are not required in a pre-AIA inventor oath or declaration:

1) a statement that the application was made or was authorized to be made
by the person executing the substitute statement (see post-AIA 37 C.F.R.
§ 1.64(b)(1) (explaining that a substitute statement must "[c]omply with
the requirements of [37 C.F.R.] § 1.63(a), identifying the inventor or joint
inventor with respect to whom a substitute statement in lieu of an oath or
declaration is executed, and stating upon information and belief the facts
which such inventor is required to state"); post-AIA 37 C.F.R. § 1.63(a)(4)
(requiring a statement that "the application was made or was authorized to
be made by the person executing the [substitute statement]"));

2) an identification of the circumstances permitting the person to execute the substitute statement in lieu of an oath or declaration under
§ 1.63, namely whether the inventor is deceased, is under a legal incapacity, cannot be found or reached after a diligent effort was made, or
has refused to execute the oath or declaration under § 1.63 (post-AIA
37 C.F.R. § 1.64(b)(3))[j]; and

---

[i]Note that pre-AIA 37 C.F.R. § 1.63(c)(2) only requires that this information be included in
an inventor oath or declaration if it is not already included in the application data sheet that
was submitted for the patent application pursuant to pre-AIA 37 C.F.R. § 1.76.

[j]Note that in pre-AIA cases where an inventor could not be found or refused to execute an
oath or declaration, the circumstances permitting the other inventors to execute an oath or
declaration on his/her behalf would be stated in a petition filed under pre-AIA 37 C.F.R.
§ 1.47(a).

3) an acknowledgment that any willful false statement made in such statement is punishable under section 1001 of title 18 by fine or imprisonment of not more than 5 years, or both (post-AIA 37 C.F.R. § 1.64(e)).

Furthermore, unlike pre-AIA inventor oaths and declarations, which were required to identify all of the inventors for the patent applications to which they pertained, a substitute statement need only identify the inventor for whom the statement is being filed so long as all of the inventors are identified in the application data sheet accompanying the patent application. *Compare* pre-AIA 37 C.F.R. § 1.63(a)(2) (requiring that an oath or declaration "[i]dentify each inventor by full name, including the family name, and at least one given name without abbreviation together with any other given name or initial.") *with* post-AIA 37 C.F.R. § 1.64(b)(4)(i) (requiring that a substitute statement identify each inventor by his or her legal name unless such information is provided in an application data sheet in accordance with 37 C.F.R. § 1.76). Also, unlike inventor oaths or declarations under the pre-AIA law, which were required to state that the person making the oath or declaration believes the named inventor to be the original and first inventor of the subject matter which is claimed and for which a patent is sought, the person submitting a substitute statement under the post-AIA law merely has to assert that he/she believes that the inventor or joint inventor with respect to whom the substitute statement is being executed is the original inventor or an original joint inventor of a claimed invention in the application. *Compare* pre-AIA 37 C.F.R. § 1.63(a)(4) (requiring that an inventor oath or declaration state that the person making the oath or declaration believes the named inventor or inventors to be the original and first inventor or inventors of the subject matter which is claimed and for which a patent is sought) *with* post-AIA 37 C.F.R. § 1.64(b)(1) (explaining that a substitute statement must "[c]omply with the requirements of [37 C.F.R.] § 1.63(a), identifying the inventor or joint inventor with respect to whom a substitute statement in lieu of an oath or declaration is executed, and stating upon information and belief the facts which such inventor is required to state") and post-AIA 37 C.F.R. § 1.63(a)(3) (requiring that post-AIA inventor oaths or declarations include a statement that the person executing the oath or declaration believes the named inventor or joint inventor to be the original inventor or an original joint inventor of a claimed invention in the application for which the oath or declaration is being submitted).

The pre- and post-AIA laws regarding the inventor oath/declaration requirements also differ in that more people are eligible to file substitute statements under the post-AIA law than were eligible to file an oath or declaration on behalf of an unavailable inventor under the pre-AIA law. Indeed, under the pre-AIA law, only a legal representative of a deceased or legally incapacitated inventor could file an oath or declaration. See pre-AIA 37 C.F.R. § 1.42 ("In case of the death of the inventor, the legal representative (executor, administrator, etc.) of the deceased inventor may make the necessary oath or declaration, and apply for and obtain the patent."); pre-AIA 37 C.F.R. § 1.43 ("In case an inventor is insane or otherwise legally incapacitated, the legal representative (guardian, conservator, etc.) of such inventor may make the necessary oath or declaration, and apply for and obtain the patent."). Furthermore, if a joint inventor refused to join in a patent application, or could not be found or reached after diligent effort, the pre-AIA law required that the other inventors execute an oath or declaration on behalf of the non-signing inventor. Pre-AIA 37 C.F.R. § 1.47(a).[k]

The post-AIA law, however, allows a substitute statement to be filed by a legal representative of a deceased or legally incapacitated inventor, an assignee of the patent application, a party to whom an unavailable inventor is obligated to assign the patent application, a party who otherwise shows sufficient proprietary interest in the patent application, or available joint inventors. See post-AIA 37 C.F.R. § 1.64(a) (explaining that "an applicant under [37 C.F.R.] § 1.43, 1.45 or 1.46 may execute a substitute statement in lieu of an oath or declaration under § 1.63 if the inventor is deceased, is under a legal incapacity, has refused to execute the oath or declaration under § 1.63, or cannot be found or reached after diligent effort"); post-AIA 37 C.F.R. § 1.43 ("If an inventor is deceased or under legal incapacity, the legal representative of the inventor may make an application for patent on behalf of the inventor."); post-AIA 37 C.F.R. § 1.45(a) ("If a joint inventor refuses to join in an application for patent or cannot be found or reached after diligent effort, the other joint inventor or inventors may make the application for patent on behalf of themselves and the omitted inventor.");

[k]Although the pre- and post-AIA laws differ with respect to the person(s) who may execute a substitute statement or oath or declaration in the circumstances where an inventor is deceased, is legally incapacitated or is a joint inventor who refuses to join in a patent application or cannot be reached or found after diligent effort, the laws do not differ for circumstances in which the *sole* inventor or *all* joint inventors refuse to execute an application for patent or cannot be found or reached after diligent effort. Indeed, under both the pre- and post-AIA law, a person to whom an inventor has assigned the invention, or agreed in writing to assign the invention, or who otherwise shows sufficient proprietary interest in the matter justifying such action, may make an application for patent on behalf of and as agent for all the inventors in such a scenario. Pre-AIA 37 C.F.R. § 1.47(b); post-AIA 37 C.F.R. § 1.45(a).

post-AIA 37 C.F.R. § 1.46(a) (explaining that a person to whom the inventor has assigned or is under an obligation to assign the invention may make an application for patent and that a person who otherwise shows sufficient proprietary interest in the matter may make an application for patent on behalf of and as agent for the inventor). By expanding the categories of people eligible to execute a substitute statement, the post-AIA law makes filing a patent application easier in instances where an inventor is unavailable.

Section II.A.ii. discusses additional differences between the U.S. PTO's practices regarding the inventor oath and declaration requirements in circumstances where the sole inventor or all joint inventors refuse to join in a patent application or where none of the inventors can be found or reached after diligent effort.

4. Post-AIA 35 U.S.C. § 115(e)—Making Required Statements in the Assignment of Record

Similar to post-AIA 35 U.S.C. § 115(d), post-AIA 35 U.S.C. § 115(e) allows patent applicants to meet the inventor oath/declaration requirement by filing a document other than an oath or declaration. In particular, post-AIA 35 U.S.C. § 115(e) states that "[a]n individual who is under an obligation of assignment of an application for patent may include the required statements [i.e., the statements required by post-AIA 35 U.S.C. § 115(b) and (c) and post-AIA 37 C.F.R. § 1.63] in the assignment executed by the individual, in lieu of filing such statements separately."

Post-AIA 37 C.F.R. § 1.63(e) implements post-AIA 35 U.S.C. § 115(e) and states that:

(1) An assignment may also serve as an oath or declaration required by [37 C.F.R. § 1.63] if the assignment as executed:
   (i) Includes the information and statements required under paragraphs (a) and (b) of [37 C.F.R. § 1.63]; and
   (ii) A copy of the assignment is recorded as provided for in part 3 of this chapter.

Before post-AIA 35 U.S.C. § 115(e) became effective, inventors who were under an obligation of assignment of an application were required to file an inventor oath or declaration and an assignment as two separate documents. 77 Fed. Reg. 48,777 (Aug. 14, 2012). Post-AIA 35 U.S.C. § 115(e) therefore streamlines the patent application process to some degree by reducing the number of documents that must be filed for patent applications that are subject to an assignment.

5. Post-AIA 35 U.S.C. § 115(f)—Postponement of the Deadline by Which the Inventor's Oath or Declaration Must be Filed

Another significant way post-AIA 35 U.S.C. § 115 changed the U.S. PTO's procedures can be seen in the timing of when inventor oaths/declarations must be filed. Indeed, prior to the AIA, the U.S. PTO considered a patent application incomplete if it did not include the inventor's oath or declaration on the date it was filed. *See* MPEP § 601.01(a)(II) (9th ed. 2015) ("For applications filed before September 16, 2012, the Office issued a Notice to File Missing Parts if an application under 37 C.F.R. 1.53(b) did not contain the basic filing fee, the search fee, or the examination fee, or the inventor's oath or declaration, and the applicant was given a time period (usually two months) within which to file the missing basic filing fee, the search fee, the examination fee, or the inventor's oath or declaration and pay the surcharge required by 37 C.F.R. 1.16(f) to avoid abandonment.").

Post-AIA 35 U.S.C. § 115(f), however, states that "[t]he applicant for patent shall provide each required oath or declaration under subsection (a), substitute statement under subsection (d), or recorded assignment meeting the requirements of subsection (e) no later than the date on which the issue fee for the patent is paid." The U.S. PTO has interpreted post-AIA 35 U.S.C. § 115(f) to allow patent applicants to postpone filing the inventor's oath or declaration until the date on which the issue fee is paid[l] so long as the applicant provides an application data sheet before examination indicating the name, residence, and mailing address of each

---

[l]Note that the version of post-AIA 35 U.S.C. § 115(f) that was passed in the AIA stated that "[a] notice of allowance under section 151 may be provided to an applicant for patent only if the applicant for patent has filed each required oath or declaration under subsection (a) or has filed a substitute statement under subsection (d) or recorded an assignment meeting the requirements of subsection (e)." Leahy-Smith America Invents Act of 2011, Pub. L. No. 112-29, § 4(a), 125 Stat. 284, 294. Post-AIA 35 U.S.C. § 115(f) was amended to postpone the deadline for filing the inventor's oath or declaration until the date on which the issue fee is paid, however, by section 1(f) of the Leahy-Smith America Invents Technical Corrections Act of 2013. Pub. L. No. 112-274, 126 Stat. 2456-57. Notably, the U.S. PTO did not amend 37 C.F.R. § 1.53(f) to reflect this change to post-AIA 35 U.S.C. § 115(f) until December 18, 2013. Accordingly, for the time period between September 16, 2012 and December 18, 2013, 37 C.F.R. § 1.53(f) only allowed an applicant to postpone filing the inventor oath or declaration for an application filed on or after September 16, 2012 until the application was "otherwise in condition for allowance." MPEP § 601.01(a)(II) (9th ed. 2015). Beginning on December 18, 2013, however, 37 C.F.R. § 1.53(f) allowed applicants whose applications were filed on or after September 16, 2012 to postpone filing the inventor's oath or declaration until the date the issue fee is paid. *Id.*

inventor.[m] Post–AIA 37 C.F.R. § 1.53(f)(3)(i) and (ii); MPEP § 601.01(a)(II) (9th ed. 2015). Accordingly, rather than being required to supply the inventor's oath or declaration within a short time after filing a new patent application, applicants now have until the date on which they pay their issue fee to supply the oath or declaration.[n]

6. Post-AIA 35 U.S.C. § 115(g)—Earlier-Filed Application Containing Required Statements or Substitute Statement

Post–AIA 35 U.S.C. § 115(g) provides certain exceptions to post–AIA 35 U.S.C. § 115(a), (d), and (e)'s requirements to file an oath or declaration, substitute statement, or assignment-statement. Post–AIA 35 U.S.C. § 115(g) reads as follows:

> (1) EXCEPTION.--The requirements under [post-AIA 35 U.S.C. § 115] shall not apply to an individual with respect to an application for patent in which the individual is named as the inventor or a joint inventor and that claims the benefit under [35 U.S.C. §§] 120, 121, 365(c), or 386(c) of the filing of an earlier-filed application, if--
> (A) an oath or declaration meeting the requirements of subsection (a) was executed by the individual and was filed in connection with the earlier-filed application;

---

[m]Applicants must provide the name of each inventor prior to examination because Patent Examiners may need to know the identity of the inventors of a claimed invention in order to determine whether one of the prior art exceptions of post–AIA 35 U.S.C. § 102(b) applies.
[n]Although the deadline by which an applicant must file the inventor's oath or declaration has been extended under the AIA, applicants entering the national stage from an international application under the Patent Cooperation Treaty (PCT) under 35 U.S.C. § 371 should file the inventor's oath or declaration at the time the national stage application is filed because failure to do so could adversely affect any patent term adjustment to which the applicant is entitled. See post–AIA 35 U.S.C. § 154(b)(1)(A)(i)(II) (explaining that the U.S. PTO's fourteen-month deadline for issuing an Office Action or notice of allowance for purposes of calculating an applicant's patent term adjustment is measured from "the date of commencement of the national stage under section 371 in an international application"); 35 U.S.C. § 371(c)(4) (requiring that an inventor oath or declaration be filed in order for the national stage to commence); see also Leahy-Smith America Invents Technical Corrections Act of 2013, Pub. L. No. 112-274, § 1(h)(1), 126 Stat. 2456, 2457 (replacing the phrase "on which an international application fulfilled the requirements of section 371 of this title" with the phrase "of commencement of the national stage under section 371 in an international application" in pre–AIA 35 U.S.C. § 154(b)(1)(A)(i)(II)). In other words, if an applicant delays filing the required inventor oath(s)/declaration(s) for a national stage application, that applicant may obtain a shorter patent term adjustment, and therefore a short patent term, than if the applicant had filed the inventor oath(s)/declaration(s) at the time the national stage application was filed.

(B) *a substitute statement meeting the requirements of subsection (d) was filed in connection with the earlier filed application with respect to the individual; or*

(C) *an assignment meeting the requirements of subsection (e) was executed with respect to the earlier-filed application by the individual and was recorded in connection with the earlier-filed application.*

(2) COPIES OF OATHS, DECLARATIONS, STATEMENTS, OR ASSIGNMENTS.-- Notwithstanding paragraph (1), the Director may require that a copy of the executed oath or declaration, the substitute statement, or the assignment filed in connection with the earlier-filed application be included in the later-filed application.

Post-AIA 35 U.S.C. § 115(g)(1)(A)-(C) specifically enable patent applicants to avoid having to file the document(s) required by post-AIA 35 U.S.C. § 115(a), (d) and (e) (i.e., an oath or declaration, substitute statement, or assignment-statement) in a continuation, divisional, or continuation-in-part patent application in certain instances where such document(s) was/ were filed in an earlier-filed application to which the continuation, divisional, or continuation-in-part patent application claims the benefit under 35 U.S.C. §§ 120, 121, 365(c), or 386(c).° For example, if an applicant filed an oath or declaration that meets the requirements of post-AIA 35 U.S.C. § 115(a) in an earlier-filed application to which a continuation application claims the benefit, post-AIA 35 U.S.C. § 115(g)(1)(A) provides that an applicant need not file an oath or declaration, substitute statement, or assignment-statement with the continuation application. Post-AIA 35 U.S.C. § 115(g)(1)(B) and (C) similarly allow patent applicants to forego filing an oath or declaration, substitute statement, or assignment-statement with a continuation, divisional, or continuation-in-part application that claims the

---

°35 U.S.C. § 120 pertains to claims for the benefit of an earlier filing date in the U.S. (e.g., a claim for the benefit of a filing date of a provisional or non-provisional application). 35 U.S.C. § 121 provides that an invention that is made the subject of a divisional application which complies with the requirements of 35 U.S.C. § 120 shall be entitled to the benefit of the filing date of the original application. 35 U.S.C. § 365(c) permits international applications that designate the United States to claim the benefit of a prior national application or the benefit of a prior international application that designates the United States, and permits national applications to claim the benefit of the filing date of a prior international application that designates the United States. Furthermore, 35 U.S.C. § 386(c) permits international design applications designating the U.S. to claim the benefit of a filing date of certain prior national and international applications and also permits a national application to claim the benefit of the filing date of a prior international design application designating the United States.

benefit of an earlier-filed application in which a substitute statement or assignment-statement was filed under post-AIA 35 U.S.C. § 115(d) or post-AIA 35 U.S.C. § 115(e), respectively.ᴾ

Before post-AIA 35 U.S.C. § 115(g) became effective, the circumstances in which a new oath or declaration was not required in a continuation or divisional patent application were set forth in pre-AIA 37 C.F.R. § 1.63(d)(1), which stated that:

> A newly executed oath or declaration is not required under [37 C.F.R.] § 1.51(b)(2) and § 1.53(f) in a continuation or divisional application, provided that:
> (i) The prior nonprovisional application contained an oath or declaration as prescribed by paragraphs (a) through (c) of this section;
> (ii) The continuation or divisional application was filed by all or by fewer than all of the inventors named in the prior application;
> (iii) The specification and drawings filed in the continuation or divisional application contain no matter that would have been new matter in the prior application; and
> (iv) A copy of the executed oath or declaration filed in the prior application, showing the signature or an indication thereon that it was signed, is submitted for the continuation or divisional application.

The pre- and post-AIA laws regarding when a new oath or declaration can be omitted in a continuation or divisional patent application are similar in many ways. For instance, both the pre- and post-AIA laws provide that a new inventor oath or declaration does not need to be filed in a continuation or divisional application where the continuation or divisional application is filed by all or by fewer than all of the inventors named in the prior application. Pre-AIA 37 C.F.R. § 1.63(d)(1)(ii); post-AIA 35 U.S.C. § 115(g)(1)(A)-(C). In addition, both the pre- and post-AIA law require that an additional inventor's oath or declaration be filed in a continuing application for an inventor for whom an oath or declaration was not submitted in an earlier-filed application. See pre-AIA 37 C.F.R. § 1.63(d)(5) ("A newly executed oath or declaration must be filed in a continuation or divisional application naming an inventor not named in the prior application."); 77 Fed. Reg. 48,801-02 (Aug. 14, 2012) ("Thus, an additional

---

ᴾNote that post-AIA 35 U.S.C. § 115(g)(2) provides that a patent applicant may be required to provide a copy of the executed oath or declaration, substitute statement, or assignment-statement that was filed in connection with the earlier-filed application, and post-AIA 37 C.F.R. § 1.63(d)(1), which implements post-AIA 35 U.S.C. § 115(g)(2), does indeed require applicants to file a copy of the earlier-filed document(s).

inventor's oath or declaration would be necessary in a continuing application only for an inventor for whom an oath or declaration was not submitted in the prior-filed application."). Both the pre- and post-AIA laws also require that a copy of the executed oath or declaration filed in the prior application be submitted in the continuation or divisional application. Pre-AIA 37 C.F.R. § 1.63(d)(1)(iv); post-AIA 35 U.S.C. § 115(g)(2); post-AIA 37 C.F.R. § 1.63(d)(1).

Despite these similarities, there are several differences between the pre- and post-AIA laws regarding the submission of oath or declaration documents in continuation, divisional, and continuation-in-part applications. The most significant difference can be seen in the content of what must be filed in an earlier-filed application in order for post-AIA 35 U.S.C. § 115(g)(1)'s exception to the inventor oath/declaration requirement to apply to an application that claims the benefit of that earlier-filed application. Indeed, the pre-AIA law does not require that a newly executed oath or declaration be filed in a continuation or divisional application provided that "[t]he prior nonprovisional application contained an oath or declaration [that complied with pre-AIA 37 C.F.R. § 1.63(a)-(c)]." Post-AIA 35 U.S.C. §§ 115(g)(1)(A)-(C), however, provide that a newly executed oath, declaration, substitute statement, or assignment-statement need not be filed in a patent application if an oath, declaration, substitute statement, or assignment-statement meeting the requirements of post-AIA 35 U.S.C. § 115(a), (d) or (e) was filed in the earlier-filed application to which the application claims the benefit. As discussed in Sections II.A.i.2. and II.A.i.3., the content of the statements that are required by a pre-AIA oath or declaration differs somewhat from that of the statements that are required in an oath, declaration, substitute statement, or assignment-statement filed under post-AIA 35 U.S.C. § 115(a), (d) and (e). Thus, if a patent applicant filed an oath or declaration in an earlier-filed patent application that was filed before post-AIA 35 U.S.C. § 115 took effect (i.e., before September 16, 2012), it is likely that the exception to the inventor oath or declaration requirement provided by post-AIA 35 U.S.C. § 115(g)(1) would not apply to any application filed on or after September 16, 2012 that claims the benefit of that earlier-filed patent application.[q] As such, a newly executed oath or declaration that

---

[q]Indeed, the U.S. PTO's final rules regarding inventor oaths and declarations explain that for ▶ applications filed prior to September 16, 2012, any oath or declaration filed before, on, or after September 16, 2012, must comply with the oath and declaration rules in effect prior to September 16, 2012. On the other hand, any oath or declaration submitted in an application

complies with post-AIA 35 U.S.C. § 115 would be required by each inventor in the later-filed application.

Yet another difference between the exception provided under pre-AIA 37 C.F.R. § 1.63(d) and post-AIA 35 U.S.C. § 115(g)(1) is that the former exception does not apply to continuation-in-part applications whereas the latter exception does apply to such applications. More specifically, under the pre-AIA law, a newly executed oath or declaration was required in any continuation-in-part application, regardless of whether the inventorship of such application was the same as the inventorship of the earlier-filed application to which the continuation-in-part claimed the benefit. *See* pre-AIA 37 C.F.R. § 1.63(e) ("A newly executed oath or declaration must be filed in any continuation-in-part application, which application may name all, more, or fewer than all of the inventors named in the prior application."). The post-AIA law, on the other hand, only requires that a newly executed oath or declaration, substitute statement, or assignment-statement be filed in a continuation-in-part application that names more inventors than the earlier-filed application to which it claims the benefit. *See* 77 Fed. Reg. 48,801 (Aug. 14, 2012) ("[Post-AIA 37 C.F.R. §] 1.63(d), as adopted in this final rule, provides for use of a copy of the inventor's oath or declaration from a prior-filed application in a continuing application, including a continuation-in-part application.").

7. Post-AIA 35 U.S.C. § 115(h)—Supplemental and Corrected Statements; Filing Additional Statements

Post-AIA 35 U.S.C. § 115(h) modifies the U.S. PTO's practices regarding inventor oaths and declarations in two major ways. First, post-AIA 35 U.S.C. § 115(h)(2) eliminates the requirement that inventors submit supplemental oaths or declarations. Second, post-AIA 35 U.S.C. § 115(h)(3) sets forth a

---

◀ filed on or after September 16, 2012 (regardless of the date of execution of the oath or declaration) must meet the requirements of post-AIA 35 U.S.C. § 115. With respect to continuing applications, 35 U.S.C. § 115(g)(1)(A) provides an exception to a newly executed oath or declaration only where the oath or declaration in the earlier-filed application meets the requirements of post-AIA 35 U.S.C. § 115(a) which must include the required statements in post-AIA 35 U.S.C. § 115(b) and (c) and post-AIA 37 C.F.R. § 1.63. Accordingly, a new oath, declaration, substitute statement, or assignment-statement would likely be required for a continuation, divisional, or continuation-in-part application that claims priority to an earlier application filed before September 16, 2012 because the oath or declaration filed in the earlier application would not meet all of the requirements of post-AIA 35 U.S.C. § 115. 77 Fed. Reg. 48,802 (Aug. 14, 2012).

savings clause that eliminates the consequences patent holders encountered under the pre-AIA law for failing to meet the inventor oath/declaration requirements. Post-AIA 35 U.S.C. § 115(h) specifically states:

> *(h) SUPPLEMENTAL AND CORRECTED STATEMENTS; FILING ADDITIONAL STATEMENTS.--*
>
> > *(1) IN GENERAL.-- Any person making a statement required under this section may withdraw, replace, or otherwise correct the statement at any time. If a change is made in the naming of the inventor requiring the filing of 1 or more additional statements under this section, the Director shall establish regulations under which such additional statements may be filed.*
> >
> > *(2) SUPPLEMENTAL STATEMENTS NOT REQUIRED.-- If an individual has executed an oath or declaration meeting the requirements of subsection (a) or an assignment meeting the requirements of subsection (e) with respect to an application for patent, the Director may not thereafter require that individual to make any additional oath, declaration, or other statement equivalent to those required by this section in connection with the application for patent or any patent issuing thereon.*
> >
> > *(3) SAVINGS CLAUSE.-- A patent shall not be invalid or unenforceable based upon the failure to comply with a requirement under this section if the failure is remedied as provided under paragraph (1).*

In addition, a comparison of pre- and post-AIA 37 C.F.R. § 1.67, which is the U.S. PTO's regulation that governs supplemental oaths and declarations, appears in the following table:

Pre-AIA 37 C.F.R. § 1.67	Post-AIA 37 C.F.R. § 1.67
(a) The Office may require, or inventors and applicants may submit, a supplemental oath or declaration meeting the requirements of § 1.63 or § 1.162 to correct any deficiencies or inaccuracies present in the earlier filed oath or declaration.	(a) The applicant may submit an inventor's oath or declaration meeting the requirements of § 1.63, § 1.64, or § 1.162 to correct any deficiencies or inaccuracies present in an earlier-filed inventor's oath or declaration. Deficiencies or inaccuracies due to the failure to meet the requirements of § 1.63(b) in an oath or declaration may be corrected with an application data sheet in accordance with § 1.76, except that any correction of inventorship must be pursuant to § 1.48.
(1) Deficiencies or inaccuracies relating to all the inventors or applicants (§§ 1.42, 1.43, or § 1.47) may be corrected with a supplemental oath or declaration signed by all the inventors or applicants.	

(*Continued on next page*)

Pre-AIA 37 C.F.R. § 1.67	Post-AIA 37 C.F.R. § 1.67
(2) Deficiencies or inaccuracies relating to fewer than all of the inventor(s) or applicant(s) (§§ 1.42, 1.43 or § 1.47) may be corrected with a supplemental oath or declaration identifying the entire inventive entity but signed only by the inventor(s) or applicant(s) to whom the error or deficiency relates.	(b) A supplemental inventor's oath or declaration under this section must be executed by the person whose inventor's oath or declaration is being withdrawn, replaced, or otherwise corrected.
(3) Deficiencies or inaccuracies due to the failure to meet the requirements of § 1.63(c) (e.g., to correct the omission of a mailing address of an inventor) in an oath or declaration may be corrected with an application data sheet in accordance with § 1.76.	(c) The Office will not require a person who has executed an oath or declaration in compliance with 35 U.S.C. 115 and § 1.63 or 1.162 for an application to provide an additional inventor's oath or declaration for the application.
(4) Submission of a supplemental oath or declaration or an application data sheet (§ 1.76), as opposed to who must sign the supplemental oath or declaration or an application data sheet, is governed by § 1.33(a)(2) and paragraph (b) of this section.	(d) No new matter may be introduced into a nonprovisional application after its filing date even if an inventor's oath or declaration is filed to correct deficiencies or inaccuracies present in the earlier-filed inventor's oath or declaration.
(b) A supplemental oath or declaration meeting the requirements of § 1.63 must be filed when a claim is presented for matter originally shown or described but not substantially embraced in the statement of invention or claims originally presented or when an oath or declaration submitted in accordance with § 1.53(f) after the filing of the specification and any required drawings specifically and improperly refers to an amendment which includes new matter. No new matter may be introduced into a nonprovisional application after its filing date even if a supplemental oath or declaration is filed. In proper situations, the oath or declaration here required may be made on information and belief by an applicant other than the inventor.	

As shown in the previous table, pre-AIA 37 C.F.R. § 1.67(a) states that the U.S. PTO may require that an inventor submit a supplemental oath to correct any deficiencies or inaccuracies in an earlier-filed oath or declaration. Furthermore, pre-AIA 37 C.F.R. § 1.67(b) mandates that inventors file a supplemental oath or declaration when: (1) a claim is presented for matter originally shown or described but not substantially embraced in the statement of invention or claims originally presented; or (2) an oath or declaration submitted in accordance with 37 C.F.R. § 1.53(f) after the filing of the specification and any required drawings specifically and improperly referred to an amendment which includes new matter.

Unlike pre-AIA 37 C.F.R. § 1.67, post-AIA 35 U.S.C. § 115(h) and 37 C.F.R. § 1.67 do not require inventors to submit supplemental oaths or declarations in any situation. *See* post-AIA 35 U.S.C. § 115(h)(2) ("SUPPLEMENTAL STATEMENTS NOT REQUIRED.— If an individual has executed an oath or declaration meeting the requirements of [post-AIA 35 U.S.C. § 115](a) or an assignment meeting the requirements of [post-AIA 35 U.S.C. § 115](e) with respect to an application for patent, the Director may not thereafter require that individual to make any additional oath, declaration, or other statement equivalent to those required by this section in connection with the application for patent or any patent issuing thereon."); post-AIA 37 C.F.R. § 1.67(a) ("The applicant *may* submit an inventor's oath or declaration meeting the requirements of § 1.63, § 1.64, or § 1.162 to correct any deficiencies or inaccuracies present in an earlier-filed inventor's oath or declaration.") (emphasis added); post-AIA 37 C.F.R. § 1.67(c) ("The Office will not require a person who has executed an oath or declaration in compliance with [post-AIA] 35 U.S.C. 115 and [37 C.F.R.]§ 1.63 or 1.162 for an application to provide an additional inventor's oath or declaration for the application.").[r]

---

[r]The U.S. PTO has also altered its inventor oath/declaration practices such that it no longer prohibits patent applicants from making changes to their patent applications after the inventor oath or declaration is executed. *See* post-AIA 37 C.F.R. § 1.52(c) ("Interlineation, erasure, cancellation, or other alteration of the application papers may be made *before or after* the signing of the inventor's oath or declaration referring to those application papers, provided that the statements in the inventor's oath or declaration pursuant to § 1.63 remain applicable to those application papers. A substitute specification (§ 1.125) may be required if the application papers do not comply with paragraphs (a) and (b) of this section.") (emphasis added); MPEP § 608.01 (9th ed. 2015) ("Effective September 16, 2012, 37 CFR 1.52(c) no longer prohibits interlineations and other alterations of the application papers from being made after the signing of the inventor's oath or declaration."). It is unclear, however, what types of changes the U.S. PTO will tolerate under post-AIA 37 C.F.R. § 1.52(c).

More significant than post-AIA 35 U.S.C. § 115(h)(2)'s abandonment of the requirement for supplemental oaths is the savings clause that is set forth in post-AIA 35 U.S.C. § 115(h)(3). Post-AIA 35 U.S.C. § 115(h)(3) specifically states that "[a] patent shall not be invalid or unenforceable based upon the failure to comply with a requirement under [post-AIA 35 U.S.C. § 115] if the failure is remedied as provided under [post-AIA 35 U.S.C. § 115(h)(1)]." Moreover, a failure to comply with the requirements of post-AIA 35 U.S.C. § 115 can be remedied at any time, i.e., either during the pendency of a patent application or after a patent has issued. *See* post-AIA 35 U.S.C. § 115(h)(1) ("Any person making a statement required under this section may withdraw, replace, or otherwise correct the statement *at any time*.") (emphasis added). Pre-AIA 35 U.S.C. § 115 does not contain a savings clause analogous to that of post-AIA 35 U.S.C. § 115(h)(3). As a result, several patents were challenged as being invalid and/or unenforceable due to a failure to meet the inventor oath/declaration requirement under the pre-AIA law. *See, e.g., Pitney Bowes, Inc. v. Hewlett-Packard Co.*, 182 F.3d 1298, 1303 (Fed. Cir. 1999) (explaining that Hewlett Packard had moved for summary judgment of invalidity of Pitney Bowes' patent on the basis that the inventors' oath that had been filed for the patent failed to comply with 35 U.S.C. § 115); *Am. Infra-Red Radiant Co., Inc. v. Lambert Indus., Inc.*, 360 F.2d 977, 994 (Fed. Cir. 1966) (noting that the defendants had argued that American Infra-Red Radiant's patent was invalid due to absence of a proper oath under 35 U.S.C. § 115); *Union Carbide Corp. v. Dow Chem. Co.*, 619 F. Supp. 1036, 1054 (D. Del. 1985) (finding that Union Carbide's intent to defraud the PTO could be presumed from the proof that it knowingly misrepresented the inventor of the claimed process and that such fraudulent intent was material in part because 35 U.S.C. § 115 requires the applicant to "make [an] oath that he believes himself to be the original and first inventor of the process, machine, manufacture or composition of matter, or improvement thereof, for which he solicits a patent."). Thus, post-AIA 35 U.S.C. § 115(h)(3) significantly changes the pre-AIA law because it strips away the consequences for failing to comply with the inventor oath/declaration requirements that existed under the pre-AIA law.

ii. AIA § 4(b)(1)'s Amendments to Pre-AIA 35 U.S.C. § 118

Pre- and post-AIA 35 U.S.C. § 118 authorize non-inventors to file patent applications with the U.S. PTO in certain situations. The following table provides a comparison of these statutes:

Pre-AIA 35 U.S.C. § 118 Filing by Other Than Inventor	Post-AIA 35 U.S.C. § 118 Filing by Other Than Inventor
Whenever an inventor refuses to execute an application for patent, or cannot be found or reached after diligent effort, a person to whom the inventor has assigned or agreed in writing to assign the invention or who otherwise shows sufficient proprietary interest in the matter justifying such action, may make application for patent on behalf of and as agent for the inventor on proof of the pertinent facts and a showing that such action is necessary to preserve the rights of the parties or to prevent irreparable damage; and the Director may grant a patent to such inventor upon such notice to him as the Director deems sufficient, and on compliance with such regulations as he prescribes.	A person to whom the inventor has assigned or is under an obligation to assign the invention may make an application for patent. A person who otherwise shows sufficient proprietary interest in the matter may make an application for patent on behalf of and as agent for the inventor on proof of the pertinent facts and a showing that such action is appropriate to preserve the rights of the parties. If the Director grants a patent on an application filed under this section by a person other than the inventor, the patent shall be granted to the real party in interest and upon such notice to the inventor as the Director considers to be sufficient.

AIA § 4(b)(1) significantly alters pre-AIA 35 U.S.C. § 118 by allowing non–inventors (i.e., assignees, obligated assignees, and persons who otherwise show sufficient proprietary interest in a matter) to file patent applications in a broader range of circumstances than was permitted under the pre-AIA law. Indeed, pre-AIA 35 U.S.C. § 118 limited the filing of a patent application by a non–inventor to circumstances where: (1) the sole inventor or all joint inventors refused to execute a patent application, or (2) the sole inventor or all joint inventors could not be found or reached after diligent effort.[s] MPEP § 409.03 (9th ed. 2015). To demonstrate that the sole inventor or all joint inventors have refused to execute a patent application or could not be found or reached after diligent effort, pre-AIA 37 C.F.R. § 1.47(b) required that a petition providing the following information be filed along with the oath or declaration:

(1) proof of the pertinent facts demonstrating that the inventor refused to execute a patent application or could not be found or reached after diligent effort;

---

[s]Note that if at least one inventor is willing to execute the patent application, the procedure for filing the patent application was governed by pre-AIA 35 U.S.C. § 116. MPEP § 409.03(b) (9th ed. 2015).

(2) a showing that filing the patent application is necessary to preserve the rights of the parties or to prevent irreparable damage; and

(3) the last known address of all of the inventors.

In addition, an applicant who filed a patent application pursuant to pre-AIA 37 C.F.R. § 1.47(b) was also required to prove that: (A) the invention has been assigned to the applicant, (B) the inventor had agreed in writing to assign the invention to the applicant, or (C) the applicant otherwise had sufficient proprietary interest in the subject matter to justify the filing of the application. MPEP § 409.03(f) (9th ed. 2015). Moreover, a patent application that was filed with a petition could not be examined until the Office of Petitions provided a written decision as to whether the patent application would be accorded status under pre-AIA 37 C.F.R. § 1.47. See MPEP § 409.03 (9th ed. 2015) ("Since an application without an oath or declaration executed by all of the inventors may be an incomplete application, an examiner should not mail an Office action in an application without a fully executed oath or declaration under pre-AIA 37 CFR 1.63 unless the application has been accorded status under pre-AIA 37 CFR 1.47 in a written decision on the petition.").

As compared to pre-AIA 35 U.S.C. § 118, post-AIA 35 U.S.C. § 118 makes it considerably easier for non-inventors to file patent applications.ᵗ For instance, under the new statute, the circumstances in which an assignee, obligated assignee, or other person who shows sufficient proprietary interest in the invention can apply for a patent are no longer limited to those where the inventor refuses to execute a patent application or cannot be found or reached after diligent effort. See AIA § 4(b)(1) (omitting the "[w]henever an inventor refuses to execute an application for patent, or cannot be found or reached after diligent effort" phrase from post-AIA 35 U.S.C. § 118); 77 Fed. Reg. 48,788 (Aug. 14, 2012) (explaining that the U.S. PTO interprets post-AIA 35 U.S.C. §§ 115 and 118 to permit an assignee to file a patent application as the applicant in situations beyond those contemplated by post-AIA 35 U.S.C. § 115(d)(2)). As such, if a person other than the inventor applies for a patent, that person is no longer required to file a petition

---

ᵗThe legislative history of the AIA also makes clear that the changes to pre-AIA 35 U.S.C. § 118 were designed to: (1) make it easier for an assignee to file a patent application; (2) allow obligated assignees (i.e., entities to which the inventor is obligated to assign the application) to file a patent application; and (3) allow a person who has a sufficient proprietary interest in the invention to file an application to preserve that person's rights and those of the inventor. H.R. REP. NO. 112-98, at 44 (2011).

under pre-AIA 37 C.F.R. § 1.47(b).[u] Instead, the applicant may apply for the patent by filing a substitute statement pursuant to post-AIA 35 U.S.C. § 115(d).[v] If the applicant is an assignee or obligated assignee, post-AIA 37 C.F.R. § 1.46(b)(1) only requires that documentary evidence of ownership of the patent application be recorded with the U.S. PTO "no later than the date the issue fee is paid in the application." MPEP § 409.05 (9th ed. 2015). Under the post-AIA law, then, assignees and obligated assignees who file patent applications will no longer have to wait for a decision from the Office of Petitions to have their patent applications examined. If, on the other hand, the applicant is a person having sufficient proprietary interest in the invention, the applicant must submit a petition that: (1) demonstrates that the applicant has sufficient proprietary interest in the matter; and (2) states that making the application for patent by a person who otherwise shows sufficient proprietary interest in the matter on behalf of and as agent for the inventor is appropriate to preserve the rights of the parties. Post-AIA 37 C.F.R. § 1.46(b)(2); MPEP § 409.05 (9th ed. 2015).

Another major difference between pre- and post-AIA 35 U.S.C. § 118 is that under the former statute, a patent is granted to the inventor(s), whereas under the latter statute the patent is granted to the "real party in interest." *Compare* pre-AIA 35 U.S.C. § 118 ("[T]he Director may grant a patent to such inventor upon such notice to him as the Director deems sufficient . . . .") *with* post-AIA 35 U.S.C. § 118 ("If the Director grants a patent on an application filed under this section by a person other than the inventor, the patent shall be granted to the real party in interest and upon such notice to the inventor as the Director considers to be sufficient."). As a result, post-AIA 37 C.F.R. § 1.46(e) now requires that a patent applicant notify the U.S. PTO of any change in the real party in interest in a reply to a notice of allowance of such patent application in cases where there has been a change in the real party in interest.

B. Legislative Goals That Were Addressed by AIA § 4

As explained by the excerpt of the final Committee Report on the AIA, AIA § 4's amendments to pre-AIA 35 U.S.C. §§ 115 and 118 were made to simplify the inventor oath and declaration requirements, align the law

---

[u] Note that post-AIA 35 U.S.C. § 115 still requires that the patent application be accompanied by an oath or declaration from each inventor, except in the limited circumstances set forth in post-AIA 35 U.S.C. § 115(d)(2).

[v] *See* Section II.A.i.3. for a discussion of what must be included in a substitute statement filed under post-AIA 35 U.S.C. § 115(d).

with today's current practices regarding the filing of patent applications, and make it easier for non-inventors to file patent applications. In particular, the final Committee Report states:

> The U.S. patent system, when first adopted in 1790, contemplated that individual inventors would file their own patent applications, or would have a patent practitioner do so on their behalf. It has become increasingly common for patent applications to be assigned to corporate entities, most commonly the employer of the inventor. In fact, many employment contracts require employees to assign their inventions to their employer.
>
> Current law still reflects the antiquated notion that it is the inventor who files the application, not the company-assignee. For example, every inventor must sign an oath as part of the patent application stating that the inventor believes he or she is the true inventor of the invention claimed in the application. By the time an application is eventually filed, however, the applicant filing as an assignee may have difficulty locating and obtaining every inventor's signature for the statutorily required oath. Although the USPTO has adopted certain regulations to allow filing of an application when the inventor's signature is unobtainable, many have advocated that the statute be modernized to facilitate the filing of applications by assignees.
>
> The Act updates the patent system by facilitating the process by which an assignee may file and prosecute patent applications. It provides similar flexibility for a person to whom the inventor is obligated to assign, but has not assigned, rights to the invention (the "obligated assignee").
>
> Section 115 of title 35 is amended to allow a substitute statement to be submitted in lieu of an inventor's oath when either the inventor (i) is unable to submit an oath, or (ii) is both unwilling to do so and under an obligation to assign the invention. If an error is discovered, the statement may later be corrected. A savings clause is included to prevent an invalidity or unenforceability challenge to the patent based on failure to comply with these requirements, provided that any error has been remedied. Willful false statements remain punishable, however, under Federal criminal laws.
>
> Section 118 of title 35 is also amended to make it easier for an assignee to file a patent application. The amendment now allows obligated assignees—entities to which the inventor is obligated to assign the application—to file applications, as well. It also allows a person who has a sufficient proprietary interest in the invention to file an application to preserve that person's rights and those of the inventor.

H.R. REP. No. 112-98, at 43-44 (2011) (citations omitted); *see also* 157 CONG. REC. S1366, 2011 (reprinting materials distributed by the Republican Policy Committee) (explaining that the changes to the inventor oath and declaration requirements were made because fulfilling such requirements under the old law can be challenging when applications are pursued by company-assignees for whom a variety of past and present employees may have played a role in developing the invention); S. Rep. No. 111-18, at 34 (2009) (explaining that the inventor oath and declaration provision of the AIA "streamlines the requirement that the inventor submit an oath as part of a patent application, and makes it easier for patent owners to file applications."); H.R Rep. No. 110-314, at 24 (2007) ("[Pre-AIA] 35 U.S.C. § 115 requires an inventor (or if relevant, joint inventors) to execute an oath stating that he believes himself to be an original inventor of the claimed invention and the country of which he is a citizen. This requirement, however, does not reflect the realities of modern research. The globalization of science and technology has led to an environment of international collaborations and migrating scientists and engineers, in which it is increasingly difficult to obtain the signatures of all the inventors when filing a patent application."); *Patent Reform Act of 2007: Hearing Before the Subcomm. on Courts, the Internet, and Intellectual Property of the H. Comm. on the Judiciary*, 110th Cong. 18 (2007) ("Section 4 [of the AIA], probably the least controversial portion of the bill, is designed to simplify the process for providing an inventor's oath.").

C. Effective Date of AIA § 4

AIA § 4(e) provides that the amendments made by AIA § 4 "shall take effect upon the expiration of the 1-year period beginning on the date of the enactment of this Act and shall apply to any patent application that is filed on or after that effective date." Leahy-Smith America Invents Act of 2011, Pub. L. No. 112-29, 125 Stat. 284, 297. The AIA was enacted on September 16, 2011. *Id.* at 284. Thus, if a patent application was filed before September 16, 2012,[w] any oath or declaration that is filed for the application must comply with pre-AIA 35 U.S.C. § 115 and pre-AIA 37 C.F.R. § 1.63. The inventor oath/declaration requirements for patent applications that are filed on or after September 16, 2012, however, are governed by post-AIA 35 U.S.C. § 115 and post-AIA 37 C.F.R. § 1.63. 77 Fed. Reg. 48,802 (Aug. 14, 2012).

---

[w]Note that the international filing date (not the date on which the application enters the national stage) of a PCT application that designates the United States will dictate whether the pre- or post-AIA inventor oath/declaration requirements apply. *See* 35 U.S.C. § 363 ("An international application designating the United States shall have the effect, from its international filing date under article 11 of the treaty, of a national application for patent regularly filed in the Patent and Trademark Office except as otherwise provided in section 102(e) of this title.")

D. Practical Implications of AIA § 4's Amendments to Pre-AIA 35 U.S.C. §§ 115 and 118

Overall, AIA § 4(a)(1)'s amendments to pre-AIA 35 U.S.C. § 115 should make it significantly easier for inventors and patent applicants to comply with the inventor oath and declaration requirements. Indeed, post-AIA 35 U.S.C. § 115(b) eliminates some of the requirements that made executing an oath or declaration somewhat burdensome under the pre-AIA law. For example, inventor oaths and declarations no longer must list all inventors for a patent application, identify the country of citizenship for the inventors, or include foreign priority information. Post-AIA 35 U.S.C. §§ 115(d) and (e) also enable inventors and patent applicants to meet the inventor oath or declaration requirement using a broader array of documents, such as substitute statements or assignment-statements in lieu of oaths or declarations. By postponing the deadline for filing an oath, declaration, substitute statement, or assignment-statement until the date on which the issue fee is paid for an allowed patent application, post-AIA 35 U.S.C. § 115(f) also eases some of the burdens inventors experience when filing their patent applications. Post-AIA 35 U.S.C. § 115(g) further assists inventors and patent applicants by expanding the pre-AIA exceptions for when a new oath or declaration must be filed in a continuation, divisional, or continuation-in-part application. Finally, post-AIA 35 U.S.C. § 115(h) removes requirements for filing supplemental oaths/declarations that were imposed under the pre-AIA law. Even more importantly, that statute eliminates the consequences that were imposed under the pre-AIA law for failing to meet the inventor oath/declaration requirement by stating that "[a] patent shall not be invalid or unenforceable based upon the failure to comply with a requirement under [post-AIA 35 U.S.C. § 115] if the failure is remedied as provided under [post-AIA 35 U.S.C. § 115(h)(1)]."

AIA § 4(b)(1)'s amendments to pre-AIA 35 U.S.C. § 118 should also make it much easier for non-inventors, such as assignees, obligated assignees, and persons who otherwise show sufficient proprietary interest in a matter, to file patent applications. In particular, post-AIA 35 U.S.C. § 118 no longer restricts filings by non-inventors to circumstances in which the sole inventor or all joint inventors refuse to execute an application or cannot be found or reached after diligent effort. In addition, assignees and obligated assignees are no longer required to submit petitions with their patent applications under the post-AIA law. As a result, applications filed by assignees or obligated assignees should be processed more quickly under the post-AIA law than under the pre-AIA law because Patent Examiners will no longer need to wait on a decision from the Office of Petitions before acting on such applications.

III) AIA § 20(a)—MODIFICATIONS TO CORRECTION OF INVENTORSHIP PROCEDURES FOR PATENT APPLICATIONS[x]

AIA § 20(a) amends pre-AIA 35 U.S.C. § 116 to allow patent applicants to more easily correct the inventorship for their patent applications.[y] As a result of AIA § 20(a)'s amendments to pre-AIA 35 U.S.C. § 116,[z] the U.S. PTO made substantial revisions to pre-AIA 37 C.F.R. § 1.48, which is the regulation that implements pre-AIA 35 U.S.C. § 116. A comparison of the pre- and post-AIA 37 C.F.R. § 1.48 appears as follows:

Pre-AIA 37 C.F.R. § 1.48	Post-AIA 37 C.F.R. § 1.48
(a) Nonprovisional application after oath/declaration filed. If the inventive entity is set forth in error in an executed § 1.63 oath or declaration in a nonprovisional application, and such error arose without any deceptive intention on the part of the person named as an inventor in	(a) Nonprovisional application: Any request to correct or change the inventorship once the inventorship has been established under § 1.41 must include:

(Continued on next page)

---

[x]Inventorship errors in issued patents must be corrected either by filing a certificate of correction under 37 C.F.R. § 1.324 or by filing a reissue application pursuant to 37 C.F.R. § 1.171. The procedures in this section only apply to corrections of inventorship in patent applications and do not apply to correcting inventorship in issued patents.

[y]Note that AIA § 20(a) works in conjunction with AIA § 4(a)(1) to ease the burdens placed on patent applicants that are associated with changing inventorship for a patent application. Indeed, AIA § 4(a)(1) amends pre-AIA 35 U.S.C. § 115, which sets forth certain guidelines for naming inventors in a patent application. First, post-AIA 35 U.S.C. § 115(a) requires that patent applications filed under 35 U.S.C. § 111(a) or that commence the national stage under 35 U.S.C. § 371 shall include, or be amended to include, the name of the inventor for any invention claimed in the application. Post-AIA 35 U.S.C. § 115(h)(1) then allows any person making a statement required under post-AIA 35 U.S.C. § 115 (e.g., a statement naming the inventorship for a patent application) to "withdraw, replace, or otherwise correct the statement at any time." Post-AIA 35 U.S.C. § 115(h)(3) further provides that "[a] patent shall not be invalid or unenforceable based upon the failure to comply with a requirement under this section if the failure is remedied as provided under [post-AIA 35 U.S.C. § 115(h)(1)]."

[z]AIA § 20(a) removed the "without any deceptive intention" language from pre-AIA 35 U.S.C. § 116. The effects of this change are discussed in Chapter 5.

Pre-AIA 37 C.F.R. § 1.48	Post-AIA 37 C.F.R. § 1.48
error or on the part of the person who through error was not named as an inventor, the inventorship of the nonprovisional application may be amended to name only the actual inventor or inventors. Amendment of the inventorship requires: (1) A request to correct the inventorship that sets forth the desired inventorship change; (2) A statement from each person being added as an inventor and from each person being deleted as an inventor that the error in inventorship occurred without deceptive intention on his or her part; (3) An oath or declaration by the actual inventor or inventors as required by § 1.63 or as permitted by §§ 1.42, 1.43 or § 1.47; (4) The processing fee set forth in § 1.17(i); and (5) If an assignment has been executed by any of the original named inventors, the written consent of the assignee (see § 3.73(b) of this chapter). (b) Nonprovisional application—fewer inventors due to amendment or cancellation of claims. If the correct inventors are named in a nonprovisional application, and the prosecution of the nonprovisional application results in the amendment or cancellation of claims so that fewer than all of the currently named inventors are the actual inventors of the invention being claimed in the nonprovisional application, an amendment must be filed requesting deletion of the name or names of the person or persons who are not inventors of the invention being claimed. Amendment of the inventorship requires:	(1) An application data sheet in accordance with § 1.76 that identifies each inventor by his or her legal name; and (2) The processing fee set forth in § 1.17(i). (b) Inventor's oath or declaration for added inventor: An oath or declaration as required by § 1.63, or a substitute statement in compliance with § 1.64, will be required for any actual inventor who has not yet executed such an oath or declaration. (c) Any request to correct or change the inventorship under paragraph (a) of this section filed after the Office action on the merits has been given or mailed in the application must also be accompanied by the fee set forth in § 1.17(d), unless the request is accompanied by a statement that the request to correct or change the inventorship is due solely to the cancelation of claims in the application. (d) Provisional application. Once a cover sheet as prescribed by § 1.51(c)(1) is filed in a provisional application, any request to correct or change the inventorship must include: (1) A request, signed by a party set forth in § 1.33(b), to correct the inventorship that identifies each inventor by his or her legal name; and (2) The processing fee set forth in § 1.17(q).

(*Continued on next page*)

Pre-AIA 37 C.F.R. § 1.48	Post-AIA 37 C.F.R. § 1.48
(1) A request, signed by a party set forth in § 1.33(b), to correct the inventorship that identifies the named inventor or inventors being deleted and acknowledges that the inventor's invention is no longer being claimed in the nonprovisional application; and (2) The processing fee set forth in § 1.17(i). (c) Nonprovisional application—inventors added for claims to previously unclaimed subject matter. If a nonprovisional application discloses unclaimed subject matter by an inventor or inventors not named in the application, the application may be amended to add claims to the subject matter and name the correct inventors for the application. Amendment of the inventorship requires: (1) A request to correct the inventorship that sets forth the desired inventorship change; (2) A statement from each person being added as an inventor that the addition is necessitated by amendment of the claims and that the inventorship error occurred without deceptive intention on his or her part; (3) An oath or declaration by the actual inventors as required by § 1.63 or as permitted by §§ 1.42, 1.43, or § 1.47; (4) The processing fee set forth in § 1.17(i); and (5) If an assignment has been executed by any of the original named inventors, the written consent of the assignee (see § 3.73(b) of this chapter).	(e) Additional information may be required. The Office may require such other information as may be deemed appropriate under the particular circumstances surrounding the correction of inventorship. (f) Correcting or updating the name of an inventor: Any request to correct or update the name of the inventor or a joint inventor, or the order of the names of joint inventors, in a nonprovisional application must include: (1) An application data sheet in accordance with § 1.76 that identifies each inventor by his or her legal name in the desired order; and (2) The processing fee set forth in § 1.17(i). (g) Reissue applications not covered. The provisions of this section do not apply to reissue applications. See §§ 1.171 and 1.175 for correction of inventorship in a patent via a reissue application. (h) Correction of inventorship in patent. See § 1.324 for correction of inventorship in a patent.

(Continued on next page)

Pre-AIA 37 C.F.R. § 1.48	Post-AIA 37 C.F.R. § 1.48

(d) Provisional application—adding omitted inventors. If the name or names of an inventor or inventors were omitted in a provisional application through error without any deceptive intention on the part of the omitted inventor or inventors, the provisional application may be amended to add the name or names of the omitted inventor or inventors. Amendment of the inventorship requires:

(1) A request, signed by a party set forth in § 1.33(b), to correct the inventorship that identifies the inventor or inventors being added and states that the inventorship error occurred without deceptive intention on the part of the omitted inventor or inventors; and

(2) The processing fee set forth in § 1.17(q).

(e) Provisional application—deleting the name or names of the inventor or inventors. If a person or persons were named as an inventor or inventors in a provisional application through error without any deceptive intention on the part of such person or persons, an amendment may be filed in the provisional application deleting the name or names of the person or persons who were erroneously named. Amendment of the inventorship requires:

(1) A request to correct the inventorship that sets forth the desired inventorship change;

(2) A statement by the person or persons whose name or names are being deleted that the inventorship error occurred without deceptive intention on the part of such person or persons;

(3) The processing fee set forth in § 1.17(q); and

(i) Correction of inventorship in an interference or contested case before the Patent Trial and Appeal Board. In an interference under part 41, subpart D, of this title, a request for correction of inventorship in an application must be in the form of a motion under § 41.121(a)(2) of this title. In a contested case under part 42, subpart D, of this title, a request for correction of inventorship in an application must be in the form of a motion under § 42.22 of this title. The motion under § 41.121(a)(2) or 42.22 of this title must comply with the requirements of paragraph (a) of this section.

(*Continued on next page*)

Pre-AIA 37 C.F.R. § 1.48	Post-AIA 37 C.F.R. § 1.48
(4) If an assignment has been executed by any of the original named inventors, the written consent of the assignee (see § 3.73(b) of this chapter).	
(f)	
(1) Nonprovisional application—filing executed oath/declaration corrects inventorship. If the correct inventor or inventors are not named on filing a nonprovisional application under § 1.53(b) without an executed oath or declaration under § 1.63 by any of the inventors, the first submission of an executed oath or declaration under § 1.63 by any of the inventors during the pendency of the application will act to correct the earlier identification of inventorship. See §§ 1.41(a)(4) and 1.497(d) and (f) for submission of an executed oath or declaration to enter the national stage under 35 U.S.C. 371 naming an inventive entity different from the inventive entity set forth in the international stage.	
(2) Provisional application—filing cover sheet corrects inventorship. If the correct inventor or inventors are not named on filing a provisional application without a cover sheet under § 1.51(c)(1), the later submission of a cover sheet under § 1.51(c)(1) during the pendency of the application will act to correct the earlier identification of inventorship.	
(g) Additional information may be required. The Office may require such other information as may be deemed appropriate under the particular circumstances surrounding the correction of inventorship.	

(*Continued on next page*)

Pre-AIA 37 C.F.R. § 1.48	Post-AIA 37 C.F.R. § 1.48
(h) Reissue applications not covered. The provisions of this section do not apply to reissue applications. See §§ 1.171 and 1.175 for correction of inventorship in a patent via a reissue application. (i) Correction of inventorship in patent. See § 1.324 for correction of inventorship in a patent. (j) Correction of inventorship in a contested case before the Board of Patent Appeals and Interferences. In a contested case under part 41, subpart D, of this title, a request for correction of an application must be in the form of a motion under § 41.121(a)(2) of this title and must comply with the requirements of this section.	

Under the U.S. PTO's practices, there are four different types of corrections that can be made to the inventorship for a patent application. The first is to correct the inventorship by adding an inventor who was omitted. The second is to correct the inventorship by deleting the name of a person who was listed as an inventor but who actually did not invent any of the claimed subject matter in a patent application. The third correction that can be made is to change the order in which multiple inventors are listed on a patent application. Lastly, an inventor's name can be corrected either because it contains a spelling or other typographical error or because the inventor's legal name has changed (e.g., a name change due to marriage). The U.S. PTO has set forth different requirements for each of the different types of correction of inventorship as well as for the different types of patent documents in which the correction is being made (e.g., a provisional patent application versus a non-provisional patent application). Sections III.A. and B. provide a comparison of the various pre- and post-AIA requirements for correcting inventorship in the aforementioned circumstances.[aa]

[aa]In addition to the changes shown in the tables throughout this section, the personnel responsible for reviewing requests for inventorship corrections has also changed. Requests for a correction of inventorship that were filed before September 16, 2012 are generally handled by a Primary Examiner. MPEP § 602.01(c)(3) (9th ed. 2015). Requests for a correction of inventorship that are filed on or after September 16, 2012, however, are handled by the Office of Patent Application Processing (OPAP). MPEP § 602.01(c)(1) (9th ed. 2015).

A. Correcting Inventorship in Provisional Patent Applications

The U.S. PTO considers the inventorship for a provisional patent application to be the inventor or joint inventors who are listed in the cover sheet prescribed by 37 C.F.R. § 1.51(c)(1) for both pre- and post-AIA provisional applications (i.e., for provisional applications filed before, on, or after September 16, 2012). MPEP § 602.01(c)(I)(A) and (B) (9th ed. 2015). Furthermore, once a cover sheet is filed in a provisional application, any correction of inventorship must be done according to the procedure outlined in 37 C.F.R. § 1.48. MPEP § 602.01(c)(I)(A) (9th ed. 2015). In other words, once an applicant has filed a cover sheet naming one or more inventors in a provisional application, the applicant may not correct the inventorship in his/her provisional application by filing another cover sheet listing the correct inventors.

The following tables provide an overview of the pre- and post-AIA requirements for correcting inventorship in a provisional patent application:

Type of Correction to Inventorship in a Provisional Patent Application	Requirements for Requests to Correct Inventorship that are Filed Before September 16, 2012 (Pre-AIA)	Requirements for Requests to Correct Inventorship that are Filed On or After September 16, 2012 (Post-AIA)
Adding an Inventor Who Was Omitted	(1) A request, signed by a party set forth in 37 C.F.R. § 1.33(b) (e.g., a practitioner of record or the applicant), to correct the inventorship that identifies the inventor or inventors being added and states that the inventorship error occurred without deceptive intention on the part of the omitted inventor or inventors; and (2) The processing fee set forth in 37 C.F.R. § 1.17(g). (Pre-AIA 37 C.F.R. § 1.48(d); MPEP § 602.01(c)(3) (9th ed. 2015))	Any request to correct or change inventorship must include: (1) A request, signed by a party set forth in 37 C.F.R. § 1.33, to correct inventorship that identifies each inventor by his or her legal name; and (2) The processing fee set forth in 37 C.F.R. § 1.17(g). (Post-AIA 37 C.F.R. § 1.48(d); MPEP § 602.01(c)(1) (9th ed. 2015))

(*Continued on next page*)

Type of Correction to Inventorship in a Provisional Patent Application	Requirements for Requests to Correct Inventorship that are Filed Before September 16, 2012 (Pre-AIA)	Requirements for Requests to Correct Inventorship that are Filed On or After September 16, 2012 (Post-AIA)
Deleting a Person Who Was Erroneously Named as an Inventor	(1) A request to correct the inventorship that sets forth the desired inventorship change; (2) A statement by the person or persons whose name or names are being deleted that the inventorship error occurred without deceptive intention on the part of such person or persons; (3) The processing fee set forth in 37 C.F.R. § 1.17(q); and (4) If an assignment has been executed by any of the original named inventors, the written consent of the assignee. (Pre-AIA 37 C.F.R. § 1.48(e))	Any request to correct or change inventorship must include: (1) A request, signed by a party set forth in 37 C.F.R. § 1.33, to correct inventorship that identifies each inventor by his or her legal name; and (2) The processing fee set forth in 37 C.F.R. § 1.17(g). (Post-AIA 37 C.F.R. § 1.48(d); MPEP § 602.01(c)(1) (9th ed. 2015))
Correcting the Order in Which the Named Inventors Occurs	(1) A petition under 37 C.F.R. § 1.182; and (2) A fee for the petition set forth in 37 C.F.R. § 1.17(f). (MPEP § 605.04(f) (8th ed. 2012))	Under the post-AIA law, the change in the order of the names of inventors is not provided for since provisional applications do not become application publications or patents. (MPEP § 602.01(c)(2) (9th ed. 2015))

(*Continued on next page*)

Type of Correction to Inventorship in a Provisional Patent Application	Requirements for Requests to Correct Inventorship that are Filed Before September 16, 2012 (Pre-AIA)	Requirements for Requests to Correct Inventorship that are Filed On or After September 16, 2012 (Post-AIA)
Correcting the Legal Name of an Inventor (i.e., because of a misspelling, typographical error, or legal name change)	(1) A petition under 37 C.F.R. § 1.182; and (2) A fee for the petition set forth in 37 C.F.R. § 1.17(f). (MPEP § 605.04(c) (8th ed. 2012))	Any request to correct or change inventorship must include: (1) A request, signed by a party set forth in 37 C.F.R. § 1.33, to correct inventorship that identifies each inventor by his or her legal name; and (2) The processing fee set forth in 37 C.F.R. § 1.17(g). (Post-AIA 37 C.F.R. § 1.48(d); MPEP § 602.01(c)(1) (9th ed. 2015))

As shown in the previous tables, the post-AIA law streamlines the procedure for correcting inventorship in a provisional application by providing the same procedure for every type of inventorship correction, i.e., adding an inventor, deleting an inventor, and correcting the legal name of an inventor, and by eliminating the need for patent applicants to correct inventorship where the only error lies in the order in which the inventors' names appear. Under the pre-AIA law, applicants were required to follow different procedures depending on the type of inventorship error they wished to correct. For instance, if an applicant wanted to correct the legal name of an inventor, he/she was required to file a petition under 37 C.F.R. § 1.182. If an applicant wanted to add an inventor, on the other hand, the pre-AIA law required that he/she file a request under 37 C.F.R. § 1.33(b). By streamlining the inventorship correction procedures, the post-AIA law should make it easier for patent applicants to correct inventorship in provisional applications and reduce the number of errors that occur in doing so.

B. Correcting Inventorship in Non-Provisional Patent Applications

The most significant changes to the U.S. PTO's procedures for correcting inventorship were made to the procedures that apply to non-provisional applications.[bb] In particular, in addition to changing the actual procedures patent applicants must use to correct inventorship for a particular patent application, the U.S. PTO has changed its own procedures in assessing inventorship for non-provisional applications where applicants have submitted inconsistent information. Specifically, if a patent applicant provided conflicting inventorship information in a non-provisional patent application that was filed before September 16, 2012, the U.S. PTO defines the inventorship as that set forth in the executed oath or declaration filed under pre-AIA 37 C.F.R. § 1.63. MPEP § 602.01(c)(I)(B) (9th ed. 2015). In contrast, for non-provisional patent applications filed on or after September 16, 2012, the U.S. PTO deems the inventors to be those listed in the application data sheet (ADS) so long as the ADS was filed before or concurrently with the inventor's oath or declaration.[cc] Post-AIA 37 C.F.R. § 1.41(b); MPEP § 602.01(c)(I)(A) (9th ed. 2015). If the ADS was not filed before or concurrently with the inventor's oath or declaration, the U.S. PTO deems the inventorship to be the inventor or joint inventors set forth in the inventor's oath or declaration (except as otherwise provided in 37 C.F.R. § 1.41(b)). *Id.*

The following tables provide an overview of the pre- and post-AIA requirements for correcting inventorship in a non-provisional patent application:

[bb]Note that besides utilizing the procedures outlined in this section, patent applicants may also correct the inventorship of a non-provisional application by filing a continuing application under 37 C.F.R. § 1.53. MPEP § 602.01(c)(III) (9th ed. 2015). This procedure can be used to correct inventorship under both the pre- and post-AIA law. If a patent applicant opts to correct inventorship via a continuing application, he/she must ensure that there is an overlap in inventorship between the continuing application and the parent application. *Id.* In other words, if inventors A and B are the only inventors listed on a parent application, an applicant cannot correct inventorship for that application by filing a continuing application naming only inventors C and D because no overlap of inventorship would exist.

[cc]Also, once an applicant files either an application data sheet or an inventor's oath or declaration, any correction of inventorship must be made in accordance with post-AIA 37 C.F.R. § 1.48(a). MPEP § 602.01(c)(I)(A) (9th ed. 2015).

Type of Correction to Inventorship Requested in a Non-Provisional Patent Application	Requirements for Requests to Correct Inventorship that are Filed Before September 16, 2012 (Pre-AIA)	Requirements for Requests to Correct Inventorship that are Filed On or After September 16, 2012 (Post-AIA)
Adding an Inventor Who Was Erroneously Omitted	(1) A request to correct the inventorship that sets forth the desired inventorship change; (2) A statement from each person being added as an inventor that the error in inventorship occurred without deceptive intent on his or her part; (3) An oath or declaration by the actual inventor or inventors as required by pre-AIA 37 C.F.R. § 1.63 or as permitted by pre-AIA 37 C.F.R. § 1.42, § 1.43 or § 1.47; (4) The processing fee set forth in pre-AIA 37 C.F.R. § 1.17(i); and (5) If an assignment has been executed by any of the original named inventors, the written consent of the assignee. (Pre-AIA 37 C.F.R. § 1.48(a); MPEP § 602.01(c)(3) (9th ed. 2015))	Any request to correct or change the inventorship must include: (1) An application data sheet (ADS) in accordance with 37 C.F.R. § 1.76 that identifies each inventor by his or her legal name; (2) The processing fee set forth in 37 C.F.R. § 1.17(i); and (3) An oath or declaration as required by 37 C.F.R. § 1.63 or a substitute statement that complies with 37 C.F.R. § 1.64 for the inventor(s) who is being added. (Post-AIA 37 C.F.R. § 1.48(a) and (b); MPEP § 602.01(c)(1) (9th ed. 2015))

(Continued on next page)

Type of Correction to Inventorship Requested in a Non-Provisional Patent Application	Requirements for Requests to Correct Inventorship that are Filed Before September 16, 2012 (Pre-AIA)	Requirements for Requests to Correct Inventorship that are Filed On or After September 16, 2012 (Post-AIA)
Adding an Inventor Who Invented Subject Matter that Was Previously Disclosed but Not Claimed in a Patent Application	(1) A request to correct the inventorship that sets forth the desired inventorship change; (2) A statement from each person being added as an inventor that the addition is necessitated by amendment of the claims and that the inventorship error occurred without deceptive intention on his or her part; (3) An oath or declaration by the actual inventor or inventors as required by pre-AIA 37 C.F.R. § 1.63 or as permitted by pre-AIA 37 C.F.R. § 1.42, § 1.43 or § 1.47; (4) The processing fee set forth in pre-AIA 37 C.F.R. § 1.17(i); and (5) If an assignment has been executed by any of the original named inventors, the written consent of the assignee. (Pre-AIA 37 C.F.R. § 1.48(c); MPEP § 602.01(c)(3) (9th ed. 2015))	Any request to correct or change the inventorship must include: (1) An application data sheet (ADS) in accordance with 37 C.F.R. § 1.76 that identifies each inventor by his or her legal name; (2) The processing fee set forth in 37 C.F.R. § 1.17(i); and (3) An oath or declaration as required by 37 C.F.R. § 1.63 or a substitute statement that complies with 37 C.F.R. § 1.64 for the inventor(s) who is being added. (Post-AIA 37 C.F.R. § 1.48(a) and (b); MPEP § 602.01(c)(1) (9th ed. 2015))

(*Continued on next page*)

Type of Correction to Inventorship Requested in a Non-Provisional Patent Application	Requirements for Requests to Correct Inventorship that Are Filed Before September 16, 2012 (Pre-AIA)	Requirements for Requests to Correct Inventorship that Are Filed On or After September 16, 2012 (Post-AIA)
Deleting a Person Who Is Erroneously Named as an Inventor	(1) A request to correct the inventorship that sets forth the desired inventorship change; (2) A statement from each person being deleted as an inventor that the error in inventorship occurred without deceptive intent on his or her part; (3) An oath or declaration by the actual inventor or inventors as required by pre-AIA 37 C.F.R. § 1.63 or as permitted by pre-AIA 37 C.F.R. § 1.42, § 1.43, or § 1.47; (4) The processing fee set forth in pre-AIA 37 C.F.R. § 1.17(i); and (5) If an assignment has been executed by any of the original named inventors, the written consent of the assignee. (Pre-AIA 37 C.F.R. § 1.48(a); MPEP § 602.01(c)(3) (9th ed. 2015))	Any request to correct or change the inventorship must include: (1) An application data sheet (ADS) in accordance with 37 C.F.R. § 1.76 that identifies each inventor by his or her legal name; and (2) The processing fee set forth in 37 C.F.R. § 1.17(i). (Post-AIA 37 C.F.R. § 1.48(a); MPEP § 602.01(c)(1) (9th ed. 2015))
Deleting a Person Named as an Inventor Due to Amendment or Cancellation of Claims	(1) A request, signed by a party set forth in 37 C.F.R. § 1.33(b) (e.g., a patent practitioner of record or the applicant), to correct the inventorship that identifies the named inventor or inventors being deleted and acknowledges that the inventor's invention is no longer being claimed in the nonprovisional application; and (2) The processing fee set forth in 37 C.F.R. § 1.17(i). (Pre-AIA 37 C.F.R. § 1.48(b); MPEP § 602.01(c)(3) (9th ed. 2015))	Any request to correct or change the inventorship must include: (1) An application data sheet (ADS) in accordance with 37 C.F.R. § 1.76 that identifies each inventor by his or her legal name; and (2) The processing fee set forth in 37 C.F.R. § 1.17(i). (Post-AIA 37 C.F.R. § 1.48(a); MPEP § 602.01(c)(1) (9th ed. 2015))

(Continued on next page)

Type of Correction to Inventorship Requested in a Non-Provisional Patent Application	Requirements for Requests to Correct Inventorship that are Filed Before September 16, 2012 (Pre-AIA)	Requirements for Requests to Correct Inventorship that are Filed On or After September 16, 2012 (Post-AIA)
Correcting the Order in Which the Named Inventors Occurs	(1) A petition under 37 C.F.R. § 1.182; and (2) A fee for the petition set forth in 37 C.F.R. § 1.17(f). (MPEP § 605.04(f) (8th ed. 2012))	Any request to correct or change the inventorship must include: (1) An application data sheet (ADS) in accordance with 37 C.F.R. § 1.76 that identifies each inventor by his or her legal name in the desired order; and (2) The processing fee set forth in 37 C.F.R. § 1.17(i). (Post-AIA 37 C.F.R. § 1.48(f); MPEP § 602.01(c)(2) (9th ed. 2015))
Correcting the Legal Name of an Inventor (i.e., because of a misspelling, typographical error, or legal name change)	(1) A petition under 37 C.F.R. § 1.182; and (2) A fee for the petition set forth in 37 C.F.R. § 1.17(f). (MPEP 605.04(c) (8th ed. 2012))	Any request to correct or change the inventorship must include: (1) An application data sheet (ADS) in accordance with 37 C.F.R. § 1.76 that identifies each inventor by his or her legal name in the desired order; and (2) The processing fee set forth in 37 C.F.R. § 1.17(i). (Post-AIA 37 C.F.R. § 1.48(f); MPEP § 602.01(c)(2) (9th ed. 2015))

As shown in the previous tables, the procedures for correcting inventorship in a non-provisional patent application have been streamlined under the post–AIA law. Indeed, for requests to correct inventorship that are filed on or after September 16, 2012, the same procedure; i.e., filing an application data sheet that identifies each inventor by his or her legal name and paying the processing fee set forth in 37 C.F.R. § 1.17(i), is used for every type of inventorship correction. The only circumstance in which a difference in this procedure exists is when an inventor is being added, in which case that inventor must file an oath, declaration, or substitute statement in addition to filing the application data sheet and paying the processing fee. Pre-AIA requests for correction of inventorship were far more complicated in that applicants were required to follow different procedures for each different type of correction being sought.

C. Effective Date of AIA § 20(a)[dd]

AIA § 20(l) states that "[t]he amendments made by [AIA § 20] shall take effect upon the expiration of the 1-year period beginning on the date of the enactment of this Act and shall apply to proceedings commenced on or after that effective date." Leahy-Smith America Invents Act of 2011, Pub. L. No. 112-29, 125 Stat. 284, 335. The AIA was enacted on September 16, 2011. *Id.* at 284. Thus, any request for correction of inventorship that is filed on or after September 16, 2012 must comply with the post-AIA 35 U.S.C. § 116 and post-AIA 37 C.F.R. § 1.48.

[dd]Note that there does not appear to be any legislative history for AIA § 20(a).

CHAPTER TEN

Miscellaneous Provisions of the AIA

Contents

I) INTRODUCTION

This final chapter is somewhat of a catch-all chapter that discusses several provisions of the AIA, which, on their own, may not warrant an entire chapter. This is not to say that the provisions discussed herein are not important. Rather, they are generally more straightforward than the other AIA provisions that have been discussed in previous chapters. As a result, the discussions of each of the AIA provisions covered in this chapter have been intentionally kept short. The AIA provisions addressed in this chapter cover a wide range of topics, including administrative changes at the U.S. PTO, jurisdictional changes, fee provisions of the AIA, U.S. PTO programs created by the AIA, and studies required by the AIA. Overall, the objective of this

America Invents Act Primer. http://dx.doi.org/10.1016/B978-0-12-812096-5.00010-7

chapter is to provide a general overview of the AIA provisions that pertain to these topics as well as resources for further information, where relevant.

II) ADMINISTRATIVE CHANGES AT THE U.S. PTO

In addition to the many substantive changes the AIA made to U.S. patent law, the legislation also brought about many changes to the manner in which the U.S. PTO operates. Indeed, the AIA created a new administrative law body known as the Patent Trial and Appeal Board to handle, among other things, the new post-grant proceedings established by AIA §§ 6 and 18. AIA § 23 also authorized the U.S. PTO to open three new satellite offices, thereby improving accessibility of the Office to patent applicants and inventors living outside of the Washington, D.C. metropolitan area. This section will provide a brief discussion of these changes, as well as a missed opportunity by AIA § 22 to correct a funding issue that the U.S. PTO has long grappled with.

A. AIA § 7(a)—Creation of the Patent Trial and Appeal Board (PTAB)

AIA § 7(a) amends pre-AIA 35 U.S.C. § 6 to establish the Patent Trial and Appeal Board (PTAB). Leahy-Smith America Invents Act of 2011, Pub. L. No. 112-29, 125 Stat. 284, 313. The PTAB, which is an administrative law body comprised of administrative patent judges (APJs), replaces the Board of Patent Appeals and Interferences (BPAI) that existed before the AIA was enacted and is charged with reviewing adverse decisions made by Patent Examiners upon a written appeal by a patent applicant, reviewing appeals of reexaminations pursuant to 35 U.S.C. § 134(b), conducting derivation proceedings, and conducting post-grant proceedings (i.e., inter partes reviews and post grant reviews). *Id.* at 313-14; H.R. REP. NO. 112-98, at 77 (2011). Unlike the BPAI, which only handled certain appeals within the U.S. PTO, the PTAB handles both appeals within the U.S. PTO and trial-type proceedings (i.e., inter partes and post grant reviews). *Id.* Furthermore, although both inter partes reexamination and interference proceedings were in the process of being phased out at the time the PTAB was created, AIA §§ 6(f)(3)(B) and 7(e)(3) provide that the PTAB shall continue handling such proceedings so long as they exist. *See* Leahy-Smith America Invents Act of 2011, Pub. L. No. 112-29, 125 Stat. 284, 311 ("For purposes of an interference that is commenced before [September 16, 2012], the Director may deem the Patent Trial and Appeal Board to be the Board of Patent Appeals and Interferences, and may allow the Patent Trial and Appeal Board to conduct any further proceedings in that interference."); *id.* at 315 ("[T]he Patent Trial and Appeal Board may be deemed to be the Board of Patent Appeals and Interferences for purposes of appeals of *inter partes* reexaminations that are requested . . . before [September 16, 2012.]").

AIA § 7(e) provides that AIA § 7(a) "shall take effect upon the expiration of the 1-year period beginning on the date of the enactment of [the AIA] and shall apply to proceedings commenced on or after that effective date[.]" *Id.* The AIA was enacted on September 16, 2011. *Id.* at 284. Accordingly, the PTAB replaced the BPAI on September 16, 2012.

B. AIA § 21—Travel Expenses and Payment of Administrative Judges

AIA § 21(a) amends pre-AIA 35 U.S.C. § 2(b)(11) to authorize the U.S. PTO to cover subsistence expenses and certain travel-related expenses incurred by administrative judges[a] who attend programs conducted by the U.S. PTO that pertain to domestic and international intellectual property law and the effectiveness of intellectual property protection domestically and throughout the world. Leahy-Smith America Invents Act of 2011, Pub. L. No. 112-29, 125 Stat. 284, 335. A comparison of pre- and post-AIA 35 U.S.C. § 2(b)(11) is shown in the following table:

Pre-AIA 35 U.S.C.§ 2(b)(11)	Post-AIA 35 U.S.C. § 2(b)(11)
(b) SPECIFIC POWERS.— The [U.S. Patent and Trademark] Office— (11) may conduct programs, studies, or exchanges of items or services regarding domestic and international intellectual property law and the effectiveness of intellectual property protection domestically and throughout the world[.]	(b) SPECIFIC POWERS.—The [U.S. Patent and Trademark] Office— (11) may conduct programs, studies, or exchanges of items or services regarding domestic and international intellectual property law and the effectiveness of intellectual property protection domestically and throughout the world, and the Office is authorized to expend funds to cover the subsistence expenses and travel-related expenses, including per diem, lodging costs, and transportation costs, of persons attending such programs who are not Federal employees[.] (Emphasis added.)

[a]When reading pre- and post-AIA 35 U.S.C. § 2(b)(11), note that administrative judges are not federal employees, but are instead federal officers of the U.S. PTO. *See* post-AIA 35 U.S.C. § 6(d) (explaining that administrative patent judges hold offices pursuant to an appointment by the Director of the U.S. PTO).

In addition, AIA § 21(b) amends pre-AIA 35 U.S.C. § 3(b) to establish the pay rate for administrative patent judges and administrative trademark judges. *Id.* at 336. In particular, the following subsection was added to the end of pre-AIA 35 U.S.C. § 3(b):

(6) ADMINISTRATIVE PATENT JUDGES AND ADMINISTRATIVE TRADEMARK JUDGES.—The Director may fix the rate of basic pay for the administrative patent judges appointed pursuant to section 6 and the administrative trademark judges appointed pursuant to section 17 of the Trademark Act of 1946 (15 U.S.C. 1067) at not greater than the rate of basic pay payable for level III of the Executive Schedule under section 5314 of title 5. The payment of a rate of basic pay under this paragraph shall not be subject to the pay limitation under section 5306(e) or 5373 of title 5.

The AIA did not provide a specific effective date for AIA § 21(a) or (b). *Id.* The Leahy-Smith America Invents Technical Corrections Act of 2013, however, provides that AIA § 21 "shall be effective as of September 16, 2011." Pub. L. No. 112-274, § 1(g), 126 Stat. 2456, 2457.

C. AIA § 22—Patent and Trademark Office Funding

AIA § 22(a)[b] amends pre-AIA 35 U.S.C. § 42(c), which pertains to the U.S. PTO's funding. Leahy-Smith America Invents Act of 2011, Pub. L. No. 112-29, 125 Stat. 284, 336; *see also* Leahy-Smith America Invents Technical Corrections Act of 2013, Pub. L. No. 112-274, § 1(j), 126 Stat. 2456, 2457 (amending post-AIA 35 U.S.C. § 42(c)(3)(A) and (B)). The following table provides a side-by-side comparison of pre- and post-AIA 35 U.S.C. § 42(c):

[b]AIA § 22(b) provides that the amendments made by AIA § 22(a) "shall take effect on October 1, 2011." Leahy-Smith America Invents Act of 2011, Pub. L. No. 112-29, 125 Stat. 284, 336.

Pre-AIA 35 U.S.C.§ 42(c)	Post-AIA 35 U.S.C. § 42(c)
(c) To the extent and in the amounts provided in advance in appropriations Acts, fees authorized in this title or any other Act to be charged or established by the Director shall be collected by and shall be available to the Director to carry out the activities of the Patent and Trademark Office. All fees available to the Director under section 31 of the Trademark Act of 1946 shall be used only for the processing of trademark registrations and for other activities, services and materials relating to trademarks and to cover a proportionate share of the administrative costs of the Patent and Trademark Office.	(c) (1) To the extent and in the amounts provided in advance in appropriations Acts, fees authorized in this title or any other Act to be charged or established by the Director shall be collected by and shall, subject to paragraph (3), be available to the Director to carry out the activities of the Patent and Trademark Office. (2) There is established in the Treasury a Patent and Trademark Fee Reserve Fund. If fee collections by the Patent and Trademark Office for a fiscal year exceed the amount appropriated to the Office for that fiscal year, fees collected in excess of the appropriated amount shall be deposited in the Patent and Trademark Fee Reserve Fund. To the extent and in the amounts provided in appropriations Acts, amounts in the Fund shall be made available until expended only for obligation and expenditure by the Office in accordance with paragraph (3). (3) (A) Any fees that are collected under this title, and any surcharges on such fees, may only be used for expenses of the Office relating to the processing of patent applications and for other activities, services, and materials relating to patents and to cover a proportionate share of the administrative costs of the Office. (B) Any fees that are collected under section 31 of the Trademark Act of 1946, and any surcharges on such fees, may only be used for expenses of the Office relating to the processing of trademark registrations and for other activities, services, and materials relating to trademarks and to cover a proportionate share of the administrative costs of the Office.

Post-AIA 35 U.S.C. § 42(c)(1) is virtually identical to the first sentence of pre-AIA 35 U.S.C. § 42(c). In particular, both pre-AIA 35 U.S.C. § 42(c) and post-AIA 35 U.S.C. § 42(c)(1) provide Congress' Appropriations Committee with the authority to allocate the fees that the U.S. PTO collects back to the U.S. PTO so that the Office can carry out its operations. Post-AIA

35 U.S.C. § 42(c)(2) modifies the pre-AIA law, however, by establishing a reserve fund for the U.S. PTO. The reserve fund captures fees collected by the U.S. PTO for a fiscal year that exceed the amount appropriated to the Office for that fiscal year. Post-AIA 35 U.S.C. § 42(c)(2). The U.S. PTO should receive authorization to spend all of the fees that are deposited in the reserve fund through its annual appropriations that are authorized by Congress. Post-AIA 35 U.S.C. § 42(c)(2); U.S. PTO, http://www.uspto.gov/patent/laws-and-regulations/america-invents-act-aia/fees-and-budgetary-issues (last visited April 28, 2016). If indeed Congress does authorize the U.S. PTO to spend the fees that have been deposited in the reserve fund, the U.S. PTO will have access to all of the patent and trademark fees it collects. *Id.*

Post-AIA 35 U.S.C. § 42(c)(3)(B) slightly modifies the second sentence of pre-AIA 35 U.S.C. § 42(c), which requires that the U.S. PTO only use trademark fees for trademark operations, including a share of administrative expenses. Post-AIA 35 U.S.C. § 42(c)(3)(B)'s change to pre-AIA 35 U.S.C. § 42(c)'s language does not appear to substantively change the law with regard to the manner in which the U.S. PTO may spend the trademark fees it collects. In addition, post-AIA 35 U.S.C. § 42(c)(3)(A) adds a clause that is analogous to post-AIA 35 U.S.C. § 42(c)(3)(B) for patent fees that are collected by the agency. The patent fee clause of post-AIA 35 U.S.C. § 42(c)(3)(A) did not exist in pre-AIA 35 U.S.C. § 42(c). Accordingly, the U.S. PTO may only use patent fees for patent operations, including the share of administrative expenses associated with such operations.[c]

Overall, AIA § 22(a) does not make any significant changes to the pre-AIA law. *See* Joe Matal, *A Guide to the Legislative History of the America Invents Act: Part II of II*, 21 Fed. Cir. B.J. 539, 644 (2012) ("[AIA § 22(a)] has no substantive effect—it simply restates pre-AIA law regarding the deposit of USPTO user fees in a U.S. Treasury account."). Indeed, the statutory amendments that were passed in AIA § 22(a) were merely leftovers from a provision that was ultimately excluded from the AIA but that had promised to make significant changes to the pre-AIA law by giving the U.S. PTO

[c] *See also* Manus Cooney, *Lame Duck Patent Reform: AIA Technical Corrections*, IPWatchdog, Dec. 2, 2012, http://www.ipwatchdog.com/2012/12/02/lame-duck-patent-reform-aia-technical-corrections/id=30778/ (explaining that the technical changes to post-AIA 35 U.S.C. § 42(c)(3)(A) and (B) that were brought about by § 1(j) of the Leahy-Smith America Invents Technical Corrections Act ensure that: (1) the rule requiring that patent fees be spent for patent purposes also applies to RCE fees and (2) all of the U.S. PTO's administrative costs will be covered by either patent or trademark fees).

direct access to its user fees and end the diversion of those fees to other federal agencies. *See id.* at 644-45 ("[AIA § 22(a)] is the vestige that remained after a revolving fund for patent fees, which would have given the USPTO direct access to its user fees and ended fee diversion, was added to the Senate bill in March 2011 via the floor managers' amendment, preserved in the House bill as introduced and through markup in April 2011, but then stripped from the final bill in June 2011 as the result of a battle with the Appropriations Committee.") (citations omitted).

Under both the pre-AIA law and current law, Congress' Appropriations Committee—not the U.S. PTO—has the authority to allocate the U.S. PTO's user fees to the agency. Pre-AIA 35 U.S.C. § 42(c); post-AIA 35 U.S.C. § 42(c)(1). As a result, an enormous amount of the fees collected by the U.S. PTO have been diverted to other agencies over the years. *See* Michael Lipkin, *Sen. Introduces Bill to Stop USPTO Fee Diversion*, LAW360, March 13, 2014, http://www.law360.com/articles/518518/sen-introduces-bill-to-stop-uspto-fee-diversion (explaining that since 1990, more than $1.1 billion has been diverted from the U.S. PTO to other federal government agencies). Many lawmakers and members of the intellectual property (IP) community have long argued that the U.S. PTO should be able to have direct access to its user fees so that the agency could use such fees to cover the expenses associated with the services it provides to the public. *See, e.g.,* 157 CONG. REC. S5417, 2011 (August 1, 2011 Letter from the U.S. PTO Director to Sen. Coburn) ("This year alone, we anticipate that the agency will collect approximately $80 million in fees paid for USPTO services that will not be available for expenditure in performing those services. Quite clearly, since the work for which these fees were paid remains pending at [the] USPTO, at some point in the future we will have to collect more money in order to actually perform the already-paid-for services . . . Further, the unpredictability of the annual appropriations cycle severely hinders [the] USPTO's ability to engage in the kind of multi-year, business-like planning that is needed to effectively manage a demand-driven, production-based organization."); Gene Quinn, *House Bill Seeks to End Diversion of Fees from the USPTO*, IPWATCHDOG, April 22, 2015, http://www.ipwatchdog.com/2015/04/22/house-bill-seeks-to-end-diversion-of-fees-from-the-uspto/id=57072/ (listing statements by Congressmen Sensenbrenner, Nadler, Franks, Collins, Deutsch, Rohrabacher, and Jeffries, as well as Congresswoman Lofgren, that support ending fee diversion in order to enable the U.S. PTO to more effectively provide its patent and trademark examination services).

Because AIA § 22(a) failed to meet its original purpose of ending Congressional oversight of the U.S. PTO's fees, the provision was a huge disappointment for many in the IP community who have long felt that the U.S. PTO should be able to retain all of the fees it collects. In the years since the AIA was passed, Congress has proposed new legislation to attempt to address the U.S. PTO's fee diversion problem. *See* Gene Quinn, *House Bill Seeks to End Diversion of Fees from the USPTO,* IPWATCHDOG, April 22, 2015, http://www.ipwatchdog.com/2015/04/22/house-bill-seeks-to-end-diversion-of-fees-from-the-uspto/id=57072/ (explaining that the Innovation Protection Act, which has been percolating in Congress over the past few years, would allow the U.S. PTO to have access to the fees it collects until such fees are expended). Whether Congress will pass the new legislation, however, remains to be seen.

D. AIA § 23[d]—U.S. PTO Satellite Offices

Since the time that the U.S. Constitution first provided authority for a patent office in 1787, the U.S. PTO's headquarters have always been located in the Washington, D.C. metropolitan area. U.S. PATENT AND TRADEMARK OFFICE, REPORT TO CONGRESS ON THE SATELLITE OFFICES 3 (2014), *available at* http://www.uspto.gov/sites/default/files/aia_implementation/ USPTO_AIASatelliteOfficesReport_2014Sept30_Online.pdf. The U.S.

---

[d]AIA § 24 also pertains to the U.S. PTO's satellite offices. Specifically, AIA § 24(a) designates the name of the Detroit, Michigan satellite office as the "Elijah J. McCoy United States Patent and Trademark Office." Leahy-Smith America Invents Act of 2011, Pub. L. No. 112-29, 125 Stat. 284, 337. The Detroit satellite office is named after Elijah J. McCoy because "[h]is life captures the spirit of Michigan ingenuity and entrepreneurship." 157 CONG. REC. S1183, 2011 (statement of Sen. Stabenow). Mr. McCoy was an African American who faced a great deal of racial prejudice throughout his life. *Id.* at S1184. Despite the hardships he faced, he secured more than 50 patents. *Id.* at S1183. He is best known for his inventions that revolutionized the functioning of heavy-duty machinery, including locomotives. *Id.* In July 1872, he invented the automatic lubricator, which kept steam engines working properly so trains could run faster and longer without stopping for service. *Id.* The term "real McCoy" was based on Mr. McCoy's inventions. *Id.* at S1184. Indeed, because of its effectiveness, many had tried to copy Mr. McCoy's engine technology, but no one succeeded so machinists began asking for the "real McCoy" technology. *Id.* at S1183-84.

PTO's first satellite office, which was located in Detroit, Michigan, was not announced until December 2010, and that office did not open until July 2012. *Id.* The Detroit satellite office was not opened as a result of the AIA. Instead, the office was established under the U.S. PTO's "Nationwide Workforce Program," which was instituted to address significant long-term challenges the U.S. PTO experienced when trying to recruit and retain Patent Examiners to work in the Washington, D.C. area. *Id.* at 2-3. In a report to Congress, the U.S. PTO explained its rationale for establishing a satellite office, stating that:

> *Through a nationwide workforce, the Office could enhance the employment candidate pool of examiners (including highly qualified candidates from industries not traditionally located in the D.C. region), reduce real estate costs associated with workforce expansion, and build the national presence of the USPTO for greater interaction with the IP community. Additionally, the Office would be positioned to hire and retain more experienced examiner recruits, such as patent attorneys and patent agents, former examiners, and skilled technologists having experience with the USPTO as inventors, who require less training and achieve higher quality production faster than inexperienced hires.*
>
> **Id. at 2.**

While the U.S. PTO's plans for the Detroit satellite office were underway, Congress embraced the idea of establishing additional satellite offices throughout the United States. *Id.* at 3; *see also* 157 CONG. REC. S957, 2011 (statement of Sen. Bennet) ("The Patent and Trademark Office has struggled to hire and retain over 6,000 examiners at a single location in Alexandria, VA. This has resulted in one-third of patent examiners having been with the U.S. Patent and Trademark Office for less than 3 years . . . The PTO recently recognized this weakness in our patent infrastructure by announcing an initial satellite pilot in Detroit, MI . . . Regional satellite offices will . . . increase outreach activities and connection to patent filers, enhance the ability of the USPTO to recruit and retain patent examiners, and improve the quality and pendency for patent applications."); 157 CONG. REC. H4486, 2011 (statement of Rep. Lujan) ("By improving access to the United States Patent and Trademark Office, satellite offices have the potential to help small businesses and independent inventors navigate the patent application process."). As a result, AIA § 23 was passed.

AIA § 23 contains four subsections, i.e., subsections (a)-(d). Leahy-Smith America Invents Act of 2011, Pub. L. No. 112-29, 125 Stat. 284,

336-37. AIA § 23(a) granted the U.S. PTO the authority to establish three or more satellite offices within three years of the enactment of the AIA (i.e., by September 2014), subject to available resources. *Id.* at 336. AIA § 23(b) lists several purposes for the satellite offices, including enhancing retention rates of Patent Examiners, improving recruitment of Patent Examiners, and improving the quality of patent examination. *Id.* at 336-37. AIA § 23(c)(1) sets forth "required considerations" for selecting the location of each satellite office to be established under AIA § 23(a). *Id.* at 337. For example, AIA § 23(c)(1) explains that in choosing a satellite office location, the U.S. PTO must consider "the availability of scientific and technically knowledgeable personnel in the region from which to draw new patent examiners at minimal recruitment cost" and "shall evaluate and consider the extent to which the purpose of the satellite offices listed under [AIA § 23(b)] will be achieved." *Id.* at § 23(c)(1)(C) and (D). AIA § 23(d) further requires the U.S. PTO to submit a report to Congress on, among other things, the rationale for selecting the location of any satellite office pursuant to AIA § 23(a), the progress of establishing such office, and whether the operation of existing satellite offices is achieving the purposes set forth in AIA § 23(b).[e] *Id.*

Since AIA § 23 was passed, the U.S. PTO has opened three satellite offices in addition to its first satellite office located in Detroit, Michigan. More specifically, the U.S. PTO opened a satellite office in Denver, Colorado on June 30, 2014, followed by a satellite office in San Jose, California, which opened on October 15, 2015, and a satellite office in Dallas, Texas, which opened on November 9, 2015. Press Release, U.S. PTO, USPTO Opens Regional Office in Dallas, Texas (Nov. 9, 2015), *available at* http://www.uspto.gov/about-us/news-updates/uspto-opens-regional-office-dallas-texas; Sharon Noguchi, *San Jose: U.S. [S]atellite Patent Office [O]pens*, MERCURY NEWS, Oct. 15, 2015, http://www.mercurynews.com/bay-area-news/ci_28975867/san-jose-u-s-satellite-patent-office-opens%29; Press Release, U.S. PTO, USPTO Announces Denver Satellite Office to Open June 30 (May 30, 2014), *available at* http://www.uspto.gov/about-us/news-updates/uspto-announces-denver-satellite-office-open-june-30.

---

[e]The U.S. PTO's report to Congress can be found at: http://www.uspto.gov/sites/default/files/aia_implementation/USPTO_AIASatelliteOfficesReport_2014Sept30_Online.pdf

III) JURISDICTIONAL CHANGES UNDER THE AIA

AIA §§ 7, 9, and 19 make several changes to the statutes that govern the jurisdiction of the Board of Patent Appeals and Interferences ("BPAI"), Patent Trial and Appeal Board ("PTAB"), district courts and Federal Circuit over certain intellectual property claims. Many of the amendments that are made by the AIA simply bring the jurisdictional statutes into conformity with other changes that were brought about by the new legislation. For example, AIA § 7(c) replaces "BPAI" in several statutes with "PTAB" so that the PTAB possesses jurisdiction over the matters that are necessary to perform its intended functions. On the other hand, AIA § 9 makes more significant changes to the pre-AIA jurisdictional statutes by changing the venue for challenging certain U.S. PTO decisions from the District for the District of Columbia (D.D.C.) to the Eastern District of Virginia (E.D. Va.), which is the district in which the U.S. PTO resides. Lastly, AIA § 19 amends several statutes, and adds a new statute, in an effort to clarify state and federal court jurisdiction over certain types of intellectual property claims. This section will provide a brief discussion of the changes made by AIA §§ 7, 9, and 19.

A. AIA § 7(c)—Appeals to the U.S. Court of Appeals for the Federal Circuit

AIA § 7(c) amends three statutes pertaining to appeals to the United States Court of Appeals for the Federal Circuit ("Federal Circuit") in patent cases. In particular, AIA § 7(c)(1) amends pre-AIA 35 U.S.C. § 141, AIA § 7(c)(2) amends pre-AIA 28 U.S.C. § 1295(a)(4)(A), and AIA § 7(c)(3) amends pre-AIA 35 U.S.C. § 143. Leahy-Smith America Invents Act of 2011, Pub. L. No. 112-29, 125 Stat. 284, 314-15.

Pre-AIA 35 U.S.C. § 141 provides certain parties (i.e., patent applicants, patent owners, and parties to an interference) with a right to appeal decisions (e.g., the decision in an appeal or on an interference) made by the BPAI to the Federal Circuit. Consistent with the AIA's establishment of the PTAB, as well as the new trial-like proceedings (i.e., inter partes reviews, post grant reviews, and derivation proceedings), AIA § 7(c)(1) amends pre-AIA 35 U.S.C. § 141 to provide a right to appeal decisions made by the PTAB, including final written decisions made in the trial-like proceedings, to the Federal Circuit. *Id.* at 314. A comparison of pre- and post-AIA 35 U.S.C. § 141 is provided in the following table:

Pre-AIA 35 U.S.C. § 141	Post-AIA 35 U.S.C. § 141
An applicant dissatisfied with the decision in an appeal to the Board of Patent Appeals and Interferences under section 134 of this title may appeal the decision to the United States Court of Appeals for the Federal Circuit. By filing such an appeal the applicant waives his or her right to proceed under section 145 of this title. A patent owner, or a third-party requester in an inter partes reexamination proceeding, who is in any reexamination proceeding dissatisfied with the final decision in an appeal to the Board of Patent Appeals and Interferences under section 134 may appeal the decision only to the United States Court of Appeals for the Federal Circuit. A party to an interference dissatisfied with the decision of the Board of Patent Appeals and Interferences on the interference may appeal the decision to the United States Court of Appeals for the Federal Circuit, but such appeal shall be dismissed if any adverse party to such interference, within twenty days after the appellant has filed notice of appeal in accordance with section 142 of this title, files notice with the Director that the party elects to have all further proceedings conducted as provided in section 146 of this title. If the appellant does not, within thirty days after filing of such notice by the adverse party, file a civil action under section 146, the decision appealed from shall govern the further proceedings in the case.	(a) EXAMINATIONS. --An applicant who is dissatisfied with the final decision in an appeal to the Patent Trial and Appeal Board under section 134(a) may appeal the Board's decision to the United States Court of Appeals for the Federal Circuit. By filing such an appeal, the applicant waives his or her right to proceed under section 145. (b) REEXAMINATIONS.-- A patent owner who is dissatisfied with the final decision in an appeal of a reexamination to the Patent Trial and Appeal Board under section 134(b) may appeal the Board's decision only to the United States Court of Appeals for the Federal Circuit. (c) POST–GRANT AND INTER PARTES REVIEWS.-- A party to an inter partes review or a post grant review who is dissatisfied with the final written decision of the Patent Trial and Appeal Board under section 318(a) or 328(a) (as the case may be) may appeal the Board's decision only to the United States Court of Appeals for the Federal Circuit. (d) DERIVATION PROCEEDINGS.-- A party to a derivation proceeding who is dissatisfied with the final decision of the Patent Trial and Appeal Board in the proceeding may appeal the decision to the United States Court of Appeals for the Federal Circuit, but such appeal shall be dismissed if any adverse party to such derivation proceeding, within 20 days after the appellant has filed notice of appeal in accordance with section 142, files notice with the Director that the party elects to have all further proceedings conducted as provided in section 146. If the appellant does not, within 30 days after the filing of such notice by the adverse party, file a civil action under section 146, the Board's decision shall govern the further proceedings in the case.

Pre-AIA 28 U.S.C. § 1295(a)(4)(A) grants the Federal Circuit exclusive jurisdiction over appeals from the BPAI's decisions in patent application appeals and interferences. Similar to AIA § 7(c)(1), AIA § 7(c)(2) amends pre-AIA 28 U.S.C. § 1295(a)(4)(A) so that it is consistent with other changes that are made by the AIA.[f] *Id.* In particular, post-AIA 28 U.S.C. § 1295(a)(4)(A) refers to the PTAB instead of the BPAI and specifically provides the Federal Circuit with exclusive jurisdiction over appeals from the PTAB's decisions in reexaminations, derivation proceedings, post grant reviews, and inter partes reviews. A comparison of pre- and post-AIA 35 U.S.C. § 1295(a)(4)(A) is set forth in the following table:

Pre-AIA 28 U.S.C.§ 1295(a)(4)(A)	Post-AIA 28 U.S.C. § 1295(a)(4)(A)
(a) The United States Court of Appeals for the Federal Circuit shall have exclusive jurisdiction … (4) of an appeal from a decision of-- (A) the Board of Patent Appeals and Interferences of the United States Patent and Trademark Office with respect to patent applications and interferences, at the instance of an applicant for a patent or any party to a patent interference, and any such appeal shall waive the right of such applicant or party to proceed under section 145 or 146 of title 35[.]	(a) The United States Court of Appeals for the Federal Circuit shall have exclusive jurisdiction … (4) of an appeal from a decision of— (A) the Patent Trial and Appeal Board of the United States Patent and Trademark Office with respect to a patent application, derivation proceeding, reexamination, post grant review, or inter partes review under title 35, at the instance of a party who exercised that party's right to participate in the applicable proceeding before or appeal to the Board, except that an applicant or a party to a derivation proceeding may also have remedy by civil action pursuant to section 145 or 146 of title 35; an appeal under this subparagraph of a decision of the Board with respect to an application or derivation proceeding shall waive the right of such applicant or party to proceed under section 145 or 146 of title 35[.]

Lastly, pre-AIA 35 U.S.C. § 143 requires the U.S. PTO to transmit to the Federal Circuit a certified list of the documents comprising the record for any U.S. PTO matter that is appealed to the Court. Similar to the changes made by AIA §§ 7(c)(1) and 7(c)(2), AIA § 7(c)(3) amends pre-AIA 35 U.S.C. § 143

[f] Senator Kyl provided an explanation of the changes to pre-AIA 28 U.S.C. § 1295(a)(4)(A) during the 2011 debates on the AIA. He stated that "section 1295(a)(4)(A) of title 28 is modified to authorize appeals of reexaminations and reviews. Interestingly, current 1295(a)(4)(A) only gives the Federal Circuit jurisdiction over appeals from applications and interferences. It appears that Congress never gave the Federal Circuit jurisdiction over appeals from reexaminations when it created those proceedings. The language of subparagraph (A) is also generalized and clarified, recognizing that the details of what is appealable will be in sections 134 and 141. Also, for logical consistency, language is added to subparagraph (A) making clear that section 145 and 146 proceedings [i.e., civil actions to obtain a patent and civil actions to challenge the PTAB's decision in an interference proceeding] are an exception to the Federal Circuit's otherwise exclusive appellate jurisdiction over applications and interferences under that subparagraph." 157 CONG. REC. 1377, 2011.

so that it is consistent with other changes that are made by the AIA. *Id.* at 314–15. For example, post-AIA 35 U.S.C. § 143 refers to the new post-grant proceedings that were created by the AIA (i.e., inter partes and post grant reviews) and references the PTAB instead of the BPAI. A comparison of pre- and post-AIA 35 U.S.C. § 143 is provided in the following table:

Pre-AIA 35 U.S.C. § 143	Post-AIA 35 U.S.C. § 143
With respect to an appeal described in section 142 of this title, the Director shall transmit to the United States Court of Appeals for the Federal Circuit a certified list of the documents comprising the record in the Patent and Trademark Office. The court may request that the Director forward the original or certified copies of such documents during the pendency of the appeal. In an ex parte case or any reexamination case, the Director shall submit to the court in writing the grounds for the decision of the Patent and Trademark Office, addressing all the issues involved in the appeal. The court shall, before hearing an appeal, give notice of the time and place of the hearing to the Director and the parties in the appeal.	With respect to an appeal described in section 142, the Director shall transmit to the United States Court of Appeals for the Federal Circuit a certified list of the documents comprising the record in the Patent and Trademark Office. The court may request that the Director forward the original or certified copies of such documents during the pendency of the appeal. In an ex parte case, the Director shall submit to the court in writing the grounds for the decision of the Patent and Trademark Office, addressing all of the issues raised in the appeal. The Director shall have the right to intervene in an appeal from a decision entered by the Patent Trial and Appeal Board in a derivation proceeding under section 135 or in an inter partes or post grant review under chapter 31 or 32.

AIA § 7(e) provides that, subject to certain exceptions,[g] the amendments made in AIA § 7(c) "shall take effect upon the expiration of the 1-year period

[g]AIA § 7(e)(1)-(4) provide certain exceptions to the September 16, 2012 effective date that is ▶ set forth in AIA § 7(e). *Id.* at 315. For instance, AIA § 7(e)(1) provides an exception to AIA § 7(e)'s effective date for "the extension of jurisdiction to the . . . Federal Circuit to entertain appeals of decisions of the Patent Trial and Appeal Board in reexaminations under the amendment made by [AIA § 7(c)(2)]." *Id.* In particular, the effective date for the extension of jurisdiction to entertain these appeals is September 16, 2011, and the Federal Circuit's jurisdiction to entertain these appeals extends "to any decision of the Board of Patent Appeals and Interferences with respect to a reexamination that is entered before, on, or after [September 16, 2011]." *Id.* In addition, AIA § 7(e)(2) provides that pre-AIA 35 U.S.C. § 6 (i.e., the statute that establishes the BPAI), pre-AIA 35 U.S.C. § 134 (i.e., the statute that allows parties to appeal reexamination decisions to the BPAI) and pre-AIA 35 U.S.C. § 141 (i.e., the statute that allows patentees to appeal the BPAI's final decisions in inter partes reexaminations to the Federal Circuit) continue to apply to inter partes reexaminations that are requested before September 16, 2012. *Id.* AIA § 7(e)(3) provides that "the Patent Trial and Appeal Board may be deemed to be the Board of Patent Appeals and Interferences for purposes of appeals of *inter partes* reexaminations that are requested . . . before [September 16, 2012]." *Id.* Lastly, AIA § 7(e)(4) gives the Director the right to intervene in appeals from decisions entered by the PTAB in inter partes reexaminations, where the inter partes reexaminations were requested

beginning on the date of the enactment of this Act [i.e., on September 16, 2012] and shall apply to proceedings commenced on or after that effective date." *Id.* at 315. Accordingly, post-AIA 35 U.S.C. §§ 141, 143 and 1295(a)(4) (A) will apply to proceedings commenced on or after September 16, 2012.

B. AIA § 9—Venue Change[h] From the District for the District of Columbia (D.D.C.) to the Eastern District of Virginia (E.D. Va.)

AIA § 9(a) amends pre-AIA 35 U.S.C. §§ 32, 145, 146, 154(b)(4)(A), and 293, as well as pre-AIA 15 U.S.C. § 1071(b)(4),[i] to provide that the Eastern

◄ before September 16, 2012. *Id.* Senator Kyl explained that in the effective-date provision of AIA § 7 ". . . various existing authorities are extended so that they may continue to apply to *inter partes* reexaminations commenced under the old system, and the apparent gap in current section 1295(a)(4)(A)'s authorization of jurisdiction is immediately filled with respect to all *inter partes* and ex parte reexaminations." 157 Cong. Rec. 1377, 2011.

[h]The AIA's venue provision originally included a section that would have codified the standard for transfers of venue that was established by the Federal Circuit in *In re TS Tech USA Corp.*, 551 F.3d 1315 (2008) and applied it to patent cases generally. 157 Cong. Rec. S1366, 2011. The standard articulated in *In re TS Tech USA Corp.* provides for transfer to the judicial district that is "clearly more convenient" for both the parties and the witnesses. *Id.* Congress initially proposed the provision governing the standard for venue transfers because the Federal Circuit's jurisprudence had held that "venue for a patent infringement defendant is proper wherever an alleged infringing product can be found." S. Rep. No. 111-18, at 19 (2009). Under the Federal Circuit's precedent, therefore, "[s]ince most products are sold nationally, a patent holder can often bring a patent infringement action in any one of the 94 judicial districts in the United States." *Id.* The Federal Circuit's standard for proper venue resulted in a great deal of forum-shopping. *See, e.g.,* 157 Cong. Rec. 1033, 2011 (statement of Sen. Coons). On March 1, 2011, however, the Senate adopted a floor manager's amendment striking the section of the AIA's venue provision that would have codified *In re TS Tech USA Corp.* 157 Cong. Rec. S1366, 2011. The summary of the floor manager's amendment explains that the provision was eliminated from the AIA because "TS Tech already applies as a matter of caselaw in the Fifth Circuit. (The Federal Circuit applies regional circuit law to procedural matters, and reads Fifth Circuit law as applying the transfer of venue rule.) Complaints about venue generally focus on [E.D. Tex.], so there is little need to apply TS Tech nationally, and it seemed odd for Congress to regulate such matters in any event.)." *Id.*

[i]Pre-AIA 35 U.S.C. § 32 pertains to U.S. PTO decisions to suspend or exclude any person, agent, or attorney from further practice before the U.S. PTO. Pre-AIA 35 U.S.C. § 145 provides patent applicants with the right to bring a civil action against the U.S. PTO if the applicant is dissatisfied with the BPAI's decision on an appeal of any patent application claims that have been twice rejected. Pre-AIA 35 U.S.C. § 146 provides a party who is dissatisfied with the BPAI's decision in an interference proceeding with the right to bring a civil action against the U.S. PTO. Pre-AIA 35 U.S.C. § 154(b)(4)(A) provides patent applicants with the right to bring a civil action against the U.S. PTO if the applicant is dissatisfied with the U.S. PTO's decision on an applicant's request for reconsideration of a Patent Term Adjustment (PTA) decision. Pre-AIA 35 U.S.C. § 293 explains who should be served process or notice of proceedings in instances where a patentee does not reside in the United States. Lastly, pre-AIA 15 U.S.C. § 1071(b)(4) pertains to jurisdiction over certain civil actions brought against the U.S. PTO to challenge decisions regarding trademark applications that are made by the Director or by the Trademark Trial and Appeal Board.

District of Virginia—not the District for the District of Columbia—shall be the venue for challenging certain decisions by the U.S. PTO. Leahy–Smith America Invents Act of 2011, Pub. L. No. 112-29, 125 Stat. 284, 316. The following text from the final Committee Report on the AIA explains the reasons for these amendments:

> In 1999, as part of the American Inventors Protection Act (AIPA), Congress established that as a general matter the venue of the USPTO is the district where it resides. The USPTO currently resides in the Eastern District of Virginia. However, Congress inadvertently failed to make this change uniformly throughout the entire patent statute. As a result, certain sections of the patent statute (and one section of the trademark statute) continue to allow challenges to USPTO decisions to be brought in the District of Columbia, a place where the USPTO has not resided in decades.
>
> Because the USPTO no longer resides in the District of Columbia, the sections that authorize venue for litigation against the USPTO are consistently changed to reflect the venue where the USPTO currently resides.
>
> **H.R. Rep. No. 112-98, at 49 (2011) (citation omitted).**

AIA § 9(b) provides that the amendments made by AIA § 9(a) "shall take effect on the date of the enactment of this Act and shall apply to any civil action commenced on or after that date." Leahy–Smith America Invents Act of 2011, Pub. L. No. 112-29, 125 Stat. 284, 316. The AIA was enacted on September 16, 2011. *Id.* at 284. Accordingly, any civil actions brought under 35 U.S.C. §§ 32, 145, 146, 154(b)(4)(A), or 293, or under 15 U.S.C. § 1071(b)(4), and that are commenced on or after September 16, 2011, should be brought in the Eastern District of Virginia, not the District for the District of Columbia.

C. AIA § 19—Jurisdictional and Procedural Matters

AIA § 19(a)-(c) amend two statutes and add a new statute in an effort to clarify state and federal court jurisdiction over certain intellectual property claims. In order to more fully comprehend the statutory changes made by AIA § 19, it is helpful to first consider the context in which the provision was enacted. Indeed, subsections (a)-(c) of AIA § 19 were enacted in response to the Supreme Court's decision in *Holmes Group, Inc. v. Vornado Air Circulation Sys., Inc.*, 535 U.S. 826 (2002). H.R. Rep. No. 109-407, at 3 (2006); *see also* H.R. Rep. No. 112-98, at 81 (2011) (reaffirming the Committee Report accompanying H.R. 2955 (i.e., H.R. Rep. No. 109-407)). The pertinent facts of the *Holmes* case, as quoted from the Supreme Court's opinion, are as follows:

> In late 1992, [Vornado Air Circulation Systems, Inc. ("Vornado")] sued a competitor, Duracraft Corp. [("Duracraft")], claiming that Duracraft's use of a "spiral

grill design" in its fans infringed [its] trade dress. The Court of Appeals for the Tenth Circuit found for Duracraft, holding that Vornado had no protectable trade-dress rights in the grill design ... [Vornado's case against Duracraft became known as the Vornado I case.]

[Despite the Tenth Circuit's unfavorable ruling, Vornado] lodged a complaint with the United States International Trade Commission against petitioner, The Holmes Group, Inc. [("Holmes")], claiming that [Holmes'] sale of fans and heaters with a spiral grill design infringed [its] patent and the same trade dress held unprotectable in Vornado I. *Several weeks later, [Holmes] filed [suit] against [Vornado] in the United States District Court for the District of Kansas, seeking, inter alia, a declaratory judgment that its products did not infringe [Vornado's] trade dress and an injunction restraining [Vornado] from accusing it of trade-dress infringement in promotional materials. [Vornado's] answer [to Holmes' complaint] asserted a compulsory counterclaim alleging patent infringement.*

The District Court granted [Holmes] the declaratory judgment and injunction it sought. The court explained that the collateral[]estoppel effect of Vornado I *precluded [Vornado] from relitigating its claim of trade-dress rights in the spiral grill design. [The District Court also] rejected [Vornado's] contention that an intervening Federal Circuit case, Midwest Industries, Inc. v. Karavan Trailers, Inc., 175 F. 3d 1356 (1999), which disagreed with the Tenth Circuit's reasoning in* Vornado I, *constituted a change in the law of trade dress that warranted relitigation of [Vornado's] trade-dress claim. The court also stayed all proceedings related to [Vornado's] counterclaim, adding that the counterclaim would be dismissed if the declaratory judgment and injunction entered in favor of [Holmes] were affirmed on appeal.*

[Vornado] appealed to the Court of Appeals for the Federal Circuit. Notwithstanding [Holmes'] challenge to its jurisdiction, the Federal Circuit vacated the District Court's judgment ... and remanded for consideration [the issue] of whether the "change in the law" exception to collateral estoppel applied in light of TrafFix Devices, Inc. v. Marketing Displays, Inc., 532 U.S. 23 (2001), a case decided after the District Court's judgment which resolved a Circuit split involving Vornado I *and* Midwest Industries.

Holmes Group, Inc., 535 U.S. at 827-829 (citations omitted).

The U.S. Supreme Court granted certiorari to determine whether the Federal Circuit had appellate jurisdiction over Vornado's appeal. *Id.* at 829. More specifically, the Supreme Court addressed the issue of whether the Federal Circuit had appellate jurisdiction over a case in which the complaint does not allege a claim arising under federal patent law but the answer contains a patent–law counterclaim. *Id.* at 827.

The Supreme Court held that the Federal Circuit did not have jurisdiction over Vornado's appeal and therefore vacated that court's judgment and remanded the case with instructions to transfer the case to the Court of Appeals for the Tenth Circuit. *Id.* at 830, 834. The Court explained that 28 U.S.C. § 1295(a)(1) vests the Federal Circuit with exclusive jurisdiction over "an appeal from a final decision of a district court of the United States ... if the jurisdiction *of that court* was based, in whole or in part, on [28 U.S.C. §] 1338 . . . ." *Id.* at 829 (quoting pre-AIA 28 U.S.C. § 1295(a)(1)). 28 U.S.C. § 1338 provides that "[t]he district courts shall have original jurisdiction of any civil action arising under any Act of Congress relating to patents . . . ." *Id.* According to the Supreme Court, therefore, "the Federal Circuit's jurisdiction is fixed with reference to that of the district court, and turns on whether the action *arises under* federal patent law." *Id.* (emphasis added).

To determine whether Vornado's case arose under federal patent law, the Supreme Court applied the "well-pleaded-complaint rule." *Id.* at 830. The Court explained that the well-pleaded-complaint rule has long been used to determine whether a case "arose under" federal law for purposes of 28 U.S.C. § 1331, a statute that uses the same operative language as 28 U.S.C. § 1338(a).[j] *Id.* In order for a case to "arise under" federal patent law, the well-pleaded-complaint rule requires that the plaintiff's well-pleaded complaint "establis[h] either that federal patent law creates the cause of action or that the plaintiff's right to relief necessarily depends on resolution of a substantial question of federal patent law . . . ." *Id.* (quoting *Christianson v. Colt Industries Operating Corp.*, 486 U.S. 800, 809 (1988)). In Vornado's case, there was no dispute that Holmes' well-pleaded complaint did not assert any claim arising under federal patent law. *Id.* Indeed, Holmes had only asserted trade dress claims. *Id.* at 828. Accordingly, the Court found that the district court's jurisdiction over Holmes' case did not "arise under" federal patent law, i.e., that the district court did not have jurisdiction over the case pursuant to 28 U.S.C. § 1338. *Id.* at 830. Thus, according to the Court, the Federal Circuit erred in asserting jurisdiction over the appeal of the district court's decision. *Id.*

In finding that the Federal Circuit did not have jurisdiction over Vornado's appeal, the Supreme Court also rejected Vornado's argument that a counterclaim that arises under federal patent law should serve as a basis for a district court's jurisdiction under 28 U.S.C. § 1338(a). *Id.* at 830. The Court explained that "[a]llowing a counterclaim to establish 'arising under' jurisdiction would . . . contravene [many] longstanding policies underlying [the Court's] precedents." *Id.* at 832. For example, by allowing jurisdiction

---

[j]More specifically, 28 U.S.C. § 1331 grants district courts "original jurisdiction of all civil actions *arising under* the Constitution, laws, or treaties of the United States." (Emphasis added.) Pre-AIA 28 U.S.C. § 1338(a), in relevant part, provides that district courts "shall have original jurisdiction of any civil action *arising under* any Act of Congress relating to patents, plant variety protection, copyrights and trademarks." (Emphasis added.) Thus, both statutes use "arising under" as their operative language.

to be established by a defendant's counterclaim, the plaintiff would no longer be "the master of the complaint." *Id*. Indeed, a plaintiff who pled only state claims in his/her complaint could be dragged into federal court (i.e., through the removal procedure set forth in 28 U.S.C. § 1441(a)) by a defendant who asserts federal counterclaims. *Id*.

Just over four years after the Supreme Court decided *Holmes Group, Inc. v. Vornado Air Circulation Systems, Inc.*, the House Committee on the Judiciary issued a report ("the Committee Report") that criticized the case for "contraven[ing] the will of Congress when it created the Federal Circuit." H.R. REP. NO. 109-407, at 5 (2006). The Committee Report explained that Congress had created the Federal Circuit in 1982 so that patent cases could be channeled "into a single appellate forum [to] create a stable, uniform law and . . . eliminate forum shopping" among litigants. *Id*. at 3. The Committee Report explained that before the Federal Circuit was created, "regional circuits were doing a poor job of developing coherent patent law" and "forum-shopping was rampant, as some circuits were regarded as 'pro-patent' and other circuits as 'anti-patent.'" *Id*. The House Committee on the Judiciary believed that *Holmes* would "induce litigants to engage in forum-shopping among the regional circuits and State courts." *Id*. at 5. The Committee was also concerned that the decision in *Holmes* would "lead to an erosion in the uniformity or coherence in patent law that ha[d] been steadily building since the [Federal] Circuit's creation in 1982." *Id*.

In 2006, in response to the Supreme Court's decision in *Holmes*, the House Committee on the Judiciary proposed a provision that was similar to the one that was ultimately passed as AIA § 19(a). *Id*. Later on in 2011, the House Committee on the Judiciary proposed the provisions that were ultimately passed as AIA § 19(a), (b), and (c) to amend pre-AIA 28 U.S.C. §§ 1338(a) and 1295(a)(1) and add new 28 U.S.C. § 1454, respectively. H.R. REP. NO. 112-98, at 54, 81, 158-60 (2011). AIA § 19(a), (b), and (c) were enacted to "clarif[y] the jurisdiction of the US district courts and stipulate[] that the US Court of Appeals for the Federal Circuit has jurisdiction over appeals involving *compulsory* patent counterclaims." *Id*. at 54 (emphasis in original).

More specifically, AIA § 19(a) amends pre-AIA 28 U.S.C. § 1338(a) to make clear that state courts have no jurisdiction over any type of claims (e.g., counterclaims or third-party claims) brought under the patent, copyright, or plant variety protection statutes. Leahy-Smith America Invents Act of 2011, Pub. L. No. 112-29, 125 Stat. 284, 331. AIA § 19(a) also amends the pre-AIA statute to clarify that certain U.S. territories qualify as states for purposes of determining whether a court located in those territories has jurisdiction over a patent, copyright, or plant variety protection claim. *Id.* A comparison of pre- and post-AIA 28 U.S.C. § 1338(a) appears in the following table:

Pre-AIA 28 U.S.C. § 1338(a)	Post-AIA 28 U.S.C. § 1338(a)
(a) The district courts shall have original jurisdiction of any civil action arising under any Act of Congress relating to patents, plant variety protection, copyrights and trademarks. Such jurisdiction shall be exclusive of the courts of the states in patent, plant variety protection and copyright cases.	(a) The district courts shall have original jurisdiction of any civil action arising under any Act of Congress relating to patents, plant variety protection, copyrights and trademarks. No State court shall have jurisdiction over any claim for relief arising under any Act of Congress relating to patents, plant variety protection, or copyrights. For purposes of this subsection, the term "State" includes any State of the United States, the District of Columbia, the Commonwealth of Puerto Rico, the United States Virgin Islands, American Samoa, Guam, and the Northern Mariana Islands.

In addition, AIA § 19(b) expressly abrogates the U.S. Supreme Court's decision in *Holmes* by amending pre-AIA 28 U.S.C. § 1295(a)(1) to state that "[t]he United States Court of Appeals for the Federal Circuit shall have exclusive jurisdiction . . . of an appeal from a final decision of a district court . . . in any civil action arising under, or in any civil action in which a party has asserted a compulsory counterclaim arising under, any Act of Congress relating to patents or plant variety protection." *Id.* at 331-32; H.R. REP. No. 112-98, at 81 (2011). In other words, contrary to what the Supreme Court held in *Holmes*, compulsory patent and plant variety protection counterclaims[k] may now serve as the basis for the Federal Circuit's jurisdiction in an appeal from a final decision of a district court. A comparison of pre- and post-AIA 28 U.S.C. § 1295(a)(1) is provided in the following table:

[k]Note that the Federal Circuit does not have exclusive jurisdiction over an appeal from a final decision of a district court in a civil action in which a party has asserted any type of patent or plant variety protection counterclaim. Instead, the Federal Circuit only has jurisdiction over *compulsory* patent or plant variety protection counterclaims. During the Senate debates in March 2011, Senator Kyl explained that "[c]ompulsory counterclaims are defined [in Fed. R. Civ. P.] 13(a) and basically consist of counterclaims that arise out of the same transaction or occurrence and that do not require the joinder of parties over whom the court would lack jurisdiction . . . Without this modification [i.e., to limit the Federal Circuit's jurisdiction to compulsory patent counterclaims as opposed to any type of patent counterclaim], it is possible that a defendant could raise unrelated and unnecessary patent counterclaims simply in order to manipulate appellate jurisdiction." Joe Matal, *A Guide to the Legislative History of the America Invents Act: Part II of II*, 21 FED. CIR. B.J. 539, 540 (2012) (quoting 157 CONG. REC. S1378-79 (daily ed. Mar. 8, 2011)).

Pre-AIA 28 U.S.C. § 1295(a)(1)	Post-AIA 28 U.S.C. § 1295(a)(1)
(a) The United States Court of Appeals for the Federal Circuit shall have exclusive jurisdiction-- (1) of an appeal from a final decision of a district court of the United States, the United States District Court for the District of the Canal Zone, the District Court of Guam, the District Court of the Virgin Islands, or the District Court for the Northern Mariana Islands, if the jurisdiction of that court was based, in whole or in part, on section 1338 of this *title [28 USCS § 1338]*, except that a case involving a claim arising under any Act of Congress relating to copyrights, exclusive rights in mask works, or trademarks and no other claims under section 1338(a) *[28 USCS § 1338(a)]* shall be governed by sections 1291, 1292, and 1294 of this *title [28 USCS §§ 1291, 1292,* and *1294]*[.]	(a) The United States Court of Appeals for the Federal Circuit shall have exclusive jurisdiction— (1) of an appeal from a final decision of a district court of the United States, the District Court of Guam, the District Court of the Virgin Islands, or the District Court of the Northern Mariana Islands, in any civil action arising under, or in any civil action in which a party has asserted a compulsory counterclaim arising under, any Act of Congress relating to patents or plant variety protection[.]

Lastly, AIA § 19(c) adds new 28 U.S.C. § 1454, which allows parties to remove from a state court to a federal court civil actions in which "any party" asserts a claim under the patent, plant variety protection, or copyright statutes. Leahy-Smith America Invents Act of 2011, Pub. L. No. 112-29, 125 Stat. 284, 332. New 28 U.S.C. § 1454 reads as follows:

(a) In General.--
    A civil action in which any party asserts a claim for relief arising under any Act of Congress relating to patents, plant variety protection, or copyrights may be removed to the district court of the United States for the district and division embracing the place where the action is pending.
(b) Special Rules.-- The removal of an action under this section shall be made in accordance with section 1446, except that if the removal is based solely on this section--
    (1) the action may be removed by any party; and
    (2) the time limitations contained in section 1446(b) may be extended at any time for cause shown.
(c) Clarification of Jurisdiction in Certain Cases.--
    The court to which a civil action is removed under this section is not precluded from hearing and determining any claim in the civil action because the State court from which the civil action is removed did not have jurisdiction over that claim.
(d) Remand.-- If a civil action is removed solely under this section, the district court--

> *(1) shall remand all claims that are neither a basis for removal under subsection (a) nor within the original or supplemental jurisdiction of the district court under any Act of Congress; and*
> *(2) may, under the circumstances specified in section 1367(c), remand any claims within the supplemental jurisdiction of the district court under section 1367.*

New 28 U.S.C. § 1454[1] clarifies jurisdiction in cases where claims relating to patents, plant variety protection, or copyrights are brought in conjunction with other claims (i.e., claims over which only state courts have jurisdiction). Indeed, 28 U.S.C. § 1454(a) and (b) permit "any party" to remove a civil action in which a claim has been brought under a patent, plant variety protection, or copyright statute from a state court to a federal district court. 28 U.S.C. § 1454(d)(1) also provides that when a civil action is removed from state court to a district court solely pursuant to 28 U.S.C. § 1454(a), the district court "shall remand all claims that are neither a basis for removal under [28 U.S.C. § 1454](a) nor within the original or supplemental jurisdiction of the district court under any Act of Congress[.]" *Id.* New 28 U.S.C. § 1454(a) should further the objectives of post-AIA 28 U.S.C. § 1338(a) by ensuring that claims brought under patent, plant variety protection, or copyright statutes are not adjudicated by state courts, which lack jurisdiction over such claims.

AIA § 19(e) provides that AIA § 19 "shall apply to any civil action commenced on or after the date of the enactment of this Act." Leahy-Smith America Invents Act of 2011, Pub. L. No. 112-29, 125 Stat. 284, 333. Accordingly, post-AIA 28 U.S.C. §§ 1338(a), 1295(a)(1), and 1454 will apply to any civil action that is filed on or after September 16, 2011.

IV) FEE PROVISIONS OF THE AIA

AIA § 10 brought about two significant changes to the administration of the U.S. PTO's fees. In particular, the first major change is seen in AIA § 10(a)(1), which expands the U.S. PTO's authority by allowing the agency to set and adjust its own fees for its patent and trademark services. AIA § 10(b)-(f) and (h)-(i) then set forth several significant restrictions with regard to the authority that is granted in AIA § 10(a)(1). The change brought about by AIA § 10(a)(1) is nevertheless still important because prior to the enactment of the provision, only Congress had the authority to set and

---

[1] Senator Kyl discussed new 28 U.S.C. § 1454 during the Senate debates in March 2011. 157 Cong. Rec. S1379. He explained that the statue instructs courts "to not remand those claims that were a basis for removal in the first place—that is, the intellectual-property counterclaims." *Id.*

adjust the agency's patent and trademark fees. The second major change to the administration of the U.S. PTO's fees can be seen in AIA § 10(g), which establishes a "micro entity" as a third type of patent applicant. Qualifying for micro entity status is crucial for many patent applicants operating on small amounts of capital because the status entitles them to a seventy-five percent reduction on most of the U.S. PTO's fees for patent services. This section will discuss the fee provisions of AIA § 10 in more depth.

A. AIA § 10(a)-(f) and (h)-(i)—U.S. PTO's Fee Setting Authority

AIA § 10(a)(1) provides the U.S. PTO with broad authority to "set or adjust by rule any fee established, authorized, or charged under title 35, United States Code, or the Trademark Act of 1946 . . . for any services performed by or materials furnished by, the Office . . . ." Leahy-Smith America Invents Act of 2011, Pub. L. No. 112-29, 125 Stat. 284, 316.[m] Before the AIA was enacted, the U.S. PTO had the ability to set certain fees, but most fees (e.g., filing fees, issuance fees, maintenance fees) were set by Congress. H.R. REP. No. 112-98, at 49 (2011). The U.S. PTO had long argued that it should be able to set its own fees in order to properly manage the agency. See id. ("History has shown that such a scheme [i.e., wherein Congress sets most of the USPTO's fees] does not allow the USPTO to respond promptly to the challenges that confront it. The USPTO has argued for years that it must have fee-setting authority to administer properly the agency and its growing workload.").[n] AIA § 10(a)(1) therefore responds to the U.S. PTO's

---

[m]Note that AIA § 11 provides the patent fee schedule as it existed on the date the AIA was enacted (i.e., on September 16, 2011). Id. at 320-25. The final Committee Report for the AIA explains that the fee schedule was included in the AIA to serve as a reference point for any future adjustments by the U.S. PTO. H.R. REP. No. 112-98, at 78 (2011).

[n]See also Letter from Gary Locke, U.S. Department of Commerce, to The Honorable Lamar Smith, Chairman of the House Committee on the Judiciary (May 31, 2011), available at http://www.patentreform.info/Legislative%20History%20PDFs/From%20USPTO/views-hr1249-america-invents-act.pdf ("Fee-setting authority, coupled with the right to use all fees paid by patent applicants without fiscal year limitation, will permit the USPTO to engage in multi-year budget planning and achieve a stable funding model that supports future investments and improvements in operations. This structure is critical to enable the USPTO to better meet the needs of America's innovators."); Letter from Gary Locke, U.S. Department of Commerce, to The Honorable Patrick J. Leahy, Chairman of the House Committee on the Judiciary and The Honorable Jefferson B. Sessions, III, Ranking Member of the House Committee on the Judiciary (Oct. 5, 2009), available at http://www.patentreform.info/Legislative%20History%20PDFs/From%20USPTO/locke-letter-oct-05-2009.pdf ("The ability to set fees to recover costs would permit the USPTO to better address operational funding needs, and provide high quality, timely examination of patent applications. This authority is critical in light of the USPTO's recent financial challenges and growing backlog.").

arguments by expanding the U.S. PTO's fee-setting authority for the fees it collects from patent and trademark applicants.

Although AIA § 10(a)(1) significantly expands the U.S. PTO's fee-setting authority, AIA §§ 10(a)(2), (b), (c), (d), (e), (f), and (h) all impose certain limitations on the agency's ability to set and adjust its fees. Leahy-Smith America Invents Act of 2011, Pub. L. No. 112-29, 125 Stat. 284, 316-19. For example, AIA § 10(a)(2) explains that the U.S. PTO may only set or adjust fees in a manner that allows the agency to recover the aggregate estimated costs to the Office for "processing, activities, services, and materials relating to patents (in the case of patent fees) and trademarks (in the case of trademark fees), including administrative costs of the Office with respect to such patent or trademark fees (as the case may be)." *Id*. at 316. AIA § 10(b) further provides that certain fees[o] that are set or adjusted under AIA § 10(a) "shall be reduced by 50 percent" when applied to any small entity that qualifies for reduced fees under 35 U.S.C. § 41(h)(1) and "shall be reduced by 75 percent" when applied to any micro entity as defined in post-AIA 35 U.S.C. § 123. *Id*. at 316-17. AIA § 10(c)(1) and (2) require the U.S. PTO to consult with the Patent Public Advisory Committee and the Trademark Public Advisory Committee[p] on the advisability of reducing any fees described in AIA § 10(a) before reducing such fees. *Id*. at 317. AIA § 10(d) then sets forth a specific procedure that the U.S. PTO must follow when consulting with the Patent Public Advisory Committee or the Trademark Public Advisory Committee. *Id*. AIA § 10(e) further requires the U.S. PTO to publish any proposed fee change under AIA § 10(a)(1) in the Federal Register and prescribes a procedure that the agency must follow when

[o]More specifically, the fees for "filing, searching, examining, issuing, appealing, and maintaining patent applications and patents" are subject to the reductions set forth in AIA § 10(b). *Id*. at 316-17.

[p]Note that 35 U.S.C. § 5 explains, among other things, how the Patent Public Advisory Committee and Trademark Public Advisory Committee are established and how members of the committees are appointed. Furthermore, 35 U.S.C. § 5 was amended by section 1(l) of the Leahy-Smith America Invents Technical Corrections Act of 2013. Pub. L. No. 112-274, 126 Stat. 2456, 2458-59; *see also* Manus Cooney, *Lame Duck Patent Reform: AIA Technical Corrections*, IPWatchdog, Dec. 2, 2012, http://www.ipwatchdog.com/2012/12/02/lame-duck-patent-reform-aia-technical-corrections/id=30778/ (explaining that section 1(l) of the Leahy-Smith America Invents Technical Corrections Act "makes the terms of [Patent Public Advisory Committee] and [Trademark Public Advisory Committee] members run for 3 years from a fixed date rather than from the date that they are appointed, and requires the Chairman and Vice Chairman to be appointed from among existing members and makes their terms run for one year. Current law designates only a Chairman and gives him a 3-year term.").

making such publication. *Id.* at 317-18. AIA § 10(f) provides that the U.S. PTO will only retain its authority to set or adjust fees under AIA § 10(a) "during such period as the Patent and Trademark Office remains an agency within the Department of Commerce." *Id.* at 318. Lastly, AIA § 10(h)(1) imposes an additional fee of $400 for each application for an original patent, except for a design, plant, or provisional application, that is not filed electronically, and specifically provides that the U.S. PTO cannot reduce this fee despite the broad fee-setting authority granted to the agency in AIA § 10(a)(1). *See id.* at 319 ("Notwithstanding any other provision of [AIA 10], an additional fee of $400 shall be established . . . .").[q]

AIA § 10(i)(1) provides that "[e]xcept as provided in [AIA § 10](h), this section and the amendments made by this section shall take effect on the date of the enactment of this Act."[r] *Id.* at 319. Thus, the U.S. PTO possessed the authority to set and adjust its fees pursuant to AIA § 10(a)(1) on September 16, 2011. AIA § 10(i)(2), however, provides that "[t]he authority of the Director to set or adjust any fee under [AIA 10](a) shall terminate upon the expiration of the 7-year period beginning on the date of the enactment of this Act." *Id.* Absent further Congressional action, therefore, the U.S. PTO will only possess the authority to set and adjust its fees pursuant to AIA § 10(a)(1) until September 16, 2018.

B. AIA § 10(g)—Establishment of the Micro Entity Status

Prior to the enactment of the AIA, the U.S. PTO recognized two types of patent applicants, namely a large entity and a small entity. AIA § 10(g) adds new 35 U.S.C. § 123[s] to the patent statutes, which changes the pre-AIA

---

[q]AIA § 10(h)(1) provides that the $400 fee for original applications that are not filed electronically "shall be reduced by 50 percent for small entities that qualify for reduced fees under [35 U.S.C. §] 41(h)(1)." *Id.* Surprisingly, AIA § 10(h)(1) does not include an analogous sentence to reduce fees by 75 percent for applicants that qualify for micro entity status under 35 U.S.C. § 123. *Id.* Despite the omission of any reference to a micro entity discount in AIA § 10(h)(1), the U.S. PTO currently provides such entities with a discount of 50 percent for non-electronically filed patent applications. U.S. PTO Fee Schedule, http://www.uspto.gov/learning-and-resources/fees-and-payment/uspto-fee-schedule (last visited April 28, 2016).

[r]The reason that AIA § 10(h) is excepted from AIA § 10(i)(1) is that AIA § 10(h)(2) provides the effective date for AIA § 10(h)(1). *Id.* at 319. In particular, AIA § 10(h)(2) states that "[AIA § 10(h)] shall take effect upon the expiration of the 60-day period beginning on the date of the enactment of this Act." *Id.* Accordingly, unlike AIA § 10(a)-(g), which took effect on September 16, 2011, AIA § 10(h) took effect on November 14, 2011.

[s]Note that the U.S. PTO has implemented new 35 U.S.C. § 123 through new 37 C.F.R. § 1.29.

law by establishing a "micro entity" as a third type of patent applicant. *Id.* at 318-19. The addition of the micro entity is significant because AIA § 10(b) enables such entities to pay only 25 percent of most of the fees that are required by the U.S. PTO for patent services. *Id.* at 316-17. In comparison, large entities are required to pay the full amount of any fee that is required by the U.S. PTO for patent services, and small entities are generally required to pay 50 percent of any fee that is required by the U.S. PTO for patent services. *Id.*

Patent applicants can qualify as a "micro entity" in two ways. *See Sometimes It's Good to Be Small…And Even Better to Be Micro! – Micro Entity Status Under The America Invents Act,* Ward and Smith, P.A., May 22, 2013, http://www.wardandsmith.com/articles/micro-entity-status-under-the-america-invents-act (explaining that entities can qualify as a micro entity on the "Gross Income Basis," the requirements of which are set forth in post-AIA 35 U.S.C. § 123(a) or on an "Institutions of Higher Education Basis," the requirements of which are set forth in post-AIA 35 U.S.C. § 123(d)).[i] Furthermore, while micro entities are a subset of small entities,[ii] there are many differences between the procedures patent applicants must follow in order to establish their status as either type of applicant. For instance, in contrast to the process for establishing small entity status, which merely requires that a patent applicant pay certain fees in the small entity amount, patent applicants who wish to be treated as micro entities must certify to the U.S. PTO that they qualify as such under either 35 U.S.C. § 123(a)

---

[i]The analysis of whether an applicant qualifies for micro entity status is complex. Indeed, the hypothetical fact patterns provided at http://www.uspto.gov/sites/default/files/aia_implementation/fitf_public_training_final3-15-2013.pdf demonstrate the challenges associated with determining whether micro entity status applies to a given patent applicant. In addition, there are reasons that an applicant may not want to certify itself as a micro entity even if the applicant qualifies for micro entity status. Accordingly, the best practice is always to consult a knowledgeable patent attorney or patent agent before determining whether to certify to micro entity status for purposes of filing and prosecuting a particular patent application.
[ii]The final Committee Report on the AIA explained that "[a]s part of the ongoing effort to nurture U.S. innovation, Congress has long recognized that certain groups, including independent inventors, small business concerns, and non-profit organizations (collectively referred to as 'small business entities') should not bear the same financial burden for filing patent applications as larger corporate interests . . . there is likely a benefit to describing—and then accommodating—a group of inventors who are even smaller [than small entities], in order to ensure that the USPTO can tailor its requirements, and its assistance, to the people with very little capital, and just a few inventions, as they are starting out. [AIA § 10(g)] defines this even smaller group—the micro-entity—that includes only true independent inventors." H.R. REP. No. 112-98, at 50 (2011).

or (d) before they are permitted to pay fees in the micro entity amounts. *Compare* 37 C.F.R. § 1.27(c)(3) ("The payment, by any party, of the exact amount of one of the small entity basic filing fees set forth in §§ 1.16(a), 1.16(b), 1.16(c), 1.16(d), 1.16(e), or the small entity basic national fee set forth in § 1.492(a), will be treated as a written assertion of entitlement to small entity status even if the type of basic filing or basic national fee is inadvertently selected in error.") *with* 37 C.F.R. § 1.29(a) and (d) (requiring that an applicant make certain certifications in order to establish micro entity status). Furthermore, micro entities must continually assess whether they qualify as such throughout both the pendency of their patent application(s) and patent terms. *See* 35 U.S.C. § 123(a)(3) and (a)(4) (requiring that a patent applicant's income level not exceed a certain amount for the calendar year preceding the calendar year in which the applicable fee is paid). Small entities, on the other hand, "need only determine continued eligibility for small entity status for issue and maintenance fee payments, but can pay intervening fees at [a] small entity rate without determining whether still entitled to small entity status." 77 Fed. Reg. 75,022 (Dec. 19, 2012); *see also* 37 C.F.R. § 1.27(g)(1) ("Once status as a small entity has been established in an application or patent, fees as a small entity may thereafter be paid in that application or patent without regard to a change in status until the issue fee is due or any maintenance fee is due.").

AIA § 10(i)(1) states that "[AIA § 10(g)] and the amendments made by [AIA § 10(g)] shall take effect on the date of the enactment of this Act." Leahy-Smith America Invents Act of 2011, Pub. L. No. 112-29, 125 Stat. 284, 319. Accordingly, new 35 U.S.C. § 123 went into effect on September 16, 2011. Note however that the 75 percent fee reduction for micro entities filing U.S. applications did not become effective until March 19, 2013 because that is the date that fees "for filing, searching, examining, issuing, appealing, and maintaining patent applications and patents" were first set or adjusted by rulemaking under the U.S. PTO's fee setting authority under AIA § 10(a).[v] MPEP § 509.04 (9th ed. 2015) (citing 78 FR 4212 (January 18, 2013)); *see also* 77 Fed. Reg. 75,021 (Dec. 19, 2012) ("However, no patent fee is currently eligible for the seventy-five percent micro entity reduction as no patent fee has yet been set or adjusted under section 10 of the Leahy-Smith America Invents Act.").

---

[v]In addition, the 75 percent reduction for certain "filing, searching [and] examining" fees for micro entities filing international applications under the PCT did not become available until January 1, 2014. MPEP § 509.04 (9th ed. 2015).

V) U.S. PTO PROGRAMS CREATED BY THE AIA

In addition to the many statutory, jurisdictional and administrative changes that were brought about by the AIA, Congress directed the U.S. PTO to establish and maintain two new programs—the Patent Ombudsman for Small Business Program and the Pro Bono Program. A brief description of each program is provided as follows.

A. AIA § 28—Patent Ombudsman for Small Business Program

AIA § 28 requires the U.S. PTO to establish and maintain a Patent Ombudsman[w] Program for Small Business Concerns using available resources.[x] Leahy-Smith America Invents Act of 2011, Pub. L. No. 112-29, 125 Stat. 284, 339. In response to AIA § 28, the U.S. PTO has established a dedicated staff to focus on the patent application filing needs of small businesses and independent inventors. U.S. PATENT AND TRADEMARK OFFICE, REPORT TO CONGRESS ON THE IMPLEMENTATION OF THE LEAHY-SMITH AMERICA INVENTS ACT 46 (2015), *available at* http://www.uspto.gov/sites/default/files/documents/Report_on_Implementation_of_the_AIA_September2015.pdf. The U.S. PTO has also developed educational materials for independent inventors and small businesses and partnered with the National Institute of Science and Technology Manufacturing Extension Partnership to develop the IP Awareness Assessment Tool, which helps small businesses and independent inventors learn about the steps they should take to protect their intellectual property. *Id.* Additionally, the U.S. PTO created an Ombudsman Program to assist pro se applicants (i.e., applicants filing a patent application without the counsel of a patent attorney or patent agent) when the normal patent application examination process seems to go awry. *Id.* Further information on the Patent Ombudsman for Small Business Program can be found at: http://www.uspto.gov/patent/laws-and-regulations/america-invents-act-aia/programs.

---

[w]An ombudsman is someone who hears and investigates complaints by private citizens.
[x]In support of the provision that was ultimately passed as AIA § 28, Senator Kirk explained that "it can ... be daunting for a small business owner or inventor to obtain a patent. In many instances, the value of a patent is what keeps that new small business afloat. It is vital for America's future competitiveness, her economic growth, and her job creation that these innovators spend their time developing new products and processes that will build our future, not wading through government redtape. [AIA § 28] would help small firms navigate the bureaucracy by establishing the U.S. Patent and Trademark Office Ombudsman Program to assist small businesses with their patent filing issues." 157 CONG. REC. S1034, 2011.

B. AIA § 32—Pro Bono Program

AIA § 32(a)[y] requires the U.S. PTO to "work with and support intellectual property law associations across the country in the establishment of pro bono programs designed to assist financially under-resourced independent inventors and small businesses." Leahy-Smith America Invents Act of 2011, Pub. L. No. 112-29, 125 Stat. 284, 340. By Executive Action taken in 2014, President Obama also required the U.S. PTO to appoint a full-time Pro Bono Coordinator and help expand pro bono coverage into all 50 states. U.S. PATENT AND TRADEMARK OFFICE, REPORT TO CONGRESS ON THE IMPLEMENTATION OF THE LEAHY-SMITH AMERICA INVENTS ACT 47 (2015), *available at* http://www.uspto.gov/sites/default/files/documents/Report_on_Implementation_of_the_AIA_September2015.pdf. In response to AIA § 32(a), the U.S. PTO launched a pilot program in Minnesota in 2011 to provide legal services to help financially under-resourced inventors and businesses obtain solid patent protection. U.S. PTO, http://www.uspto.gov/patent/laws-and-regulations/america-invents-act-aia/programs (last visited April 28, 2016). In addition, in response to President Obama's Executive Action, the U.S. PTO appointed a full-time Pro Bono Coordinator in 2014 and hired additional patent attorneys to assist with the Patent Pro Bono Program. U.S. PATENT AND TRADEMARK OFFICE, REPORT TO CONGRESS ON THE IMPLEMENTATION OF THE LEAHY-SMITH AMERICA INVENTS ACT 47 (2015), *available at* http://www.uspto.gov/sites/default/files/documents/Report_on_Implementation_of_the_AIA_September2015.pdf.

Using the Minnesota pilot program as a model, the U.S. PTO reached out to numerous legal associations across the U.S. to try to assist the associations with establishing new pro bono programs. U.S. PTO, http://www.uspto.gov/patent/laws-and-regulations/america-invents-act-aia/programs (last visited April 28, 2016). As a result of the U.S. PTO Pro Bono Team's efforts, the Office was able to secure nationwide pro bono coverage in June 2015 so that under-resourced independent inventors and small businesses in every state would have access to free patent legal help. U.S. PATENT AND TRADEMARK OFFICE, REPORT TO CONGRESS ON THE IMPLEMENTATION OF THE LEAHY-SMITH AMERICA INVENTS ACT 47 (2015), *available at* http://www.

[y]AIA § 32(b) states that AIA § 32(a) "shall take effect on the date of the enactment of [the AIA]." Leahy-Smith America Invents Act of 2011, Pub. L. No. 112-29, 125 Stat. 284, 340. Accordingly, the U.S. PTO was required to begin working to establish pro bono programs to assist under-resourced independent inventors and small businesses on September 16, 2011.

uspto.gov/sites/default/files/documents/Report_on_Implementation_of_ the_AIA_September2015.pdf. In addition, on October 25, 2013, several patent practitioners and pro bono experts signed the first-ever charter of the newly formed Pro Bono Advisory Council (PBAC) to oversee the pro bono programs that the U.S. PTO helped establish. *Id.* Information for the pro bono programs in each state can be found at: http://www.uspto.gov/ learning-and-resources/patents-help/united-states-map.

VI) STUDIES REQUIRED BY THE AIA

Several AIA provisions require the U.S. PTO and/or others to conduct studies on certain topics. For example, AIA § 3(l)(2)(A) requires the Chief Counsel of the U.S. PTO to consult with the General Counsel to conduct a study on the effects of eliminating the use of dates of invention in order to determine whether an applicant is entitled to a patent. Leahy-Smith America Invents Act of 2011, Pub. L. No. 112-29, 125 Stat. 284, 291. In other words, AIA § 3(l)(2)(A) requires that a study on the effects of the first-inventor-to-file system (i.e., the "first-to-file study") be conducted. In addition, AIA § 3(m)(1) requires the U.S. PTO to provide a report on prior user rights as the defense is used in select countries in the industrialized world. *Id.* at 292. Also, AIA § 26(a) requires the U.S. PTO to conduct a study on the manner in which the AIA is being implemented by the Office. *Id.* at 338. AIA § 27(a) requires the U.S. PTO to "conduct a study on effective ways to provide independent, confirming genetic diagnostic test activity where gene patents and exclusive licensing for primary genetic diagnostic tests exist." *Id.* Furthermore, AIA § 29 requires the U.S. PTO to establish methods for studying the diversity of patent applicants. *Id.* at 339. AIA § 31(a)(1) requires the U.S. PTO to conduct a study to determine how the Office, in coordination with other federal departments and agencies, can help small businesses with international patent protection. *Id.* Finally, AIA § 34(a) requires the Comptroller General of the United States to study the consequences of litigation brought by non-practicing entities or by patent assertion entities. *Id.* at 340. Reports on the studies required by AIA §§ 3(m)(1), 26(a), 27(a), 31(a)(1), and 34(a) can be found at: http:// www.uspto.gov/patent/laws-and-regulations/america-invents-act-aia/aia-studies-and-reports#heading-5. A report on the first-inventor-to-file study required by AIA § 3(l)(2)(A) is still forthcoming. *Id.*

CASE INDEX

STANDARD INDEX

Printed in the United States
By Bookmasters